A New Conservation Politics

Dedication

In memory of Frederick Douglas and for Nelson Mandela. Two who acted against enormous odds and who did not fear to follow their vision of justice with the fierceness of a mother Grizzly. They fought for justice among humans. May they inspire those who fight for a greater justice – for the living Earth and all of her species. Truly "In Wildness in the Preservation of the World." For John le Carré, master storyteller. And for Denzel and Carol.

A New Conservation Politics:

Power, Organization Building, and Effectiveness

David Johns

WILEY-BLACKWELL

A John Wiley & Sons, Ltd., Publication

This edition first published 2009, © 2009 by David Johns

Blackwell Publishing was acquired by John Wiley & Sons in February 2007. Blackwell's publishing program has been merged with Wiley's global Scientific, Technical and Medical business to form Wiley-Blackwell.

Registered office: John Wiley & Sons Ltd, The Atrium, Southern Gate, Chichester,
West Sussex, PO19 8SQ, UK

Editorial offices: 9600 Garsington Road, Oxford, OX4 2DQ, UK
The Atrium, Southern Gate, Chichester, West Sussex, PO19 8SQ, UK
111 River Street, Hoboken, NJ 07030-5774, USA

For details of our global editorial offices, for customer services and for information about how to apply for permission to reuse the copyright material in this book please see our website at www.wiley.com/wiley-blackwell

Library of Congress Cataloging-in-Publication Data:

Johns, David, 1951–
  A new conservation politics : power, organization building, and effectiveness / David Johns.,
     p. cm.
  Includes bibliographical references and index.
  ISBN 978-1-4051-9014-5 (pbk. : alk. paper) – ISBN 978-1-4051-9013-8 (hardcover : alk. paper)
1. Nature conservation–Political aspects–United States. 2. Environmental protection–Political aspects–United States. 3. Conservationists–United States–Political activity. I. Title.

  QH76.J64 2009
  333.720973 – dc22

                                                                    2008045481

ISBN: 9781405190138 (hardback) and 9781405190145 (paperback)

A catalogue record for this book is available from the British Library.

Set in 10/13 RotisSemiSans by Laserwords Private Limited, Chennai, India
Printed and bound in Malaysia by Vivar Priting Sdn Bhd

1   2009

Contents

Acknowledgments

My debts are enormous as the bibliography attests. The use to which I have put the work of those I owe is not their responsibility, but I have endeavored to present their contributions accurately and in context. I owe a larger debt to those I have learned from directly in the struggle to ensure the integrity of the Earth. These include advocates and scientists whose courage, innovation, perseverance, toughness and capacity for empathy with the natural world are exemplary: Rob Ament, "Nitro" Ament, Mario Boza, Tom Butler, Romain Cooper, Ol'ga Chernayagina, John Davis, Dominick DellaSala, Bob Ekey, Libby Ellis, Brock Evans, Dave Foreman, Wendy Francis, Mitch Friedman, Alan Watson Featherstone, Mary Granskou, Karsten Heuer, Webster Heuer, Doug Honnold, Malcolm Hunter, Peter Ilyn, Andy Kerr, Harvey Locke, Rurik List, Curt Meine, Oscar Moctezuma, Jerry Mander, Brendan Mackey, Bill Marlett, Vance Martin, Bill Meadows, Murphy Marobe, Alec Marr, Margo McKnight, Brian Miller, Reed Noss, Elliot Norse, Doug Peacock, Juri Peepre, Ian Player, Bart Robinson, Rich Reading, Conrad Reining, John Robinson, Michael Soulé, Bittu Sahgal, Wayne Sawchuk, Jim Strittholt, John Terborgh, Kim Vacariu, Kelpie Wilson, Louisa Willcox, L.G. Willcox, and Virginia Young. There are many, many others too numerous to mention. They do not live for anything so transient as the acknowledgment of history but for the transcendent – the wild, self-willed and undomesticated. The ocean flows in their veins and a fierce green fire burns in their hearts.

It has been a privilege to work with a handful of visionary funders who understand the world can't be changed without taking risks. Among them are Doug Tompkins, Rose Letwin, Tim Greyhavens, Denise Joines, Gary Tabor, Bill Lazar, Don Weeden, Yvon Choinard, Rick Ridgeway, and Ted Smith.

Not least among those to whom I am indebted are storytellers, writers, songsters, film makers, photographers, poets, painters and poster makers: Jonathan Cobb, Barbara Dean, Lou Gold, Barry Lopez, Cecelia Ostrow, Doug Peacock, Paul Shepard, Gary Snyder, Walkin' Jim Stoltz, and Terry Tempest Williams.

Special thanks to Ellen Main for her unfailingly sound criticism and editing, and to Phyllis Ray and Margaret Flagg who proofread the manuscript. The editorial and production team at Blackwell have been superb, including editor Ward Cooper, Delia Sandford, Rosie Hayden, and Gopika Sasidharan.

Introduction: hard times

The wind will not cease, though the trees wish to rest.

Early Chinese aphorism

Conservationists have made enormous strides. In the high tide of the early 1970s, grassroots activism and lobbying in the United States prevailed on a Republican president and Democratic congress to adopt the Endangered Species Act, Marine Mammal Protection Act, and National Environmental Policy Act, and to ratify the Convention on Trade in Endangered Species of Wild Flora and Fauna. Since that time, new parks, wilderness and other protected areas have been established, the ivory trade banned, and much else achieved. Despite these successes the extinction crisis is accelerating. The political influence of conservationists continues to fall short of what is necessary.

Within a decade of the disgrace of the president who signed many of these laws, his political party declared war on Nature and has since then been attacking these laws and what they seek to protect. Agencies stealthily announce administrative policy changes – such as announcing the release of 200 million acres of wilderness-quality Bureau of Land Management lands for development at 5 p.m. on Friday afternoon, knowing that Saturday papers and news shows are little read and watched. The US experience is not unique. Despite electoral promises to the contrary, Brazilian President Lula has publicly allowed the assault on the Amazon to continue. China's Three Gorges Dam project and other such projects throughout Asia attest to the political weakness of conservationists.

How has this seeming decline from the high tide of the 1970s come about? Leaving aside authoritarian regimes for the moment, how is it that in democracies, conservation goals remain unrealized when the overwhelming majority of the people consistently say that they value Nature and want to protect species and natural systems? Although some countries are doing better than others, the global trajectory is not good. Deborah Guber (2003: 105–6), noted similar findings among US researchers: conservationists had little affect on the votes of national legislators. One observer noted that no other lobby was so routinely ignored. Why have conservationists been unable to build on the values espoused by so many and create the political force needed for better policy? This is especially vexing in the United States because as the world's biggest economy and preeminent military power, its behavior affects the entire globe as does no other country.

Explanations and excuses for this sorry situation are many. It is true that conservationists are up against enormous obstacles. Economic and political institutions are mostly in the hands of those dedicated to preserving their privileged positions, positions that depend on the transformation of Nature into commodities. It is true that political opportunities for change are often lacking because they are cyclical. But conservationists do have control over some things and that includes themselves. Change to make conservation more effective must begin here.

What can conservationists do differently to build a much more effective political force? One that is capable of sustained action over decades. One that can take and keep leadership away from the self-serving, and transform institutions into servants instead of masters. Creating a political force requires mobilizing people. It is *action* that makes a difference – action in support of conservation policy and conservation-compatible personal decisions. Mobilization is the process wherein people commit their time, energy, money, and other resources to collective action for a common purpose. Instead of buying a new toy or sitting in front of the tube and occasionally sending a check to an NGO, they vote on conservation issues, write letters to officials and the media, attend hearings, and take to the streets on behalf of the wild Earth. In the United States, 80% of the population agree with conservation goals, but conservation is neither central to them nor do they feel strongly about it (Deborah Guber, 2003: 3, 38). That's why political leaders know they can safely ignore the views of most people on conservation issues; views don't translate into action.

To mobilize people, conservationists need to touch more people much more deeply than they have. This also presents formidable obstacles because most means of reaching people – cultural institutions – are also businesses and they are not in the hands of conservationists. Notwithstanding, the cultural arena is not confined to the mass media and its few owners but remains a contested terrain. Indeed, even the corporate media is divided between its political agenda – protecting its privilege – and short-term profit that may cause it to carry messages it doesn't agree with. This cultural space provides the opening for conservationists to mobilize people to act politically and economically to ensure that the integrity of the natural world is respected and protected.

Taking advantage of the opportunities for mobilization begins with an unflinching look at the current situation. This means putting aside the upbeat pabulum that is routinely cranked out by most NGOs on the advice of public relations and fund-raising consultants. Virtually all existing human societies are destructive of the natural world because they are dependent on massively transforming it. The grim numbers are all too familiar: humans take or destroy 40% of terrestrial net primary product and at least 30% of marine net primary product (Stuart Pimm,

2001). In 2006, the US gross domestic product – a close approximation of the market value of Earth transforming economic activity – was about $13.2 *trillion*. US charitable giving in 2006 to conservation, environmental and animal rights groups was $6.6 billion (USA Foundation, 2007). The problem is obvious.

Reforming human societies so that they become less destructive of the natural world is a long-term effort. Extinction is a much more immediate problem, and in the near term, conservationists can more consistently win legislative, policy, and electoral battles if they can mobilize powerful constituencies. Conservationists will never match the enormous financial and political resources of their opponents and all that big money can buy and enormous power can coerce. They can make up for this by being smarter, more creative, and more agile than their opponents. And that entails becoming more realistic about the political process, its institutions, and people.

Many, perhaps most, conservationists are trained in the natural sciences and have a high regard for rationality – careful thought and analysis, and measured consideration of alternatives and consequences. They want others to share these traits and come to the right conclusions by means of the right process; they want people to do the right thing for the right reasons. But all the rational arguments in the world for protecting shrinking grizzly bear or tiger habitat, for example, won't save the day if people don't *care* about grizzlies or tigers. Caring is not a matter of what is usually denoted by the term rational, that is, the cognitive or the mental. Caring is about emotion and need. *Emotion* and *motivate* come from the same root – to move. Thoughts, beliefs, and values are anchored in how people feel, in what they are connected to, in what they long for, and in what they require to sustain their lives and make them meaningful. Critical thought, and especially reflection, are important in examining the authenticity of emotions, but thoughts do not move people. Conservationists who love Nature and people in love with money and toys are both motivated by caring, by desire, by the connections (which are always emotional) to other living things or to inert matter like a car. Thought helps to sort out which connections serve human well-being and which are compensatory, but cognitive arguments by themselves do not persuade people to reject the compensatory; rather, a lack of emotional satisfaction does.

I use the term rational in this book to refer to both cognitive and emotive aspects of the personality that are healthy, authentic and grounded, as distinguished from those aspects that are compensatory, self-destructive, or contrary to needs and biological well-being over time. Anthropologist Roy Rappaport (1976: 65) argued similarly that survival is at root biological, and institutions that are destructive of biological well-being are maladaptive. Thus, an emotion such as fear may be rational or irrational, depending on the circumstances; similarly, thoughts may be rational or irrational depending on internal consistency, approximation to reality, and so on. In this view, the irrational is a major part of politics, as philosopher Herbert Marcuse (1964) noted when he described modernity as the rational pursuit of irrational goals, that is, the methodical way in which humans created and deployed weapons of mass (self) destruction. Writer D.H. Lawrence (1936: 284–5), who spent much of his life seeking to free people from sexual and cultural repressiveness, nonetheless agreed with Dostoevsky's Grand Inquisitor: most people require and demand "miracle, mystery and authority". He also observed that in the 20th century at least, people found their miracle and mystery more in the products of science than in religion. Political effectiveness depends on attending to these observations. That ir/rational emotion and need (and structural factors) move people, rather than reason, is not a weakness – it's just how we are. It will be catastrophic for life on Earth if conservationists confuse how they want people to be with how people are. Such confusion *is* a weakness.

Because so many conservationists do not have training in the social sciences nor the experience of long involvement with social movements, electoral campaigns or political infighting outside of NGOs, they are often engaged in reinventing the wheel. This is a waste of precious energy. Many of the answers to a more effective conservation movement lie with learning from other social movements. There is a vast body of practitioner and academic literature that documents important and hard-won lessons. It is not reasonable to expect conservationists to sort through and absorb this enormous body of knowledge. The purpose of this book is to make available, in summary form, the critical research findings and practitioner experience from a range of movements and nonmovement politics across the globe. The nature of this literature and my own experience is unavoidably skewed to the developed world, but many of the lessons are transferable among regime types. I have provided extensive citations, not to back up statements, but to make more detailed discussions of topics easily accessible to readers.

This book is organized around the elements of one central question: how can conservationists mobilize influential segments of the population to change policies and institutions and keep them on track to achieve conservation goals. Although I will present examples of success, lessons from failure, and even some truisms, the real answers are found only in practice, in doing.

The first part of the book examines the main obstacles to conservation, both within the movement and external to it. It also includes an outline of the main factors that motivate human political action. These factors may be thought of as levers that conservationists must learn to operate simultaneously rather than one at a time.

Building an effective movement has been likened to forging a hammer. Along with an uncompromising vision, flexibility of means, and the ability to convince opponents and decision makers that one will never tire or fade away, and being able to reward and punish (the hammer) is essential to political success. Part 2A is devoted to the many elements involved in forging the hammer and Part 2B to the maintenance of the hammer over time. Although the chapters build on each other in important ways, readers can usefully access them individually and in an order that best suits their purpose. Where something critical appears in another chapter, I have made reference to its location.

If conservation remains a sideshow or social afterthought it cannot ensure the survival of wild (self-willed) creatures in "natural patterns of abundance and distribution" (Reed Noss, 1993: 11) and of wild places. While it is probably true there will never be a world free of greed, hubris, myopia, and stupidity, a world in which the most powerful institutions are not constitutionally enslaved to endless material growth and accumulation of material wealth is possible. It is also desirable and necessary. The well-being of humans as well as the natural world depends on the former becoming, in Aldo Leopold's (1987 [1948]: 204) words, "citizens of the land community rather than its temporary conquerors."

A note on terminology

Some conservationists object to military-sounding terminology because they feel it reinforces political antagonism: it paints political *opponents* and even the uninformed as *enemies*. What is needed, they argue, is language that better reflects the need to bring people together and to achieve social change by means that transcend the violence of those who have tried to subjugate Nature. I understand and respect this position, but do not fully agree with it and so will use the objectionable terminology now and then. First, because it fits, and second because there are often no alternatives that so clearly convey by *metaphor* the meaning intended. As Evan Cornog (2004: 163), the chronicler of presidential stories observes, the metaphor of war is instantly understandable. Second, human societies *are* waging *war* on Nature and have been doing so for millennia. The violence is horrendous. Many institutions are also waging war on conservationists. To not acknowledge this, and to refrain from language that makes this clear, is a disservice, akin to the militaryese used to obscure mass murder, for example, as collateral damage. It also obscures the need to act intelligently by taking this violence seriously and recognizing the need for self-defense. Abolitionists from Frederick Douglass to Nelson Mandela had no illusions. Their lack of illusions, their refusal to confuse the way they wanted the world to be with the way it actually was, helped guarantee their day-to-day survival and eventual triumph. Conservationists need to heed Douglass' and Mandela's example.

Throughout these chapters I use the term "resources" extensively when speaking about mobilization. I also frequently refer to people as the objects of mobilization. As objects of mobilization, *resources* and *people* or *groups of people* are more or less interchangeable. Political resources are mostly to be found in people.

Sometimes it makes sense to specify what particular qualities are being sought: money, certain skills, bodies on the line, access to leaders, control of institutions, and so on. At other times, when talking about political clout, it makes more sense to talk about people or groups of people.

And there are nonhuman resources, and it can make sense to talk of mobilizing them. Wolves in the Lamar Valley of Yellowstone are a political resource – conservationists call them flagship species for that reason. They can be photographed or shown to potential members, donors, or legislators. Watching wolves creates a bond, and some people thereby become devoted to protecting them. To label something a political resource is not to reduce it to that. An ecological disaster like a mudslide resulting from deforestation is real – it kills fish and causes stream damage. Such an event can be used as a resource to make dramatically clear the costs of roadbuilding and logging. A freak 1999 hurricane that struck France undermined French faith in the human control of Nature (Gary Bryner, 2001: 96). Severe disease outbreaks have had similar political effects (Hays, 2000). In some cases, disease outbreaks are closely tied to human intrusion into and disruption of ecosystems, and to long-distance trade. In some instances, diseases have been directly mobilized as an instrument of politics, as when plague-infected corpses were catapulted over the walls of fortified cities in medieval Europe. Biowarfare programs and use of antibiotics in factory farming are more recent examples.

Usually, however, *resources* refers to a short list of those things important to achieving political goals:

- self-consciousness
- an understanding of the social system including the levers of political control and the systemic dynamism that generates opportunities and threats
- a grounded ideology
- control of important institutions
- control of the instruments of force
- access to decision makers
- organizing and communication skills
- access to or ownership of media for mass communication, and possession of internal media for communication with members
- charismatic spokespeople and good stories, and
- a number of members and supporters, and their degree of commitment, unity, and internal division.

Part 1
The Gauntlet:
we have met the
enemy and they are
both us and them

1   Us

History is shaped . . . not by the cunning of Reason but by the cunning of Desire.

Norman O. Brown, *Life Against Death* (1959: 16)

The destruction of creatures with whom we have long shared the Earth is accelerating, despite the efforts of conservationists to slow or stop it. Whole ecosystems are being buried under asphalt, concrete, subdivisions and domesticated monocultures at an exponential rate just as surely as great tides of Hollywood lava once consumed whole cinematic villages of noble savages – only this is real, not a movie. Despite efforts rooted in rural communities, in centers of world power, and everywhere in between, conservationists have been unable to stem the cancer that is inexorably devouring grizzlies, wetlands, dry forests, raptors, butterflies, tropical forests, boreal forests, tundras, and caribou.

What can conservationists do differently that will make them more effective? This book is written for those who care enough about the natural world to examine their assumption and the current way of doing things. It is written for those who understand that extinction is irreversible and that alongside the tyranny of swelling human numbers and demand for even more stuff, are conservationists who are wedded to business as usual. This book is not for those who think things will somehow work out in the end, or for those who think they can magically have Nature and the equivalent of 6.5 billion American consumers. It is for those who are willing to look squarely at current practices and to dump approaches that aren't working for more promising approaches. It is for those for whom Nature is more important than cherished ideas or the need for recognition from other humans.

Effective political and social change begins with those who seek to make change and ground themselves in what works.

What's not Working

Thomas Patterson (2002: 13) echoes the observations of many social scientists when he states that if all those eligible had voted in 2000 the electoral outcome would have been very different. Having the presidency and both houses of Congress in other hands would have not halted human-caused extinctions in their tracks, but it would have been far better for the natural world than the actual outcome. The point is that a lack of action on the part of potential voters made a difference for conservation. Action is what counts. Action changes outcomes. One action damages Nature, another nurtures it.

There are many reasons conservationists have not been effective in getting people to act. Some of the most salient are related to their assumptions about what motivates people to act. Conservationists know that what people think and feel counts, but not how these thoughts and feelings generate action. Why do some thoughts and feelings move people to action and others don't? What happens when people's hearts tug in one direction, but their calculations in another? What causes people to publicly espouse one view but act contrarily? The relationship between emotions, values, and views of the world on one hand and action on the other, are complex and not always obvious. It is difficult to gain insight, however, when one already possesses insight.

Conservationists are fond of quoting Margaret Mead on how small groups of committed people drive change in the world, but Mead lacked a full understanding. A small group can start a snowball rolling down a hill, but the group needs a hill, a way to get to the launch place on the hill, the right kind of snow, and much more. For the conservation snowball to become a daunting boulder a good understanding of the sources of political action are needed. Such knowledge is not innately mysterious or hard to come by, but much gets in the way. Conservationists too often:

- Focus not on generating action but on the precursors to action.
- Focus too much on the cognitive and on education as transmission of knowledge.
- Fail to follow through when they have emotionally energized people by involving them in a community or organizational structure that can nurture their energy and sustain it.
- Do not understand or do not want to understand what causes decision makers to act.

Zeno and Conservation

Conservationists often aim not at instigating action, but at some intermediate point in a process that is supposed to lead, in some vague way, to action. If loving Nature leads to action, and experiencing Nature causes people to love Nature, then conservationists focus on hiking programs. The other elements that determine whether or not people act – the need for constant encouragement, the overwhelming importance of collective reinforcement in sustaining action, the role of organization – are never addressed. Similarly, conservationists sometimes treat lobbying like the unmet demand for contraception. If one provides information, states a preferred outcome along with some poll numbers, things will work out. In the case of contraception there is often a preexisting motivation, but for most objects of lobbying the motivation must be provided.

In each of these instances conservationists become trapped like Zeno's hypothetical arrow; they only get half of the rest of the way toward their goal. In the case of instigating action, there is often no understanding of the need for creating a conservation community – something which has empowered many other movements.

Conservationists may limit their activities to those short of directly generating political action because it is less risky. Action can create controversy. Action draws the attention of sometimes powerful and violent opponents. Whatever the reasons, There is a lack of recognition that half-steps will not stop extinctions. Action can run contrary to some countries' laws that limit the political activities of non-governmental organizations (NGOs) that have charitable tax status – a status many rely on to raise funds.

Education and Emotion

The focus of conservation NGOs on education as provision of knowledge is not just a reflection of tax laws but of a pervasive belief that if people are given good information they will do the right thing; they will act rationally in their long-term interests and with generosity toward the natural world. Those conservationists trained in the natural sciences seem particularly susceptible to this Enlightenment predisposition. But the predisposition often goes beyond faith in the cognitive and the notion that people reason through issues and act accordingly. Perhaps they regard appeals to emotion as inherently manipulative – the province of the wealth and power driven. Perhaps it reflects living in one's own head too much. Ted Brader (2005: 21–3) reports that political scientists share this problem: they acknowledge the importance of emotion in political behavior but don't study it, whereas political practitioners hold political action to be essentially emotional and operate on that basis. Certainly many scientists have an abhorrence of irrational behavior, a reticence about delving into the emotional which is often identified as irrational, and a faith that people learn from their mistakes. Yet there is little support for such faith. Often enough people's emotions are irrational. In the United States, for example, 19% of Americans believe they are among the richest 1% of the population and another 20% believe they will be in that 1% some day (John Micklethwait and Adrian Wooldridge, 2004: 307–8). Most who believed the Bush administration's misrepresentations about Iraq's involvement in the 9/11 attacks and possession of weapons of mass destruction, still believed them three years later, despite incontrovertible evidence to the contrary (Steven Kull, 2004: 3–6). Just as some people fall in love with those who abuse them, sometimes repeatedly, so do people embrace and make excuses for political leaders who ill-serve them. If people can't get basic economic self-interest or security issues right, are they likely to get ecology right based on emotion?

Good information might be of help if bad information caused irrational emotions or bad political choices. But it does not; bad information, emotions ungrounded in reality, and bad decisions have a common cause. This is why proffering and even instilling more accurate views of the human impact on the natural world has not paid off with commitment and action as anticipated.

As we will see a range of views about reality and a range of values can contribute to the actions conservationists desire and conservationists are most successful when they can work across a range of beliefs. People may act to save a species, protect a wilderness area, have small families, and limit consumption because they are ecocentric, theocentric, and for a variety of anthropocentric reasons including aesthetics, quality of life, humility, a love of wilderness solitude or recreation, and so on. Few people, however, who chose to have one or no children do so because they have "correctly" reasoned that another human life in a developed country adds yet another straw to the camel's back. Reproductive decisions are usually associated with the quality of one's own childhood experience, peer pressure, calculations concerning the cost of children, the existence and desirability of other options besides motherhood, which are related to self-esteem, selfishness, and attitudes toward birth control (Laurie Mazur, 1994: 111–299; Alan Durning and Christopher Crowther, 1997).

Thus, although the natural world is real and operates as it does regardless of what we believe to be true, people's views of how the world works need not be strictly accurate in order to give rise to desirable action. It is not necessary to understand evolutionary biology to act to protect a species, though clearly it informs how to achieve protection. People often work hard to conserve a place without understanding its biological value. Indeed, humans are seldom in possession of complete knowledge about the world around them and so fill in the gaps. Whether one avoids lightning because it's lethal or one is frightened by Zeus's anger makes little practical difference if people are similarly motivated to stay off hilltops.

Not all views similarly motivate, however. Elizabeth Barber and Paul Barber (2004: 13) observed that people are predisposed to explanations for events that are both deterministic and purposive. Such explanations are more economical than the usually more complicated reality. Deborah Keleman (1999: 283–9) noted that both children and adults use their own intentionality as a model to explain causality in the larger world. Most observers of human thought and behavior, she goes on to say, see this approach as so widespread because it is adaptive. Though literally mistaken, it organizes experience in a functionally successful way. A central feature in most religions is a purposeful god(s) or Nature and this has caused some believers to reject biological evolution because of its random elements. But recall that Einstein dismissed quantum mechanics with the statement that god does not play dice with the universe.

Of course mistaken views do create problems for people and society, but successful conservation depends on right action, not purity of motivation. Indeed, many of those who cannot accept that the universe lacks purpose increasingly evince strong support for conservation on the basis of their religious views. Conservationists who cut their teeth on Lynn White's (1967) essay on Christianity and conservation may have a tough time with this, but if conservation is really the priority, then both the religious and nonreligious need to focus on conservation goals and not on perceived imperfections in each other's world views and motives if they are immaterial to outcomes. Conservationists would do well to heed anthropologist Roy Rappaport (1974: 56, 1999) who observed that "(i)t is not merely that adaptive behavior may be associated with understandings which do not accurately reflect material conditions, but that some adaptive behavior may be elicited only by such understandings." Marvin Harris (1974: 11–32) documents many instances of this as well.

Most people will reject factual information that seems to contradict their values or what they find meaningful, and that is worth keeping in mind. Neither conservationists nor scientists are immune from the effect of values, beliefs, and their emotional underpinnings on the acceptance of knowledge. Values, beliefs, and emotions do change, but less as a result of contrary knowledge than their failure in the face of generational change or values and emotional orientations that are more functional. In the context of mobilization to address near-term issues it is inescapable that a wide variety of views must serve conservation. Grizzly habitat, for example, will more likely be protected for clean water and for fish than for the great bear. Beliefs and rational thought are for the most part after the fact justifications or strictly utilitarian, that is, in the service of securing emotionally determined goals. Critical thinking is rare.

The Russian psychologist Luria Vygotsky (1962: 150) wrote many decades ago that "Thought itself is engendered by motivation, that is, by our desires and needs, our interests and emotions. Behind every thought there is an affective-volitional tendency, which holds the answer to the last 'why' in the analysis of thinking." More recently neurobiologists such as Antonio Damasio (1994, 1999) have demonstrated not only that human needs and emotions guide people, but that reason cannot function without them. The options people face, especially when making complex social decisions, are too great. Emotional filters, shaped by genome and experience, whittle the universe of options down to a few that conscious intellect can manage (Damasio 1994: 165–201). Even then emotions influence choice. That's why successful advertising is aimed at the heart, not the cortex.

The conservation reticence to fully engage emotion in mobilization is ironic given that conservationists are so plainly motivated by their own passions. The conservation literature is full of passion and emotional epiphanies. Aldo Leopold's (1987 [1949]: 130) is a moving example: having shot an old wolf and her pups because fewer wolves meant more deer, he approaches to finish the job, only to see a fierce green fire fade in the dying wolf's eyes. He realized in that wolf's eyes "something known only to her and the mountain." That something encompassed a more profound and larger view of life than more deer for human hunters. For Mike Harcourt, the former British Columbian Premier who created many protected areas, the moment came when he visited an enormous clear-cut in the heart of Vancouver Island. It looked to him like a massive bomb had exploded, leveling the great forest for miles around. Harcourt saw the wrongness of this intentional destruction and knew he had to try and stop it. Peter Illyn, the founder of Restoring Eden had a gentler epiphany: rising early one morning he saw through the mist an elk grazing near his tent. It is perhaps trite to say that Illyn saw the "miracle of the ordinary": another creature breathing, eating, living; a creature so much like him, yet different, and no less remarkable. People see most deeply with their hearts.

The eminent biologist David Ehrenfeld (1979: 142, 224) noted that emotions have been around for many millions of years in the mammalian line and have been long tested; our "higher" cortical functions are much more recent and still an evolutionary experiment. Rachel Carson (1984: 24) argued that "it is not half so important to know as to feel." It is our emotions that connect us to others, and to our selves. Our needs – for survival (food, shelter, and sex), for love and belonging, for making sense of the world – impel us to meet them. Our reflexes, our pleasure/pain responses and our emotions fit us to the world in ways most likely to meet our needs based on evolutionary experience. Only by touching people at this level will they be moved to act on behalf of the Earth and all of its life.

When conservationists do use emotion in their campaigns it is often to good effect. But frequently campaigns demonstrate a superficial understanding of emotions. Fear is a powerful motivator, as governments and political candidates know. Both regularly and successfully use fear to mobilize support or draw attention away from their own weaknesses and misdeeds. Conservation has been less successful for a number of reasons: they have sometimes overstated or exaggerated threats and industry has pounced on the slightest error or misprognostication, undermining conservationist credibility; threats to biodiversity, unlike threats to human health from polluted air and water, are not experienced by most people as salient; fatigue sets in, especially if the threat is distant in time or place or emotional ties to the natural world are weak or absent.

Conservationists often appeal to people's concerns for their progeny. Certainly most people express great concern for the lives of their children and grandchildren, but psychiatrist Harold Searles (1979) has questioned reliance on these statements. He argues that given the level of human inaction and apathy in the face of biological meltdown, people do not hold their children in such high regard. They are not willing to sacrifice to give them a better world, in part because they resent their own parents for passing on a world of problems. Certainly there is some corroboration for this view in the United States where increasingly education and similar services are underfunded.

In later chapters we will devote significant space to how conservationists can make better use of emotion to reconnect people with the natural world and each other and motivate more committed action.

Conservation as a Tease

It is the rare conservationist who has not attended a conference or other meeting where some inspiring speaker thoroughly excited hundreds of people, priming them to act. Invariably members of the audience ask the speaker what they can do, only to hear vague and general answers. Write a check. Fill out this form, so we can inundate you with pleas for money and the occasional request to send a postcard or letter to an official. The crowd goes home, and in the noise and distraction of day-to-day life, the positive energy dissipates. Meanwhile the subdivision and strip mall developers, and oil drillers are highly energized and organized. Conservationists have failed to *involve* people in a setting that sustains sympathizers' energy and commitment. Doing so requires creating a home for the whole person. It requires facilitating the creation of a conservation community and organization that involves people on a regular basis over time.

Except for a small portion of conservationists – professional staff and committed volunteers – there is no conservation community. Most NGO members live lives in which their social networks have little or nothing to do with conservation, important events and rituals have little or nothing to do with conservation, conservation is not routinely celebrated, nor its value routinely experienced. There exists no conservation equivalent of Black churches or White universities that provided the substrate for the thick webs of friendship and mutual support that sustained commitment in the US civil rights movement. Within the movement people made friends, met their spouses, socialized and relaxed together, shared risks, disappointments and euphoria, and found common meaning. The web of relationships contributes enormously to making a cause central in people's lives. That's because relationships and the venues that support them meet people's needs, just as conservation meets the need for meaning. When these needs are intertwined the ties that bind are strengthened.

Organization is also critical to sustaining, building, and deepening commitment to political action. Apartheid was not brought down with organizations of check writers and postcard signers. Nor did such people bring down the Berlin Wall, create strong unions, achieve suffrage, or topple corrupt leaders. Organizations of check writers and postcard signers will neither halt the extinction crisis nor slow climate change. Broad-based and strong political organizations will do that. Organizations that have a place for all of those who become excited hearing a speech for the first time and want to do more. Organizations that involve people regularly and nurture their involvement. When people are left hanging, when their contact is a quarterly newsletter or even a glossy monthly, they are not drawn to greater involvement and they are not available down the road when they are needed.

The reality is that getting involved in politics is not most people's idea of a good time. Most of them need strong encouragement to act politically (or to act to restore an ecosystem or boycott a product) and consistent reinforcement to sustain action. By involving people in regular activity and making an organizational place for them encouragement and reinforcement can be provided, relationships built, and reticence about taking action overcome. People gain experience and mutual trust. The activities in which people are involved do not need to be directly political, for example, monthly visits to the state capital or quarterly rallies, or regular visits to a local wetland to cull exotics. They can be social or educational activities. The purpose is that they involve and make people part of the organization in a way writing a check usually does not. People are then available for political action when the time comes.

One approach to organization is to organize the already organized – those who are self-conscious and have political experience and clout. They bring more to the table than the unorganized although more of an investment is required to bring them to conservation. On balance the investment pays off more than organizing the merely sympathetic.

Science and Crassness

Many conservationists believe or want to believe that decision makers, at least elected officials in democracies, are responsive to scientific findings and otherwise persuadable by reasoned arguments. They are shocked when non-science (non-sense) or pseudoscience holds sway (Paul Ehrlich and Anne Ehrlich, 1997; Todd Wilkinson, 1998). They should not be shocked. Politics operates by a different rationality – one that is focused on getting and keeping power (Johns, 2000:

226–8). Politicos *are* concerned with the substance of policy in a few areas that are priorities for them, but their position in these areas is usually consonant with their more powerful supporters – that is often the basis for successful campaigns. When scientific findings support a policy position being backed they will certainly be cited; if not, they will be ignored, denigrated, or the "tobacco company doctors" rolled out in support. Science does find a receptive audience when elected officials genuinely care about problem solving. As with other audiences, if a legislator values the natural world the science important to protecting it will be valued. Science can also play a pivotal role in a crisis or when decision makers are closely divided. Scientific findings or lobbying by prominent and high-profile scientists can provide one more hook to put one's position over the top. Scientific findings are more often probative before courts and before agencies in some circumstances.

An old proverb provides useful direction to conservationists: good does not triumph over evil because it is good but because it is strong. That's the reason for this book.

There are other problems within the conservation movement that limit its effectiveness and are within the power of NGOs to change. Dependence on foundation largess limits organizations in many ways: many foundations are conservative and action-averse or seek to set recipients' agendas; being a tax-exempt entity limits political action; and foundation support in total is inadequate to support a movement of the size and strength needed. Ultimately the most reliable support is self-funding. That's a different book.

Increased effectiveness also depends on conservationists' understanding of what they are up against. In politics illusions about opponents can be mortal. We turn to that now.

2 Them: inertia, inequality, and propaganda

In the affairs of this world men are saved not by faith, but by the want of it.

Benjamin Franklin, *Poor Richard's Almanac* (1987 [1732]: 37)

It must be remembered of course that good will . . . can be preserved in the long run only by those whose actions warrant it. But this does not prevent those who do not deserve good will from winning it and holding onto it long enough to do a lot of damage.

Edward Bernays, *The Engineering of Consent* (1947: 116)

In his 1999 documentary *Free Speech for Sale*, Bill Moyers tells the story of a North Carolina state representative who took on the hog industry. Businesswoman Cindy Watson was the first Republican elected to the state legislature from her district in decades. A supporter of economic development, she also took seriously the concerns of her ordinary constituents when they told her that huge hog farms were making their lives miserable. Massive factory farms were spraying hog waste on crops, poisoning groundwater used for drinking and causing high levels of airborne ammonia that made breathing difficult. To address this she sought to change a state law that had years before removed the power of counties to zone hog operations, and to have air and water quality laws enforced. The industrial hog farmers reacted. Wendell Murphy – owner of Murphy Family Farms (a billion-dollar operation) and a former state legislator who had pushed the hog farm zoning exemption into law – along with several cohorts created a political front group called "Farmers for Fairness". They did this because they knew messages directly identified with the hog farmers would be discounted as self-serving. The industrial hog farmers launched a saturation media campaign attacking Watson's integrity, defeating her in the Republican primary.

Many, many years before Cindy Watson confronted the marriage of campaign cash and mass media, Rachael Carson experienced a concerted effort by industry to suppress her book, *Silent Spring*. When the chemical industry learned of the book, they sought to stop its publication by threatening the publisher with economic injury. When that failed, the industry trade group doubled its public relations (PR) budget, formed front groups (like Farmers for Fairness), and recruited supposedly independent scientists and doctors to counter her claims. *Silent Spring* was published nonetheless, and changed the world by helping to create the modern environmental movement. In other cases, industry has been more successful, such as when Dupont persuaded Book-of-the-Month Club to break its contract and rescind its selection of Gerard Zilg's history of the family and its corporation. Industry-hired PR firms have infiltrated groups to spy on and disrupt critics by spreading rumors, sowing dissension among them, and making threats (John Stauber and Sheldon Rampton, 1995). They have undertaken attacks on character, put pressure – sometimes successfully – on people's employers, and sabotaged the efforts of people to promote views contrary to the interests of the wealthy and powerful. Powerful interests routinely create phony "grassroots" groups (insiders call their creations astro-turf groups) to mislead the public and hide who is funding and supporting the campaigns. They spend huge amounts of money not to compete on the merits but to dominate the debate and drown the opposition.

Campaigns aimed at manipulating the public agenda do not stop with spending huge sums and mild forms of coercion. We will deal with the exercise of power in the next chapter. Here our focus is on the obstacles conservationists face in the cultural arena: the vastly greater resources opponents of conservation can employ to neutralize or overcome conservation mobilization and the capacity and willingness of opponents to use nondemocratic techniques to shape how important groups in society see the world and interpret their experience. A focus on the cultural arena does not limit us to the media or worldviews; the cultural is entwined with economic and political power. Most media operations are businesses; governments routinely engage in controlling and spinning information, issuing disinformation, and much more. Even though powerful interests are able to influence or dictate the rules of the game, they seldom hesitate to break the law when it suits them.

Cultural Inertia

Human social arrangements tend to reproduce themselves and this includes people's view of the world. Change is incremental most of the time and existing relationships with Nature appear as the only possible way of relating to it. Overcoming this usually nonconscious acceptance takes great energy; a systemic crisis is often required to open opportunities for change.

Over the last 12,000 years, human population growth and growth in per capita consumption – both of which are directly related to loss of biodiversity and wildlands – have been the norm and are mostly seen as desirable. At the beginning of the Neolithic age, there were about 10 million people on Earth. It took from our emergence as a species to about 1820 to reach 1 billion. Today, at 6.6 billion, it takes us only 14 years to add a billion. Only 20% of the Earth's terrestrial surface outside of Antarctica is relatively untouched (Eric Sanderson et al., 2002) and only 4% of the ocean (Benjamin Halpern et al., 2008). Even though many recognize that human-caused changes are at the root of many problems, the pervasive human presence is taken for granted. Although many developing countries in the most recent century have sought to slow growth, the United States issued a major report in the early 1970s calling for population stabilization at 200 million (Commission on Population and the American Future, 1972), and there are many nongovernmental organizations (NGOs) trying to address the issue, there is no strong awareness that reducing human numbers significantly is likely to be necessary to maintain ecological health and biodiversity or to ensure healthy human populations.

Human reproductive urges and security needs certainly are important drivers of population, but population growth has strong cultural and social components. Decisions about fertility are usually made for personal reasons (influenced by the cultural and social); policy decisions likewise tend to focus on the parochial rather than the ecological. High fertility is closely associated with high levels of childhood mortality and other forms of insecurity, which produce cultural prohibitions against birth control, divorce, abortion, or women undertaking roles other than mother (Pippa Norris and Ronald Inglehart, 2004: 233). Economic motives become enshrined in cultural directives ensuring, for example, that there are children to provide agricultural labor and old age support for parents. Children help allay parents' fear of death by providing a kind of immortality (Sheldon Solomon et al., 2004) (Kasser et al., 2004). Children also provide needy parents with objects for control and vicarious living (Christopher Lasch, 1978). The economically powerful have often favored a rapidly growing population because it keeps the price of labor down while allowing for expanded production. Migration can fuel fertility. Ethnic and political competition can generate high birth rates, especially when leaders encourage high fertility or reward it. Ancient and modern states have encouraged high fertility because armies need soldiers and economies need laborers (Bernard Van Praag, 2003; Partha Dasgupta, 2003; Peter McDonald, 2003).

Growth in the human footprint is also due to increasing consumption per capita, which has skyrocketed with the capture of much higher energy subsidies. Personal striving plays a major role here as well. The Neolithic brought a marked decline in nutrition and health from hunting and gathering days (Marshall Sahlins, 1972; Marvin Harris, 1975: 233–55; Richard Steckel and Jerome Rose, 2002); not until well into the modern period did nonelites recover pre-Neolithic stature and nutrition. Since the Neolithic, the vast majority have struggled to escape poverty and billions still do. Much of the world wants to consume like Americans. Consumption is also status driven (Grahame Clark, 1986; Andre Gorz, 1980). For some, there is never enough. Adam Smith (1976 [1759]: 50–1) noted that although wealth gives great power of many sorts – control over goods, over the labor of others, over political and military machines – "(t)he rich man glories in his riches because he feels they naturally draw upon him the attention of the world. At the thought of this, his heart seems to swell and dilate itself within him, and he is fonder of his wealth on this account, than for all the other advantages it procures him." Status is not just an elite phenomenon, however. Consumption is also strongly linked to fear of mortality, and if available is preferred to children and religion (Sheldon Solomon et al., 2004) (Kasser et al., 2004). Consumption, like belief in immortality, building monuments, or having children, is used to assuage the "terror" associated with awareness of death. Their research shows that consumption increases with awareness of mortality. This relationship is not modern, as the story of Gilgamesh, King of Uruk, attests.

The acquisition of goods and so-called wealth is not just personal, but built into the organization of societies. The institutions that organize the division of labor, allocate resources, govern, maintain order, and interact with other societies in war, trade, and peace must be fed. To be stronger than other groups in society or stronger than another society not only gives access to more wealth but also requires more wealth. Probably no form of social organization in human history has been primarily organized around getting wealth than modern capitalism and its imitators (communism or state capitalism as some called it). It is now the globally dominant system, based on the European conquest of much of the world, and does not tolerate alternatives (David Abernathy, 2000; Robert Beil, 2000; Sharon Beder, 2006). It has produced and sustained a series of cultural beliefs embraced by many. John Locke (2003 [1689]: 111–9) believed that human happiness depended on the domination of Nature. David Ehrenfeld (1979) has ably documented the faith that emerged with modern capitalism: endless progress based on the human ability to solve all problems with technology or changes in social organization and the limitlessness of the material world to supply human wants. Progress as growth has found support in religious doctrine (Carolyn Merchant, 1980; David Noble, 1997). Modern economics holds that economies are closed systems, that is, not meaningfully connected to the natural world and physical law; this is metaphysical hocus-pocus under a cover of complex math, not science (Robert Nadeau, 2006: 102–23, 135–45). These views both rationalize the pursuit of wealth and serve as a cultural impetus, providing support for its goodness and rightness. Such views reflect the military power and wealth that growth has conveyed on some groups and societies. For those victimized by this system, growth seems the only way out.

Countervailing views persist. Lao-tzu, William Blake, Henry Thoreau, and count-less others who have not attained a dominant historical voice have remained con-nected to the pulse of life and inspire others. Some societies reject or seek to tame cowboy capitalism. The culture ministers from the developed countries – hardly a subversive group – have expressed major concerns that language, art, and other aspects of distinctive cultures could be destroyed by unrestrained markets (Anthony DePalma, 1998: E1). Only a small minority, albeit articulate and growing, has voiced such concerns about the effects of growth and markets on biodiversity (Brian Czech, 2000; Richard Tucker, 2000; Herman Daly and Joshua Farley, 2004; Robert Nadeau, 2006). The great irony of the last 12,000 years of biologically destructive growth is that "early hunters and plant collectors enjoyed luxuries that only the richest" enjoy today, and "worked far fewer hours for their sustenance" (Marvin Harris, 1977: x).

David and Goliath

Enormous political and economic inequality marks virtually all extant societies. Some groups command enormous resources; these resources enable them, over long periods of time, to mold mythology, inculcate values and control information directly or through the state. Powerful groups in the United States, for example, have helped to make pervasive the story that Americans are a chosen people, destined to impose civilization on unenlightened peoples (Richard Hughes, 2003). In the short term, the resources of these groups enable them to frame and reframe public debates for their benefit. The Bush administration's Healthy Forest Restoration Act became law in part because its supporters had the resources to overwhelm the voices of critics. Supporters' false claims that the bill didn't involve ecologically destructive activities such as logging of old growth or in roadless areas could not be effectively challenged (Jacqueline Vaughn and Hannah Cortner, 2005).

There are many other examples of this imbalance in resources. Political scientist Ronald Libby's (1999) analysis of the defeat of Big Green in California and a humane farming initiative in Massachusetts demonstrates that the capacity to frame or reframe the issues is determinative of outcomes. If the issue is defined as whether states should regulate for sustainability or whether farm animals should be treated humanely, people come down on the side of sustainability and humane treatment. If the issue is defined as whether to impose regulations that will burden business and cost consumers exorbitantly or that will allegedly drive "family farmers" out of business, people will oppose regulation. The tobacco industry spent over $40 million on television advertising portraying a popular US Senate bill to regulate tobacco as a tax bill aimed at lower income Americans (*Free Speech for Sale*, 1999). The ads included toll-free numbers so voters could call their senators without cost or much effort. The industry took another step, promising members of the Senate that if they voted against the bill, the industry would continue the PR spending in their states through their fall reelection campaigns. Bipartisan support for the popular bill was turned around, and it was defeated.

It is not only well-bankrolled campaigns that rely heavily on purchased media that present obstacles for conservationists. Media ownership is highly concentrated. In the United States (and other countries) a handful of global conglomerates own the television networks, cable companies, film studios, music labels, book publishers, and thousands of TV and radio stations, newspapers, news magazines, and other outlets (Bagdikian, 2004). These companies are not just media giants interested in maintaining the growth-oriented social system they benefit from. Many have far flung interests that profoundly affect biodiversity. Can we expect these companies to cover or investigate their many interests such as weapon development and sales, media concentration, nuclear power plants, or massive land developments? General Electric, the owner of NBC, is a felon and a major polluter, and they have pulled no punches in trying to escape responsibility for polluting the Hudson River (John Schwartz, 2003: C1). If these enormous companies don't or won't cover their own interests and crimes, who will?

Media companies are not just looking after in-house interests; most are heavily dependent on ad revenue. Can we expect news outlets dependent on auto and oil company advertising to fairly cover political battles over mileage standards, climate change, or alternative energy? The cable companies that took the $40 million in ad revenue from the tobacco companies did not report their windfall or the lobbying campaign by tobacco companies as a story. The link between cancer and cigarettes was largely ignored by the media until tobacco ads were banned (Bagdikian, 2004: 251–5).

Yes, there are other news sources, but in the United States, for example, these alternative sources are usually print and often have small circulations. About 80% of the US population say they get their "news" from television; only about 30% read a newspaper on a daily basis and 99% of dailies have no competition in their home town. Other nations are more literate than the United States, many are less so, but the dominance of the electronic media has profound effects in the United States and other countries where reading is increasingly a minority affair. It is perhaps ironic that the state-funded electronic media of many European countries do a much better job at their watchdog role than the so-called free US media. In many US localities, the local news outlets are owned by the same company, making it easier for them to avoid controversial issues or to further their own political agenda. Increasingly in the United States, media companies – large and small – are cutting back on news, substituting the much cheaper and more entertaining talking heads who convey very little hard information.

The media are subject to pressure, threats, and manipulation from government as well as advertisers and powerful friends. The US media's failure at the highest levels in recent years is legend (W. Lance Bennett et al., 2007; *Buying the War*, 2007) and extends to coverage of the ecological costs of war throughout southwest Asia and of obtaining oil globally. Journalists often avoid criticism of powerful sources because they are afraid they will be cut off, hurting their careers. Some journalists, either busy or lazy, have grown reliant on PR firms or government to provide tips, ideas and angles, news copy, video, photos, and interviews with VIPs. Too frequently, the information provided is not checked or evaluated. A formulaic approach of presenting "both sides" results in "he said–she said" news, without any independent evaluation of actual behavior. Exxon knew what it was doing when it spent more on PR related to the Exxon Valdez oil spill, than on cleaning up the spill (Aaron Freeman and Craig Forcese, 1994: A31). It's also no surprise that in the United States, there are more PR professionals (about 150,000 in 1995) feeding the media than there are news people (about 130,000) (John Stauber and Sheldon Rampton, 1995: 2–3).

Reporters and editors, who are otherwise conscientious, practice self-censorship. They recognize the sources of most media revenue are advertising or corporate sponsorships. They know that stories critical of these revenue sources, whether actually printed or broadcast, are likely to be viewed by higher ups as evidence of poor judgment. Eighty percent of US reporters polled indicated they had felt such pressure or knew of stories killed as a result (Ronald Bettig and Jeanne Hall, 2003: 30).

Clearly there are many aspects of biodiversity loss and ecological ruin that concern powerful interests and that the media is willing to cover and cover sympathetically. But to the degree that conservation calls for policies that could damage powerful interests, and if they lack powerful allies, they face the same obstacles as other movements: being outspent, ignored, or portrayed in a distorted fashion. There are media outlets that still do hold to high standards of new coverage and recognize their responsibility is not to feed people what they want to hear. But in many parts of the world, they are a shrinking minority.

Technique

Once upon a time the US CIA secretly purchased the film rights to Orwell's *Animal Farm* and *1984* (Frances Saunders, 1999; Laurence Zuckerman, 2000: A15). They bought the rights to make sure the films ended with their own views, not Orwell's. In *Animal Farm*, Orwell's view of capitalists and communists as both oppressive is replaced with a benign view of capitalism. In *1984* the hero is portrayed as defiant to the end, rather than as crushed by the state. Today such actions seem mild, although cultural politics has long since been dominated by tactics other than efforts to persuade on the merits. The combination of resource inequality and ever more manipulative techniques of communication is insidious and presents significant obstacles for conservation.

The practice of PR, as we noted above, extends far beyond spin and cinematic deceit. It includes the often surreptitious business and government collection of information on target audiences to determine their emotional and mental states and vulnerabilities, what symbols, stories, and themes resonate with them, and other information that is important in crafting efforts to obtain desired actions such as buying a product, candidate, economic system, or war. As a Standard Oil company executive put it, we want people to know we share their values, that we want what they want, and that they should join our crowd (Ewen, 1996: 380). It includes illegal data mining and the use of magnetic resonance imaging (MRI) machines for gauging responses to messages. And it goes beyond simple manipulation (this beer, car, candidate will make you, happy, sexy, free or advertisers that will only sponsor programming with themes that encourage consumption). It includes espionage, infiltration, disruption, and harassment of groups, creation of astro-turf groups, outright deceit (lies, withholding information), character attacks, and much else. It may be combined with intimidation and economic or political coercion.

PR has long been practiced, but as a profession it has its roots in the early 20th century and was a response to those mass media outlets performing their watchdog role (Ewen, 1996: 400–1). For all of its effectiveness PR can be successfully fought and conservationists have done so despite being unevenly matched in resources. Much more problematic for conservation and any effort at reform is propaganda. Propaganda is not simply a nastier PR on a grander scale, although it utilizes many of the same tools. If PR says "Buy this product and you will be admired," or "If you vote for Jack, you will feel safer or more virile," propaganda says "Identify yourself with this system, this way of life – it makes you who you are. Without this identification you are nothing."

The Nazi propaganda minister Joseph Goebbels summed up the distinction when he wrote that "Propaganda has no policy, it has a purpose," and "We do not talk to say something. We talk to obtain a certain effect." (James Combs and Dan Nimmo, 1993: 70–1). Propaganda aims not to instill specific beliefs, but to make people dependent on the source of propaganda for their views and values and actions. Propaganda is distinctive in its incessant, long-term, and comprehensive (addressing the whole person) qualities, aimed at establishing and maintaining a dependent relationship between target audience and propaganda source. It surrounds targeted populations with a ready-made worldview that touches profound feelings and greatly enhances the predisposition for reflexive rather than conscious action (Jacques Ellul, 1972: 266). The "certain effect" is supported by target groups for the propagandists' cause in getting and keeping power.

To promulgate, propaganda requires enormous resources. Resources that only governments, ruling political parties or similar entities, or large businesses usually possess. During World War I the US government employed 75,000 orators who delivered 3- to 4-min speeches over the course of two years to 300 million

Americans. These pep talks in support of the war utilized speakers selected to fit with the audiences, varying by ethnicity, geographic region, and so on (James Combs and Dan Nimmo, 1993: 130). The American Medical Association has long beaten back efforts at universal health care in the United States with a variety of tactics, ensuring during one campaign that every waiting room and clinic in the country had colorful posters of a doctor helping a child with language that urged people to "keep politics out of this picture" (James Combs and Dan Nimmo, 1993: 154). In both cases, these efforts went beyond furthering a particular goal and were aimed at fostering deference: we will show you the way.

Propaganda operates at both the psychological and social levels. Jacques Ellul (1972: 146, 180) argues that its seductiveness lies in making the complex simple and not asking people to think. By providing a ready-made worldview – often by taking vague notions targeted audience possess and transforming them into something focused – propaganda reinforces intellectual laziness and discourages reflection. Propaganda also seduces by offering individuals the "opportunity" to experience life vicariously through a leader or cause; great historical events can be embraced without having to take on the responsibilities of decision making and its consequences (at least immediately, anyway).

The dependence forming quality of propaganda acts like other drugs by assuaging anxiety, insecurity, and uncertainty about meaning (Jacques Ellul, 1972: 91–2, 100–1, 159, 176). Like other drugs, it does so without addressing the underlying causes of these feelings, which a grounded worldview would do. Thus, *Advertising Age* editorialized that although poverty, racism, and despair may have contributed to the Los Angeles riots of 1992, the root cause was a lack of community values and voices of restraint (Ewen, 1996: 32). Advertising's role, according to the editors, was to help bring these voices into existence; in short, to generate beliefs that would keep the peace without having to implement the changes needed to alleviate poverty and racism. Like tranquilizer habit, propaganda acts to mask reality and further disconnect people from the world.

Over time propaganda changes people's personalities (Jacques Ellul, 1972: 59, 140, 166–7, 183, 186, 205; Bruce Mazlish, 1990: 249–66). Dependence on propaganda diminishes the capacity for complex and critical thought and for the autonomous recognition and evaluation of patterns in events and information. The capacity for genuine communication is short-circuited. Together with its seductive qualities, erosion of the capacity for independent thought reinforces dependence on propaganda. At some point, any information or message from other sources comes to be regarded as propaganda.

Propaganda is often identified with lies, but effective propagandists only rely on deceit when there is little chance of discovery or it seems less risky than trying to spin the truth (Leonard Doob, 1950: 428). Propaganda at heart is not about disseminating disinformation but destroying the capacity for independent thought and feeling and making people dependent on the balm of familiar voices, symbols, and rhythms. Then inconsistencies don't matter. One day war is about retaliation for an enemy attack, the next day it is about preempting a threat based on an opponent's possession of weapons of mass destruction, and on the third day it is about a heroic mission to build democracy.

Effective propaganda operates at the social as well as psychological level. It depends on similarities among people and reinforcement among people within groups. It thus depends on mass society, the technologies of mass society, and the alienation generated by mass society. Only in recent centuries have large numbers of people become similarly situated by such an extensive division of labor and resulting inequality and differences in experience (Emile Durkheim, 1995 [1912]; Marvin Harris, 1975). To manage mass society, the modern state (and corporations) must further impose uniformity on populations (James Scott, 1998: 9–84). Policy cannot deal with idiosyncrasies. By categorizing and dealing with people based on those categories, powerful political and economic actors reinforce the attributes they use to create the categories. The interactions based on these categories are as real as class, ethnicity, gender, and history or language. Propagandists both create and rely on categories created by others, focusing on some attributes while ignoring others. They are not unique in this regard – political campaigners, advertisers, and pollsters understand the need for categorization and, for example, the use of a group's myths and rituals.

Large-scale social groups provide propagandists many access points, including messengers that enjoy existing reputations with a group (Karen Johnson-Cartee and Gary Copeland, 2004: 113–9). Shared attributes – particular fears, insecurities, desires – allow common messaging aimed at large numbers. The group provides a setting in which individuals can be observed and acted upon by group leaders and peers (Jacques Ellul, 1972: 81). If an individual's commitment starts to wane, they can be encouraged to keep up and not let down their cohorts, rewarded for devotion, criticized for failing the cause, and punished with the loss of material and symbolic perquisites. It has worked well in communist factories, in capitalist corporations, and in political systems of stripes.

Propaganda exacerbates and exploits the strong human tendency to hold more firmly to beliefs they have acted upon (Jacques Ellul, 1972: 28–9; Carol Tavris and Elliot Aronson, 2007: 32–3). By involving people in action, propagandists can strengthen people's reliance on propaganda-supplied justifications. Action also reinforces identification with the group. This cycle of action and justification enables those using propaganda to eventually generate actions that people would otherwise not undertake, such as renouncing their rights, turning over property, killing, and sacrificing their own lives. In the early stages much of the action propaganda instigates may not be particularly important in terms of achieving goals, but it furthers propagandists' purposes: dependence and obedience.

Propaganda is most effective when it coincides with a group's existing perceived material interests or values. No better example of successful anticonservationist propaganda combined with a myopic sense of self-interest exists than the so-called wise-use movement in the United States. Once extractive industries figured out that most Americans had little sympathy for them and probably could not be made to feel much sympathy, they started using their employees and rural communities to gain support. They created groups to portray conservation as anti-people and of trying to destroy rural communities and traditions (David Helvarg, 1994: 15–165). Industry glossed over jobs lost to automation, moving mills and mines off shore, and their long record of poor safety. Many industry-funded groups took on lives of their own, simultaneously providing steady employment to the petty demagogues and giving effective populist cover to industry and their political representatives, intimidating agencies, and threatening violence against opponents. Although many in rural communities and elsewhere have seen through the facade or found wise-use arguments absurd or simple-minded, these astro-turf groups have undercut the credibility of conservationists.

Effective propaganda is made easier where its purveyors enjoy a monopoly on the institutions of communication and can avoid competing messages. As the run up to the 2003 US war against Iraq indicates, media concentration can be just as effective. One does not have to posit a media conspiracy; it is simply that large corporations share common interests which they seek to perpetuate, notably the system in which they hold power and wealth. Moreover the mass media, especially television, provides the repetition and uniformity of message content necessary for effectiveness (Elisabeth Noelle-Neuman, 1973). Network news broadcasts, for example, report the same stories, give similar amounts of time to them, and follow a pattern of problem description, interview with authorities, and end with reassuring statements that action is being taken (David Paletz and Robert Entman, 1981). Cable channels now permit targeting audiences more precisely.

Propaganda relies heavily on the manipulation of symbols that can evoke strong emotion with great economy (Doris Fleischman and Howard Cutler, 1955). Such symbols are the quintessential "picture worth a thousand words." With enough resources symbols can be stripped of their moorings and put to use as doublespeak. A speech justifying government secrecy is made with the Jefferson Memorial as backdrop. The Interior Secretary announces that wildlands will be opened to oil and mineral development with mountains as a backdrop and invokes the government's commitment to stewardship and progress. Goebbels invented the "national moment," in which all citizens were expected to stop what they were doing at a particular time and listen to a short speech, a song, or an announcement (James Combs and Dan Nimmo, 1993: 71).

In a leaked document (Environmental Working Group, 2003) Bush propagandist Frank Luntz offered the following advice on the use of symbols.

- First, put audience suspicions to rest; show them that you share their values.
- Assure your audience that you are committed to "preserving and protecting" the environment, but that "it can be done more wisely and effectively." Tell a personal story. Helpfully, Luntz provides sample speeches, including one that takes responsibility for protecting "sacred places."
- Use words like *safer, cleaner, healthier, accountability*, and *responsibility*. Stay away from *risk assessment, cost–benefit analysis, roll back*, and similar terms.
- Stress a fair balance between economy and environment when making economic arguments.
- Stress local control and the undesirability of faraway bureaucrats making the rules for local communities – describe it as about freedom versus faceless federal bureaucrats.
- Call for solutions that rely on American creativity and technological progress and point out that regulations can stifle this.

This advice avoids almost any discussion of substance and emphasizes use of technique: dangle symbolic generalities before the audience after establishing a rapport by means of a personal story and espousing the audience's values. Anticonservationists know that clarity about their values, goals, and the substance of their policies will not find broad support among Americans. Anticonservationists can only use words that mean one thing to most Americans and another to themselves and try to link their policies to widely shared symbols.

Luntz's techniques include deceit. He advises clients to state there is no consensus on global warming, while acknowledging elsewhere in the briefing book that the scientific debate "is closing against us" but the public doesn't know the debate is closing. The oil–auto–Bush administration effort to stall climate change action worked. Only the action of an international body with global political backing forced a belated recognition of the problem.

Karen Johnson-Cartee and Gary Copeland (2004: 164–76) provide a list of propaganda techniques, including name calling; accusing opponents of propaganda so as not to have to deal with the substance of their arguments; hate speech; attaching an idea or proposal to some authority figure to give it credibility, or conversely attaching an idea to be denigrated to a negative authority; positive and negative testimonials given by respected or well-known persons; overwhelming audiences by providing enormous volumes of purported evidence; manipulation of statistics; false analogies; reliance on secret sources; appeal to audience's belief that they are superior or have superior knowledge. The list goes on. A. Robert Ginsburgh (1955: 214–36) examines the techniques of propaganda delivery and their relative usefulness with particular audiences.

One propaganda technique that leaves conservationists almost too exasperated to respond involves accusing opponents of the very misdeeds the propagandists' cause is perpetrating. Thus, businesses that despoil the natural world for profit accuse conservationists of being in it for the money. The first step in responding is to stop being surprised at the lack of principle in politics. We will discuss effective responses to this technique and others in Part 2, including many that are common to PR and even to legitimate efforts to persuade.

Propaganda is usually intentional, but Jacques Ellul (1972: 69) recognized it is pervasive in all modern societies. Some long-term PR spills over into propaganda as does entertainment made for profit and without intent to create a power base. Over decades Hollywood churned out simple-minded morality tales of conflict between men in white hats and men in black hats that were easily co-opted by cold war propagandists. A press report summarizing negative Scottish attitudes toward nuclear power found that *The Simpsons* cartoon series, along with industry secrecy, significantly contributed (Murdo MacLeod, 2002). Perhaps, some of *The Simpsons* writers had an agenda, but it's safe to assume that generating antinuclear attitudes was not on Fox network owner Rupert Murdoch's mind.

Propaganda can take extreme forms as when it is combined with coercion. Brainwashing, military basic training, and initiation into some religious groups all involve isolating an individual or small group, placing them under extreme duress, often including deprivation of food and sleep, thereby breaking down personality, and enabling its reconstruction in a controlled collective setting. Very severe social crises – invasion, pandemics, famine – can have similar effects, causing personalities to disintegrate and making people particularly susceptible to manipulation.

Fortunately conservationists have some built-in allies in battling propaganda that do not include relying on propaganda themselves. First, conservationists lack the enormous resources needed to engage in propaganda. Only states and other large and wealthy entities such as businesses or religious organizations control such resources. Second, and more important, propaganda fosters the incapacity of people to connect directly with reality; it does not generate greater clarity any more than being punched from the right undoes being punched from the left (Jacques Ellul, 1972: 181). More than anything else, conservation success depends on people genuinely connecting with the world – the living, respiring Earth with its great green forests, stark red-brown deserts, and wild gray-blue oceans.

It is precisely this connection or grounding that is conservation's primary ally. When individuals and groups are grounded in and connected with their emotions and experience, when they define themselves on those bases rather than being defined by others, they are not susceptible to propaganda. Swiss psychiatrist Alice Miller (1983: 43) observed that "our capacity to resist has nothing to do with our intelligence but with the degree of access to the true self. Indeed, intelligence is capable of innumerable rationalizations when it comes to the matter of adaptation."

It is widespread rootlessness that leaves people vulnerable to propaganda and the peddlers of endless consumption (Christopher Lasch, 1978; Roger Rosenblatt, 1999). To the degree that real communities exist – and it is never all or nothing – with their own stories, music, and places, propaganda is ineffective.

Another ally against propaganda is the awareness of its purveyors that it compromises their own survival. Propaganda undermines a good grasp of reality on which survival depends. Universities are an interesting example. Although very much linked to two primary purveyors of propaganda – business and the state – their usefulness to these two requires that they maintain some capacity for critical perspective. The maintenance of this critical function provides real, autonomous space that nurtures grounding and reflection. Thus, a study of Catholic students found that after four years of college they no longer accepted church teaching as dogma, holding instead to more independent views (Tamar Lewin, 2003). This was true even when they attended Catholic universities, but was more pronounced when they attended nonreligious universities. Effective propaganda requires that the propagandists maintain the ability to read their audiences. When regimes are not smart enough to recognize these limits – and often they are not – it undermines their power.

With this in mind, it is a mistake for conservationists to dismiss propagandists as corporate shills or government hacks, though some certainly are. The most effective propagandists believe in their cause (Jacques Ellul, 1972: 24, 241). They do not, however, confuse belief in the cause with belief in what they say, which must constantly change. Propagandists have different tolerances for how far they will go in the service of a "higher truth," although rationalization once embarked upon knows few limits as we noted just above. Thus, Edward Bernays (1928: 9, 1947: 113–4), who made his living as an influential propagandist, embraced his work enthusiastically, arguing it was essential to functional democracy. He said it was the only means by which intelligent men could control disorder and chaos, and guide the many who cannot guide themselves. It was a "transhistoric" " . . . requirement, for those people in power, to shape the attitudes of the general population." (Stuart Ewen, 1996: 11). The engineering of consent was essential to ruling (Edward Bernays, 1955: 4). An odd view of democracy.

Jacques Ellul (1972: 126–7, 135) was a fierce critic of propaganda but agreed with Bernays in one sense. He recognized that government leaders, whatever their personal predisposition, have little choice but to use propaganda. In the absence of genuine and deeply held values that unify society and with public opinion frequently ill-informed and always shifting, leaders must use propaganda if they are to hold a mass-based society together. Leaders with democratic predispositions may feel inclined to limit its use, but self-restraint has seldom proved an adequate safeguard against abuse. It is no protection against the damage propaganda does to personality. Conservationists must be both realistic and guarded about what they can expect even from political leaders who are friends.

The obstacles conservation faces in the cultural arena are significant. Propaganda, unequal resources, and people's tendency to conform their views to long-standing social relationships and relations with Nature put conservation at a serious disadvantage. To think that the truth of experience or the desirability of protecting the biological basis of life will somehow automatically win out over propaganda is naive and probably delusional. The Vatican took 300 years to apologize for its persecution of Galileo; he spent the last years of his life under house arrest after having been threatened with torture and forced to recant his blasphemous heliocentric views. Giodorno Bruno was not so fortunate and was burned at the stake. General Motors (GM) was convicted in 1949 – along with other corporations – of secretly buying and dismantling over 100 public transit systems in order to replace electric trolleys with GM buses and encourage automobile use over public transit (Jack Doyle, 2000). But GM's conviction did nothing to reverse the damage done. For all of this, the mobilization of people starts in the cultural arena and the opportunities for mobilization are many.

3 Them: power

Men ought either to be indulged or utterly destroyed, for if you merely offend them they take vengeance... (I)t is not reasonable that he who is armed should yield obedience willingly to him who is unarmed, or that the unarmed man should be secure among armed servants.

Niccolo Machiavelli, *The Prince* (1984 [1532])

What country can preserve its liberties, if its rulers are not warned from time to time that people preserve the spirit of resistance? Let them take arms... The tree of liberty must be refreshed from time to time, with the blood of patriots and tyrants.

Thomas Jefferson, letter to William S. Smith (1955 [1787])

When the runoff from a foreign mining company's operation in Papua New Guinea poisoned drinking water, killed fish, and took land with paltry compensation, it sparked protests. When protests didn't change things rebellion ensued. The government – heavily dependent on royalties from the mine – sent in the army and put down the rebellion (Michael Klare, 2001: 197). French President Francois Mitterrand approved the bombing of the Greenpeace ship *Rainbow Warrior* while in a New Zealand port to prevent it being used to protest nuclear tests in the Pacific. A crew member was killed in the blast. After failing to hide its involvement France successfully brought the threat of an European Union (EU) wool boycott against New Zealand to gain the agent's transfer to a French jail from where he was quietly released (David Lange, 1995). For millennia Egypt has "protected" the upstream Nile for its own use, using threats against upstream countries that propose dams and supporting rebellions to discourage the investment needed for such dams (Michael Klare, 2001: 148–54). Several Thai environmentalists, including monks, have been murdered for protesting illegal logging and both businessmen and officials have been implicated (Asian Human Rights Commission, 2005). In 2007 four policemen in Honduras were charged with murdering conservationists who were protesting illegal logging (Amnesty International 2007). Murders in the Brazilian rain forest over poaching and illegal logging have occurred all too often and many remain unsolved. Shell Oil paid for Nigerian soldiers and police used to lethally repress those protesting toxic pollution and other abuses (Madelaine Drohan, 2003: 163–88). Although the US government has not been implicated in killings of US conservationists – as it has members of the Black Panther Party and American Indian Movement (Jules Boykoff, 2007) – it has routinely violated civil liberties, used intimidation, harassment, provocation, and tolerated illegal activities against conservationists by third parties (David Helvarg, 1994, 2004; Michael Isikoff, 2005: 6). The United States and other democratic governments have a long history of repression and violence against movements for labor and civil rights, suffrage, and others that have challenged the status quo (see Chapter 19). These same governments have overthrown democratically elected governments, assassinated foreign leaders, launched wars directly and through proxies, engaged in kidnapping, torture, and political disruption and much else in order to gain and keep access to resources and markets (Chalmers Johnson, 2001, 2004, 2007; Kevin Phillips, 2006). One-time Wall Street investment banker and Under Secretary of State George Ball stated with unusual candor that the purpose of American foreign policy was to make sure there were no obstacles to America doing business around the globe (Controlling Interest, 1978). Authoritarian regimes are under even fewer constraints when it comes to the use of power. Despite the high material and political costs of violence, both democratic and authoritarian regimes rely on it when they have felt their position seriously challenged.

Inertia, being outspent, and competing with propaganda are clearly not the only obstacles confronting conservationists. The most powerful opponents of conservation – those whose power and wealth rests on the transformation of Nature into commodities – have ready access to the tools of violence and the threat of violence. They may hold state power directly, exercise extraordinary leverage over it, or act directly where states refuse to act or are otherwise weak. Direct action is most common in weak states and those with "lootable" resources, but by no means is it confined to them (Collier and Hoeffler, 1999; Michael Klare, 2001; Madelaine Drohan, 2003; Ann Hironaka, 2005). Often the only difference between private and state perpetrators of violence is a change of clothes – the police pull a sheet over their uniforms or soldiers change into death squad garb.

That resource wars are common should not be a surprise to anyone conversant with politics. As Michael Klare (2001) has documented, states battle over oil, minerals, water, cheap labor, markets, and much else. Energy is particularly important because it acts like a growth hormone for economies and political power. Much of human history is understandable as the search for and fight over energy subsidies – domesticated animals, slaves, hydropower, good soils naturally renewed, coal, oil, and uranium (David Johns, 2002: 12–3). The capture of energy enhances the capacity to remake the Earth and to capture more energy, materials, and markets. Resource wars are not just fought between countries, however, but within countries – between groups seeking access to and control of resources, and between such groups and others seeking to preserve "resources."

Conservation's opponents

Is violence against conservationists inevitable? The work of a few out-of-control operatives? What can conservationists expect?

Conservation nongovernmental organizations (NGOs) usually aim to influence society's decision makers concerning the exploitation or protection of wild places or creatures. Most often the state is their target although societies have other important centers of power such as big businesses. These power centers can be usefully thought of as part of the state (states are not monolithic), whether the relationship is formal as with the Chinese Communist Party or less formal as with what former US President Eisenhower called "the military-industrial complex." In instances where states are weak conservationists may aim to influence groups that act or try to act in its stead. Whatever the path of influence – insider lobbying, outsider pressure, the selection of officials, boycotts – conservationists are in the business of bringing pressure, though some don't see it this way.

The most important opponents conservationists confront are not other nonelite interest groups, but many of the very decision-making institutions they are trying to influence. Institutions may be hostile to particular conservation projects, to conservation generally, or to any NGO seeking to increase its capacity for influence, even though the NGO does not seek to take power. The power and wealth of decision-making institutions, whether state, business, or another sort, rests ultimately on subduing and exploiting the natural world, much of which conservationists want to protect from exploitation. An institution's direct dependence on resource exploitation is usually an important factor in the strength and centrality of its opposition to conservation claims.

The decision-making institutions that conservationists seek to influence, then, are not neutral. Even democratic states are not waiting on competing interests to persuade them on the merits of issues. Officials are responsive to their own interests, to the particular coalition of interests that put them in power and on which they depend to keep power, and to the requirements of maintaining the system of power relationships from which they benefit. These institutions – state, business, and theocratic entities – take sides and they have an unmatched capacity for coercion and violence. Confronting such institutions is unavoidable if conservationists seek to protect what is needed to avoid mass extinctions. Some conservationists do not see an essential contradiction between continued growth and conservation, either because they have limited goals or think "ecological modernization" is the future. The circumstances that have allowed rich countries to protect some reserves and clean up pollution are not repeatable, that is, by transferring their dirty industries and surplus population elsewhere or by protecting forests at home by cutting them down elsewhere. Moreover, the damage caused by rich countries with these transfers is largely irreversible. Even if, for argument's sake, growth and conservation are considered compatible overall, humans usually seek the most biologically valuable lands and waters for exploitation and they are the ones most important for conservation. To avoid being overwhelmed by much more powerful actors requires practical intelligence and knowledge of the specifics of how institutions and whole societies are dependent on transforming Nature, and how they use violence.

The sources of conflict

States act to contain internal conflict, legitimate themselves and the social order, defend themselves from other powers, and they seek to secure the wherewithal to achieve these goals. For some time modern states have understood that maintaining their strength relative to other states requires a growing economy. A growing economy also enables domestic peace. If a state fails to act effectively in any of these areas, disorder and serious challenges are likely. Although economic and political leaders sometime behave like looters, states usually attend to these tasks. Although observers generally regard democracy as better for the "environmental" cause (not the same as the conservation cause), John Dryzek et al. (2003: 3–6, 79) note that when any group interferes with one or more of these tasks, a state will treat it as a threat regardless of its form of government. States have little choice. Thus, when conservationists pursue policies that seek to limit access to lands, waters, animals, or plants *necessary* for economic activity states will oppose these actions. When alternatives exist, for example, importing foreign logs to keep markets supplied, conflict may be deferred. States have also created Nature reserves because factions of the elite have wanted to do so based on their own values. Although some interests may have wanted to exploit the reserves, their exploitation was not deemed systemically necessary. When resources are seen as economically necessary for growth and other central state functions they are taken, as demonstrated by Hetch Hetchy in Yosemite Park and tropical forests. Norway committed to a big reduction in greenhouse gases but backed off when the policy threatened to harm its industry (John Dryzek et al., 2003: 172).

Nature protection that constrains growth is not the only activity that gives state leaders' anxiety. NGO actions that undermine a state's legitimacy and generate internal or international conflict will be seen as threats and trigger state responses. On the other hand, if meeting conservation claims are constructed to enhance a state's legitimacy, they are more likely to receive a sympathetic hearing so long as the growth impact is minimal.

Although states' actions in fostering material consumption, obtaining energy resources, and defending themselves result over time in destroying their biological underpinnings, they are nonetheless captive to them. States must be fed and the international state political-economic system is such that if they do not grow their relative power declines. Decline is to be avoided at all costs.

Business entities – whether privately or state-owned – are also captive to imperatives given existing forms of social organization. These imperatives include profitability, which is tied closely to growth, access to resources to enable growth, and the autonomy to make decisions that serve both growth and profit. Growth is also linked to being able to utilize profitable technologies that are scale dependent and to compete with or fend off takeovers from other firms. Firms compete internationally for markets and resources directly and rely on their states – and increasingly the international trading system created by the most powerful states – to run interference for them. (Recall George Ball's statement above.)

Although much profit and growth occurs in the financial sector such as banking or trading in currency, futures or "structured investment vehicles," the material economy remains foundational and is rapidly growing. People do not eat or drive money; they eat meat, and drive steel, rubber, plastic, and glass. Profits and growth depend on consuming ever greater amounts of resources – minerals, water, food, energy . . . and habitat. Both population and per capita consumption are expanding. In 2008 China had 7000 steel mills, a doubling since 2002 (Clifford Krauss, 2008: C1). Metals and grain consumption have been growing rapidly not because of typical boom–bust cycles but because of Chinese and Indian industrialization and growing consumption in many other countries. Per capita meat consumption has doubled since 1961, with absolute production rising from 71 to 284 million tons in 2007 (Mark Bittman, 2008: Wk1). Factory farming has made this possible, which is energy, water, and grain intensive, resulting in destruction of tropical forests, extensive water pollution and greenhouse gas emissions.

We will examine the economic dynamics of growth more in Chapter 4; for now we need only note that businesses drive and are driven by growth. It is not a matter of choice for a firm, and it puts them at odds with many of the goals of conservationists. Economic actors, like states, have significant direct coercive resources at their disposal to deal with obstacles to growth that they encounter (Madelaine Drohan, 2003; Sharon Beder, 2006). Whether enterprises are state-owned or privately owned *as a whole* they enjoy unprecedented access to state's coercive resources and in this sense can be considered functionally part of the state. State and business, each for their own reasons and for shared reasons, seek the transformation of the natural world into commodities and utilize not just persuasion and propaganda, but coercion and force in that pursuit. Conservationists are frequently in the way.

The United States as a case

A brief look at the way in which US economic and political institutions are integrated in the pursuit of growth is instructive. Although the United States is not typical of developed countries in some respects – for example, it has no strong social democratic tradition – it is in others. It is relatively open with a well-organized conservation movement and explicitly pursues growth in material consumption. Its decision makers are not neutral on growth. Despite the US conservation movement's important achievements and claim of broad support from the public, it has made little dent in consumption or population growth.

The US Constitution, which defines the country's political structure, reflected the values and interests of its authors and was designed to aid growth. It leaves the economy in the hands of private entrepreneurs, gives the central government power to aid accumulation, restricts government at both levels from imposing certain impediments to growth, and protects private property (Article 1 §§ 8, 9, and 10, and Amendment 5). The charter reflects the view that increasing material consumption, access to resources, and a growing population were important to prosperity, human well-being, and national power.

These views have been reflected in constitutional interpretation (David Kairys, 1998: 445–637) and in the political culture of elites and much of the rest of the country. The pursuit of growth and its necessaries have been cloaked in various missionary rationalizations from Manifest Destiny and the Open Door policy to democratization, anticommunism, and new world orders (Richard Hughes, 2003). Obtaining access to markets and resources deemed vital to US interests was conveniently seen as inseparable from bringing order, enlightenment, and democracy to peoples in need such as Native Americans, Mexicans, Filipinos, Chinese, Vietnamese, Nicaraguans, and most recently Iraqis (Richard Barnet, 1990; Kevin Phillips, 2006). It was obvious that "the wealth that for untold ages has lain hidden beneath the snow-capped summits has been placed there by Providence to reward the brave spirits whose lot it is to compose the advance-guard of civilization. The Indians must stand aside or be overwhelmed" (Bighorn Association, 1870). Think tanks, trade associations, and countless ad hoc groups espouse the centrality of growth and rationalize policies pursuing it (G. William Domhoff, 1998: 71–123).

The centrality of growth to US decision makers (who control massive coercive resources) is revealed by looking at who they are and how sectors of decision makers are related. Political economist Thomas Dye (2002: 11) identified 7300 people who lead the institutions that make major policy decisions. Less than 300 people are top government officials, among them the President, Vice-President, key cabinet officials, the Congressional leadership and committee leaders, justices of the Supreme Court, the joint chiefs of staff, the Federal Reserve's Board of

Governors, and key White House advisors (Thomas Dye, 2002: 56–7, 73–96). These elected or appointed *public* officials are far outnumbered by the 7000 decision makers who are unelected and sit on corporate boards, are partners in powerful law forms, lobbying firms, or brokerage houses, or who lead prestigious think tanks and major research universities (Thomas Dye, 2002: 11 et seq). The vast majority of the important decisions that affect people's lives and Nature are made by large, profit- and growth-driven businesses: where to build factories, what to produce, whether to pollute, whether to pay living wages, how to manage land and natural "resources", and which political policies to pursue and leaders to support. Of these 7000, 4300 people sit on the boards of corporations that control two-thirds of banking and insurance assets, one-half of industrial assets, and one-half of communication assets and utilities. Of the 200,000 industrial corporations in the United States, the 100 largest firms control almost 75% of all of industrial assets; the top five control 28%. Nearly half of all wealth and nearly 60% of all business interests are owned by one-half of 1% of the US population. Of 5 million corporations that file US tax returns, 22,000 receive 70% of all revenue; 500 receive 60% of all revenue; 500 corporations account for 70% of world trade.

Not only does big business directly make decisions affecting Nature and people, but also it exercises enormous influence over public decisions. It does so by way of its social position, through its role *inside* the state, and through funding candidates and political parties.

Concentrated ownership and control of business gives its leaders unparalleled bargaining power with government, regardless of who holds office (Charles Lindblom, 1977), although influence is greater for businesses with office holders they have helped place in power. Big business alone, Linblom concluded, could say to both political parties, to Congress and the President, give us the policies we want or we will move operations off shore; we will tell the people you are hurting economic growth or we will sabotage your programs. Only they could put their arm around an elected official's shoulders and say, "Look dear friend, we all want prosperity, and in order to get it we must have things this way." Senator John McCain noted that the threats are never voiced, but the implication is clear (Moyer, 1999). Big business not only has more resources than opponents but it is easier for them to mobilize their resources (Michael Kraft and Sheldon Kamieniecki, 2007: 13; Robert Duffy, 2007: 83). They have the resources to attend every committee hearing, to draft major tax and spending bills, and to be present and involved with the rulemaking process that follows the passage of legislation (Gary Bryner, 2007: 143–4). Laws usually give agencies significant discretion and businesses are masters at turning agency rules that apply the law to their benefit. Business also controls much information that Congress and agencies need to govern (Gary Cogkianese, 2007: 204).

Big business's bargaining power with officials is reinforced by the widely held view that government, not business, is responsible for economic well-being.

Political leaders well know that voters hold them accountable for economic performance. Business leaders understand political leaders need their cooperation to meet expectations. US President Kennedy summed the result up nicely: "This country cannot prosper unless business prospers. . . . We cannot succeed unless they succeed" (William Grover, 1989: 1).

Powerful interests do not just lobby from outside government. They are inside the government. Many agencies, like the US Forest Service or the National Park Service, were the result of lobbying by powerful private interests who wrote the mission of agencies, influenced their developing culture, established close working relationships, and shared personnel (the revolving door). Pinchot's vision of "practical forestry" (industrial forestry for long-term profit) triumphed over both the "scientific forestry" of botanists and the Muir's vision of wilderness, due in large part to the influence of industry (George Gonzalez, 2001: 30, 36). The Forest Service kept public forests mostly off the block prior to World War II, keeping timber prices higher; following World War II, when demand for lumber increased sharply and with private forests depleted, the public forests were opened up to industry. The creation of National Parks and the National Park Service in the United States converted otherwise economically worthless lands into money-making enterprises for railroads and other developers (George Gonzalez, 2001: 52, 58). The National Park Service's professionalism served the interests of "responsible" tourist development, and was opposed to both preservation and allowing local business interests to impose development willy-nilly.

Most elected and top appointed federal officials are from wealthy backgrounds or have a close association with the powerful private institutions, that is, big business, major law firms, top think tanks, and universities (G. William Domhoff, 1998: 241–96; Thomas Dye, 2002: 55–96). Those who have made long careers in government service also have long and intimate working relationships with the business community (Thomas Dye, 1995: 88–94; Thomas Dye, 2002: 81).

The US electoral system provides another major point of access for powerful private interests. Although elected officials need the support of the voters, they need money to obtain the support of voters. The irony of party reforms in the 1970s is that money has come to play a bigger role in elections because of greater reliance on media and experts. Over $4.2 billion was spent on the 2004 US federal elections and that amount was surpassed in the 2008 election cycle with weeks yet to go; estimates are that $5.3 billion will be spent in 2008, including $2.4 billion on the presidential race (Center for Responsive Politics 2008). About half of eligible Americans regularly vote at the ballot box, but only about 4% contribute to campaigns, with most giving less than $200. The 0.25% of voters that give over $200 account for 80% of political contributions (G. William Domhoff, 1998: 217–8). Thus, from 1992 to 2004 environmental and conservation political action committees (PACs) contributed just under $6 million to candidates. Forestry, mining, chemical, oil, gas, utilities, and auto interests contributed over $156 million (Robert Duffy, 2007: 70–4). This number doubles if contributions to the parties are counted.

It is not so much that money changes candidates' minds, although it certainly does on issues of lesser importance to them. Rather, money is important in obtaining victory for candidates most sympathetic to the donor and in maintaining access to all officials. Many of the biggest contributors give to both parties and even both candidates if a race is close. There are candidates and elected officials with integrity, but even the best learn to go along to get along on some issues; they are trapped by the tyranny of small decisions and backscratching, group think, the need to define partisan distinctions, and avoid needlessly taking on powerful opponents. Some officials thrive on a good fight with a tough opponent, but all generally want to hold onto power in a system where growth is a central value and interest.

What happens when the interests of those with the money and voters collide? That is best answered by asking several questions: Who watches the performance of officials most closely? Who has good sources of information about official performance in real time (remember where most people get their "news")? Who is in a position to bargain with candidates and officials? Who has the best access to officials? Who has the most influence over defining the issues? How long are elite memories versus voter memories? How close is the next election? Voters, and especially unorganized voters and nonvoters, usually come in second (Murray Edelman, 1964; William Greider, 1981). Sheldon Kamieniecki (2006: 207–9, 254) finds that industry has been extremely successful at blocking conservation through leveraging economic power and hefty campaign contributions, not just successfully framing the issues. The auto industry and energy industries have kept clean air laws weak because of their enormous power (Walter Rosenbaum, 1998: 189–203; Jack Doyle, 2000; George Gonzalez, 2001: 111–3).

When conservationists win

The conditions I have just described suggest that conservationists don't stand a chance, yet the United States has parks, the Endangered Species Act, and a political culture that is supportive of conservation so long as it does not require much material sacrifice. There are a range of political conditions that have contributed to conservation successes.

Conservation victories occur more easily when powerful interests do not oppose them. Not all lands or waters conservationists want protected are of current or near-term interest to business. Sheldon Kamieniecki (2006: 250, 257) argues that business ignores about 80% of legislation with environmental consequences.

Some sectors of the business community favor protection when it benefits them. Wealthy businessmen supported the creation of a Redwood National Park, but also limited its size to keep adjacent property for economic uses (George Gonzalez, 2001: 79–94). Conservationists failed to gain a larger park. The resulting park was (and remains) too small to be biologically viable; moreover, it included extensively cut over areas purchased from timber companies at above market prices.

Some wealthy individuals support conservation projects for personal reasons. Included in these is Doug Tompkins, who, through his Foundation for Deep Ecology, supported path-breaking NGOs that transformed conservation in North America and globally. Tompkins, Yvon Choinard, and others who have a profound commitment to the wild have been instrumental in purchasing large tracts of ancient forest in Chile and Argentina. Others have been less visionary, supporting projects that do not stir conflict or run contrary to important economic interests.

Opponents of conservation are sometimes caught napping, as with the passage of the Endangered Species Act or National Environmental Policy Act. Given their efforts to weaken or dismantle these laws since their passage it's safe to assume that had they realized the consequence of these laws they would have vigorously opposed them.

Neither the state, nor big business, nor the two considered together are monoliths. Not only is the US state quite fragmented but parties, legislators, and presidents are backed by differing sectors of the elite. Although they are usually united on protecting their wealth, power, and position, they differ on many policies and priorities because of their differing interests. When the elite is divided, and especially when one sector is backing a conservation project or policy, it is easier for conservationists to prevail. These divisions have helped create a shift in climate change policy, pass wilderness legislation, and much more. When the elite is united they generally prevail (George Gonzalez, 2001: 95–114; Judith Layzer, 2007: 120).

Crises tend to provide states with greater autonomy from powerful interests, as when Franklin Roosevelt was able to achieve major economic reform as a result of a deepening depression. Crises generate this autonomy because they often divide the elite and the state needs greater cooperation of nonelite groups – either because they are mobilized and making claims that cannot be ignored, or because their sacrifice is needed as in war. (The powerful do not fight their wars.) The capacity to exercise greater autonomy and think longer term does not mean it will be exercised.

Other cases

In authoritarian regimes powerful interests may so thoroughly dominate the state that it rules at their behest rather than simply on their behalf, requiring significant resources be devoted to repression and the likelihood of high levels of corruption. Conservation is likely to be a low priority unless unusual circumstances obtain – it's more of a money maker than resource extraction or the autocrat has a soft-spot for wildlife. Neither are good bets.

In many western European states nonelite interests are much better organized than in the United States so that the power of business has a heavier counterweight (e.g., much higher levels of unionization, International Labor Organization, 1997). Money is also less important in campaigns. These circumstances, combined with parliamentary systems of government that enable smaller parties to play a signifi-cant role, have provided political opportunities not available in the United States. Environmentalists more than conservationists have been able to take advantage (Neil Carter, 2007: 48–142). John Dryzek et al. (2003) point out, however, that states vary in their openness to conservation concerns and inclusiveness of NGOs, with a mix of results. Inclusion makes achieving some goals easier, but tends to co-opt groups, blunting their ability to achieve larger goals. Exclusion can cut both ways as well, and parliamentary systems can also result in being shut out of influence for long periods, for example, Thatcherite Britain.

The exercise of power

Protecting species and wild places brings conservationists into conflict with those who want them as fodder or resources. These interests include the most powerful actors in society, such as big business and the state – the very decision makers conservationists must influence. Should conservationists gain serious traction or otherwise be seen as a threat to growth they will become targets for coercion. The exercise of power – the threat of violence or its use – takes many forms.

Economic violence and coercion

Depending on the country economic entities have varying authority and power to make decisions that have a profound effect on people and the natural world and this includes violence. In the normal course of business habitat is destroyed, people and other creatures are displaced or destroyed; the course of life is unilaterally changed through investment and management decisions, and decisions about technology and pollution. Cost-cutting and poor maintenance at Union Carbide's Bhopal plant lethally poisoned thousands. The single-hulled tanker Exxon Valdez ran aground and spilled oil that killed thousands of animals and continues to negatively affect the ecosystem. Businesses cripple entire communities when they close plants or move them, often to escape environmental regulation.

Businesses do not just collaterally destroy and injure places, creatures, and humans in the pursuit of profit and growth; they inflict injury and destruction deliberately on those deemed obstacles. Efforts to coerce, punish, or reward people by controlling jobs or effective access to food, medical care, shelter, electricity, and transport, by raising prices, withholding and awarding contracts, bribery, and blacklisting, are common and effective. So is conditioning employment and other economic benefits on acquiescence to biological degradation and threats to health. Businesses use their ability to bestow and withhold economic benefits to obtain deference to their political views and the accumulation of centralized economic decision making in their hands as well.

States also rely on economic coercion. Government-owned or -controlled enterprises use the same tools that private enterprises do to reward, punish, and otherwise discipline to gain compliance from individuals, groups, communities, and even entire nations. They have other mechanisms as well, such as the ability to cut off or manipulate terms of trade, loans, access to resources and technology, and so on. The crude brutality of gunboat imperialism is far from dead, but much international economic coercion is exercised through the International Monetary Fund, World Bank, World Trade Organization, G8 decisions and policies, UN sanctions, and via bilateral economic pressure. These economic hammers are used to pry open access to raw materials, enforce unequal terms of trade, privatize publicly owned natural and human-created capital, buy off leaders, and construct ecologically disastrous projects from dams to roads (Robert Beil, 2000; Richard Tucker, 2000; Michael Goldman, 2005; Sharon Beder, 2006). Through these means pollution is transferred from rich to poor countries and carrying capacity from poor to rich ones.

Some efforts to protect the natural world rely on some of these mechanisms, for example, debt for Nature swaps. The danger in using such tools is blowback.

Political violence and coercion

When most people think of coercion it is the political forms and institutions that come to mind: physical intimidation, beatings, arrest, imprisonment, torture, maiming, and individual or mass killing by police, armies, mercenaries and other hired thugs, and death squads.

Although political in form, the use or threat of armed force is a tool of economic entities not just states or other political actors such as rebel groups, warlords, and the like. Indeed, the lines can blurred, as when businesses work hand in glove with political actors. Often large businesses employ their own substantial security forces that function like police or military forces and act with the same or greater impunity, as multinationals have done in Africa (Madelaine Drohan, 2003;

Singer, 2008) and elsewhere. Smaller businesses may rely on more ad hoc forces, as Amazonian ranchers and loggers have in intimidating or killing those resisting deforestation such as Chico Mendez and Sister Dorothy Stang. Businesses in much of the world have a long history of violence against labor and the same strategies and tactics are applicable to conservation. That includes instigating violence by others, as when companies point the finger at conservationists for causing job losses (David Helvarg, 1994, 2004).

Private companies generally prefer to rely on governments or those with claims to government authority to bring force to bear, though in some parts of the world authorities are effectively absent. Governments are preferable because they provide a veneer of legitimacy that direct action by business lacks. States generally claim only they can use violence legitimately, but other entities such as those in rebellion against states also claim the right to use violence. Nothing can compare with the capacity of modern states to inflict violence on human populations and Nature. They control the world's armies, navies, air forces, police forces, prisons, and courts, nuclear arsenals, and employ most mercenaries. They have generally not hesitated to use violence, despite its cost, to gain access to resources or deny competitors access be they other humans or other species. Imperialism's roots go back to the first urban center's efforts to dominate adjacent hinterlands and towns. Modern imperial wars imposing direct colonial rule or economic inequality, and wars among imperial powers themselves, continue down to the present. Other tools in the repertoire of states seeking to guarantee growth include coups, civil wars, state terror, destabilization, electoral and other political interference, and destruction of opponents' capacity to utilize resources, for example, destroying dams, power stations, species (e.g., the bison), or making the resources unusable.

Not all wars and violence are growth- and resource-driven, of course. Often the rationalizations for economic violence – religion, nationalism, civilizing crusades, fear – take on lives of their own. Violence breeds blowback and countries can become trapped in feuds. Leaders placed in power by economic elites can turn out to be uncontrollable, with self-aggrandizing fantasies that lead to violence. Incompetence on the part of those using violence, or the resistance of its victims, can frustrate the purposes and goals of the former in using violence. In other cases economic and political actors get at cross purposes as when the Marxist of government of Angola sent Cuban troops to defend a US oil company it had granted a concession to, from rebels armed and funded by the United States (Madelaine Drohan, 2003: 195).

The point is that political and economic elites in democratic and authoritarian regimes alike have long been willing to use violence against each other and segments of their own population to gain control and use of resources deemed necessary to their interests. Conservationists have not been immune to such violence. Neither have they been a major target. That may be because their demands are mostly too timid to be considered much of a threat or that they can usually be defeated without violence. Certainly the social position of many conservationists – middle class with ties to elites – provides some protection. The mobility of capital also provides a safety valve *in some cases*. In the face of opposition it's often easier to obtain resources elsewhere, at least for the time being. What would occur if conservationists effectively demanded that remaining healthy forests be protected *across the globe*? What if the last grizzly habitat was wanted by energy companies and conservationists gave an emphatic no? Would those willing to kill in pursuit of "full spectrum dominance" just walk away? What would happen if conservationists no longer tolerated the slaughter of wildlife and extinction the way countless South Africans decided to no longer tolerate apartheid?

The issue for conservationists is not whether violence will be threatened and used against them, or against anyone who might threaten what Gary Snyder (1990: 5) called "the growth monster." Rather, the matter before conservationists is understanding how violence is used, when it might be expected, and what have been successful strategies to avoid or counter it without compromising goals. (More on this in Chapter 19.) The problem of growth's destruction of life on Earth and the willingness of states and businesses to use violence in its defense is not about individuals, but a political-economic *system* organized on the basis of unlimited growth and profit. Individual leaders make a difference, but they do not rule in a vacuum. They aim to ensure the continuation of the societies they govern within a framework – one that protects their interests and those who put them there. This raises two important questions: To what degree can conservationists turn decision makers to their goals? And, to the degree that conservationists gain influence with the state, to what degree are they going to be willing to use its tools on behalf of the natural world?

4 Why we act – from the double helix to world systems and sunspots

The body is the tree of perfect wisdom.

Zen aphorism

... (T)he fundamental problem ... is not why a hungry man steals but the exact opposite: Why doesn't he steal?

Wilhelm Reich (1972 [1934]: 294)

Men make their own history, but they do not make it just as they please; they do not make it under circumstances chosen by themselves, but under circumstances directly encountered, given and transmitted from the past. The tradition of all the dead generations weighs like a nightmare on the brain of the living.

Karl Marx, *The 18th Brumaire of Louis Bonaparte* (1979 [1869]: 103)

Hikers in the Rocky Mountains from Yellowstone Park north now and then encounter a grizzly bear. When a hiker or bear is surprised anxious moments ensue, especially in the fall when bears are trying to store up fat and can be a bit short-tempered. Grizzly country generates an alertness and engagement with the world that is lacking in other places, and some hikers thrive on the intensity. Only rarely does an encounter result in injury to a bear or hiker. It's not the same when hunters encounter grizzlies. A dead bear is often the result. What accounts for the difference? The presence of firearms in one case? Different attitudes toward Nature or bears? A disposition to see a serious threat when it is slight?

On the other side of the planet people and tigers encounter each other and sometimes people become prey. Despite these lethal encounters in many parts of India a tolerance for other animals precludes the typical North American reaction – that any creature threatening a human must be killed. Indian sensibilities do not run in the direction of cowboys taming the wild, though historically Mogul rulers and Indian princes regarded tigers as trophies of manliness.

In the United States grizzlies are relatively safe in protected areas. One of the strongest laws in the world protecting rare plants and animals is in place and enforcement, though underfunded, is somewhat reliable. Should a bear find itself somewhere it's not wanted, however, histrionic citizens, zealous police, and wildlife officials all spring into action as if Western civilization were threatened. The bear will be pursued, trapped, relocated, or killed. Despite this populations of grizzlies are recovering a bit. In India tiger population is plummeting. Laws are good, but law enforcement is uncoordinated, spotty, and less than rigorous in many places. Prosecution is also hit and miss. Protected-area boundaries offer no obstacles to poachers, the current primary culprits in the loss of tigers.

India and the United States are strikingly similar in one way, however. The decline of grizzlies and tigers over the longer term is the result of a cancer eating away at their habitat: industrial ranching, agriculture, oil and gas development, and recreational settlements in the United States and expanding subsistence agriculture and commercial logging in India. People have invaded, stolen, and transformed the habitat grizzlies and tigers need.

What accounts for these differences in individual, group, and institutional behaviors? What explains the gaps between the laws on the books and enforcement? What explains enforcement priorities? And what explains the similar overarching societal trends of more people and fewer bears and tigers?

It is behavior that counts. It is behavior destructive of biodiversity that conservationists must change. First by mobilizing people to act on behalf of life, and then by bringing that action effectively to bear on institutions to change institutional behavior. To achieve this conservationists need a good understanding of the factors that shape behavior at all levels. This is not the understanding of the medical assistant who knows how to give a shot but that of the doctor who understands host and disease organisms, the dynamics of disease transmission, and how the vaccine works on disease organisms and host individuals and populations. Typically advocates ignore the need to address the range of factors that bring about mobilization and institutional change. They focus on the psychological factors and ignore the structural factors. They focus on interests and assume behavior rests on a calculation of interests and ignore the emotional factors.

Human social and political behavior is shaped by a number of factors that can be thought of as "mechanisms" that regulate behavior. These factors interact in complex ways. Humans have a shared genetic endowment that impels behavior: a need for water, food, warmth, to belong to a group, to love, be loved, and be recognized, to make sense of the world and who we are, to find a purpose. Needs are associated with pain–pleasure sensations and emotions – bodily experiences that motivate. The expression of the genetic endowment – which varies somewhat from person to person – is shaped by numerous forces during the course of our growth: family, community, schools, church, government, ritual, the media, and others. As we grow into adulthood our behavior is shaped by the opportunities and constraints of the social networks to which we belong and by economic, technological, and political structures and processes that impose rules, create possibilities, and limit options – mostly without regard to our motivations and inclinations. These larger structures – essentially patterned interactions among hierarchically related institutions – are subject to their own limits, imposed by the natural world (disease, climate, finite resources, and the rare large meteorite) and by ever shifting relationships of cooperation and conflict with each other. Biological and cultural (the rules governing the behavioral flexibility human biology provides) factors interpenetrate as individuals adapt to groups, groups to societies, and societies to each other and the natural world. Larger-scale factors like societies must also accommodate smaller scale factors, such as human motivations.

These factors do not operate as a seamless web, but are often at odds with each other. Even within the individual impulses may clash with each other or with our sense of propriety.

The discussion of the application of these factors to mobilization in Part 2 relies on the overview which follows.

The human heritage: from needs to myth

Nothing living is self-sufficient. Needs are part of humans' evolutionary heritage, biologically wired, and at the root of behavior. If needs are not met, people suffer or die. Deprivation gives rise to compensatory behaviors, which can be quite destructive. The relationships entered into to meet needs – and relationships of many kinds are needs themselves – give rise to social structures and processes. *The success of the conservation movement depends on its ability to link its goals to human needs and associated emotions, sensations, and cognitively expressed values and myths.*

Needs

Despite differences of emphasis or definitions, there is substantial agreement about what human beings need and about how those needs generate behavior (Abraham Maslow, 1968, 1970; Wilhelm Reich, 1949, 1972; Alexander Lowen, 1975; Antonio Damasio, 1994, 1999; Craig in Sandra Blakeslee, 2003: D1).

- Physiological needs are associated with growth and maintaining the organism's integrity. They include objects such as food, water, air, shelter, and the avoidance or removal of things threatening bodily safety.

- Tactile, sensual, and sexual needs are important to behavior and well-being throughout life for physical and emotional well-being. Interpersonal and social bonds are dependent on and are strengthened via the expression and satiation of this cluster of needs.

- Belonging needs motivate much human sociality and underlie the capacity to adapt to the world as cooperative groups. These impulses are flexible enough to extend beyond the family and pre-Neolithic band to larger social networks, class, ethnicity, nonethnic nationalism, regionalism, humanism, and so on. Individual identity is defined in part by group memberships.

- Love and esteem needs express themselves in our pursuit of affection and respect from others and to have others in our lives we love and respect. These needs underpin strong intimate bonds but can extend to groups. Group cohesion in turn serves to improve the chances of meeting all needs including intimate love objects.

- Creativity, spontaneity, generativity, and freedom needs are the impetus to the play and work that foster development of our faculties and abilities, important to successful adaptation. Creativity and spontaneity, tempered by belonging, are the substrates for resisting social domination. Generativity is the basis for nurturing the next generation.

- Cognitive and meaning needs arise from human behavioral flexibility. With so much behavior not hardwired, humans have evolved mental structures for navigating the world that include understandings of how the world works, values, and meanings. Like language, the particulars (which noises to make in what order, what the noises mean) must be filled in. These maps often take the form of story. Humans also evolved a strong sense of propriety – a need to believe their understandings and values are right – in order to stabilize the content of these structures in the face of alternatives.

All human needs are to some degree plastic – people can learn to feel bad about them as well as good, to deny some needs or repress their expression; hence cultural and subcultural as well as individual differences. Those needs most closely tied to basic survival are least pliable; one can live longer without love than without food or water. Abraham Maslow's (1970: 97–104) notion that needs are hierarchical, with some emerging only when others are sated is not born out by the evidence. Even the hungry desire self-expression; the poor want Nature protected and enjoy art.

In considering how these needs relate to conservation it is worth recalling that hunting, and gathering peoples considered many other creatures and elements of Nature such as rivers, part of their community; they were bound to them by love, respect, and a sense of belonging as well as physiological dependence (Joseph Campbell, 1959; Paul Shepard, 1973).

Pain and pleasure

Action to meet needs is generated primarily by pleasure and pain responses and related reflexes, and by emotions. Pain–pleasure responses may be entirely nonconscious (as distinct from unconscious, that which is repressed) or have substantial conscious aspects; they may be innate or learned or a combination. A prick from a thorn causes recoil without thinking. So do putrid smells because they usually mean the source of the smell is unhealthy to eat, drink, or be exposed to. The pain felt when someone disappoints may involve conscious consideration, but denial may kick in quickly and even nonconsciously (Elliot Aronson, 1997: 131). Warmth, good smells, sweets, and fats attract. But conscious considerations often intrude. By a wide margin people prefer Pepsi over Coke based on taste (Clive Thompson, 2003: 54–7). When the labels are attached the results are reversed because affective memories associated with Coke override the brain's taste reward center. Marketers have long understood that associating a product with a deeply held value, sensation, or emotion sells; this is branding.

Pleasure and pain responses are also learned. Paying attention to traffic helps avoid the pain of being hit. Driving fast for pleasure is also acquired. Ideologies, candidates for office, and groups of people may become identified with pain or pleasure, either by direct association or via a metaphorical link, for example, an ideology or group is rotten. Such reactions can become second nature, operating nonconsciously.

Pain–pleasure responses are triggered not just by encountering some object, but by a memory or anticipation of an encounter. Memory of a great meal and looking forward to seeing one's lover are examples. So is the memory of painful medical procedures. The pleasurable tension of anticipating a meal or sexual encounter can turn to painful frustration if gratification is delayed for too long.

Objects and memories we identify with pleasure draw us to them. The experience of pleasure – whether from memory, anticipation, or contemporaneous experience – makes people psychologically open to connection with other living things (Wilhelm Reich, 1949; Alexander Lowen, 1975). Pain or its anticipation causes withdrawal, as when tissue is damaged or a heart broken and people turn away from contact (with the exception of caregivers), concentrating energies on healing. Associating pleasure with the natural world is more likely to generate empathy with it and protective behavior toward it. Although pain generally closes people off, when some action causes pain to something one takes pleasure in and feels empathy for, it is most likely to generate action against the cause of the pain than a retreat.

Emotion

People act when they care, when they are touched emotionally. Emotions are reactions to objects in the world (or memories or imagination) that cause us to pay attention to objects that may meet our needs or that threaten us (Antonio Damasio, 1994: 115; Antonio Damasio, 1999: 282–5). Encountering certain objects causes changes in the brain, musculature, and chemistry, which lead to action. Emotions are usually observable through changes in facial musculature and movements.

Understanding the differences between emotion and mood and between primary and social emotions is important when speaking to people on an emotional level. Moods (Damasio calls them background emotions) are responses to basic body states, such as excitement, fatigue, wellness, sickness, discord, and harmony (Dylan Evans, 2001). They reflect biological predispositions shaped by experience, i.e., fundamental orientations toward life such as a sense of initiative or guilt that body states reflect (Erik Erikson, 1968: 94; James Masterson, 1985; Stephen Johnson, 1985, 1987). This includes how people are oriented to the natural world – an orientation closely linked to the experience of body (Harold Searles, 1960: 101–2, 120, 137, 386).

Moods are relatively long lasting and predisposition to certain moods lifelong, for example, some people are excitable. Although easily overshadowed by emotion, moods, unlike emotions, are always present and influence the evocation of emotion and otherwise directly affect behavior (Antonio Damasio, 1994: 150–2; Antonio Damasio, 1999: 52, 286). Moods are the body states that say: "I'd don't feel like it." "Damn, yes, let's do it." Moods are expressed throughout the musculature, not just facially, affecting how people walk, breathe, and hold themselves. Those who lack the ability to read body language are handicapped politically and socially.

Primary emotions (happiness, sadness, fear, anger, surprise, and disgust) are innate and often acted on without awareness (Antonio Damasio, 1999: 50–1; Dylan Evans, 2001: 6–7, 28). Indeed, most of the information we sense, including that which triggers emotion, enters without conscious notice. Moreover, we do not always tend to our emotions by feeling them or by admitting them into consciousness. Because primary emotions arise automatically and quickly an observer may see the emotion's expression in the person experiencing it even though that person is unaware they are experiencing it. Lack of self-awareness opens people to political manipulation.

Socialization and acculturation only minimally shape the experience and expression of primary emotions. The objects that stimulate them, however, are affected experience. Thus, disgust, which is innately triggered by the smell of decomposing flesh, may come to be associated with sexuality in puritanical cultures, or spontaneity in extremely authoritarian orders. Stimulating primary emotions makes significant energy available for action. Fear is probably the most important primary emotion to politics, but ungrounded hope is a close second.

Social emotions (pride, envy, love, embarrassment, shame, guilt, and jealousy) are strongly affected by socialization and acculturation and contemporaneous social context (Antonio Damasio, 1999: 50–1; Dylan Evans, 2001: 28–30). Some social emotions, like shame, may be absent when cultures exclude them from childcare practices. The experience and expression of social emotions is much more variable than primary emotions and reflects differences between and within societies. Urban settings and rural settings impose different kinds of familiarity; industrial societies are regimented in different ways than agricultural societies. At the same time differences in history can lead similar societies, for example, capitalist-industrial Japan and the United States, to use different emotional regimes to achieve similar goals. Just as industrial warfare requires estrangement from the "enemy," so the conquest of Nature necessitates estrangement, but it can take different emotional forms – an important point for mobilization.

Unlike primary emotions, which arise automatically and share the same hard-to-hide facial expressions, social emotions tend to arise and dissipate more and vary in their facial expression. They are also easier to mask.

That emotions have been tested much longer by evolution than our reasoning centers (David Ehrenfeld, 1979: 142; Dylan Evans, 2001: 33–4) is reflected in their power. They are not trouble-free, but do fit us to the world remarkably well as human daily, monthly, seasonal, and life cycle rhythms attest (Russell Foster and Leon Kreitzman, 2004).

Cognition, consciousness, and the sacred

Cognition.   Much of conservation outreach – especially that designed by natural scientists – is aimed at the consciously cognitive, appealing to people with information and argument conveyed in words. A program of outreach must include cognitive components – emotion is not sufficient. But neither is the cognitive sufficient, nor is it well understood. Most conservationists are correct that the cognitive is about thinking, analysis, reflection, categorization, frameworks for understanding the world, and values. Frequently they conflate the cognitive with consciousness, although thinking is like emotion – overwhelmingly (95%) nonconscious (George Lakoff and Mark Johnson, 1999: 13). Although language enhances thinking (and communication) enormously, it is not synonymous with it; most thinking is in images.

Effective use of the cognitive depends on an understanding of its varied functions. Much cognitive processing, for example, is merely utilitarian. It serves emotions and need states by providing an enhanced capacity to obtain emotional and need objects, but does not independently motivate. Most thinking is habitual and habits are hard to overcome. Images or concepts are formed by neurons firing in a particular pattern (Antonio Damasio, 1994: 89–90, 112; George Lakoff and Mark Johnson, 1999: 20–2). Once a pattern is created by an initial encounter it is strengthened with each additional encounter or recollection like a path worn in the forest floor. Over time, concepts and categories are integrated creating a structure or representational map of ourselves and the world. This mapmaking function provides a basis for interpreting new information, comparing it with existing information, and considering how the terrain might change or be changed in the future (Antonio Damasio, 1999: 21). Understanding the mental habits of those one seeks to communicate with is important.

Understanding two other aspects of thinking is equally important to effective communication. First, people's self-images are rooted in nonconscious neural patterns standing for the body (Antonio Damasio, 1999: 134, 225–6; Craig in Sandra Blakeslee, 2003: D1). These images rely on the body as their ground, for example, top political leaders are heads of state. Images and concepts of the external world emerge from neural structures that are adapted to the world and tend to correspond to "natural" categories in the world (George Lakoff and Mark Johnson, 1999: 17–37), i.e., categories that have proven adaptive. Thus, Mayr and Wilson, in Michael Soulé (1995: 151–3) found that hunter–gatherer categorizations of the natural world are very similar to those of modern biologists. Paul Shepard (1995: 26) thinks animal behavior suggests they see a common reality despite species differences.

Second, most thinking and decision making (how to act), especially social decisions, are shaped by emotion (Antonio Damasio, 1994: 173–200). The "Ben Franklin" model of reasoning – defining a problem, developing options, and considering the pros and cons of the possible options – is not the way things work. Instead, very early on in the reasoning process, information from innate biological preferences and preferences generated from experience with similar situations and associated body states (emotional and need states) are also evaluated. The linking of somatic states (e.g., fear associated with challenging authority, discomfort from knowledge that runs contrary to desire) with certain options occurs nonconsciously and vets the universe of options, selecting a few for final judgment. The criteria for vetting are provided by the body states associated with preferences both innate and acquired. Morality is less about learning rules than about developing certain emotional capacities (Dylan Evans, 2001: 67). Thus, a decision about whether to engage in political activity will be heavily influenced by body states previously associated with political participation – perhaps fear and loathing, frustration, or a strong sense of efficacy and achievement. Damasio refers to these associations as somatic markers. In addition, those parts of the brain that regulate metabolism and other basic functions heavily influence "higher" centers thereby ensuring primary needs are taken into account by these centers (Antonio Damasio, 1994: 111; Antonio Damasio, 1999: 21). If conservationists attempt to mobilize people in the face of negative previous experiences it will be very difficult – unless they understand what is going on.

Consciousness.   Self-awareness of needs, emotions, and cognition adds an important additional level of regulation to human behavior. Although most thinking is nonconscious and much that cannot be made conscious, some can be brought into the light of reflection. Most of the time humans speak and write without consciously considering grammar, but they can. It is possible to control body states associated with emotion and to consider whether and how to act on emotions and even whether an emotional response is healthy or the result of past experience that has distorted emotional processing and expression.

Consciousness confers additional behavioral flexibility, and enhances problem solving and innovation beyond that permitted by nonconscious mental processing, allowing humans to consider options more thoroughly and to respond in new ways based on that consideration (Richard Dunbar, 2001: 189–91). The capacity for choosing among many possible behaviors – although constrained by habit and emotion – requires a framework for making choices. Such frameworks are themselves not hardwired in most respects and need to be created and adhered to in order to be adaptive and serve as guidance mechanisms.

Anthropologist Roy Rappaport (1999: 21) argues that humans have long grappled with the tension between the human propensity for innovation and the presence of alternative frameworks and maintaining the need to stabilize the cultural order, i.e., frameworks of meaning, purpose, and understanding. He wrote that humanity ". . . can only live by meanings and understandings it itself must construct in a world devoid of intrinsic meaning. . . . It is . . . a world in which the . . . 'reality' or 'truth' of key elements, like gods and values and social orders, not only have to be invented but maintained in the face of . . . ever-burgeoning alternative possibilities . . . ."

Several factors predispose individual frameworks or worldviews and the cultural order to stability while allowing for innovation and change. Each generation is acculturated to a particular order as it matures; it becomes ingrained in neural pathways and otherwise familiar. Humans are disposed to emotionally invest in their worldviews and culture, making them resistant to easy change. Peer pressure and aspects of self-interest militate toward adherence because it does no good to speak a language no one else understands. These have not been adequate from an evolutionary perspective, however, because another regulatory mechanism is in place: humans possess a strongly felt need to believe their individual and group view of the world is the right one, the proper one, ordained by the very order of things.

The need to explain self and world and give it purpose is accompanied by the need to believe the explanation and purpose is right, that it is in Rappaport's terms, sacred or unquestioned. Thus, it is the sacred

> ...in the absence of genetic specification of behavior (that) stabilizes the conventions of particular societies by certifying directives, authorities who may issue directives, and all of the mythic discourse that connects the present to the beginning, establishing as correct particular meanings from among the great range of meaning available to the genetically unbounded human imagination.
>
> Rappaport (1999: 321)

To speak to people at the level of the cognitive and the conscious and gain active support for conservation requires understanding what people hold sacred. It is the sacred which gives authority to views about specifics, such as how people ought to behave in particular circumstances. Reaching people emotionally and addressing their needs is not sufficient. Their understanding of themselves and the world must also be addressed. Cognitive appeals that are not consonant with an individual's or group's sense of the sacred will not be effective (Eddie Harmon-Jones and Judson Mills, 1999).

The sacred is one of four levels of views comprising a framework or worldview (Roy Rappaport, 1999: 263–7). Understanding the specifics of a group's world-view at each of these levels and how the levels are related is important because at each level, different sorts of behaviors important to conservation are regulated. This is true whether mobilization seeks to build on extant views and rules or to change them. The other three levels are cosmological axioms, views about and rules that govern daily behavior, and views related to interpreting important events in the world and guiding responses to them.

The sacred.   Sacred views are by definition unquestioned and held to be unchanging. They constitute the most fundamental assumptions about the order of the world and the purpose of life. The sacred is the word on which other words are based. The sacred often invokes the gods or a creator (David Wilson, 2002: 49), but it need not. The gods may bring order out of chaos, or Nature may be seen to be inherently orderly and meaningful. Gods are often metaphors for natural forces. Whether divinities are part of sacred views or not, sacred propositions are usually untestable because they are not about the material world. Their authority rests on the need to construct a view of self and world that makes sense and the subsequent acceptance of a specific order (Roy Rappaport, 1999: 283). It is acceptance – to act as if there are no doubts - that transforms the culturally created into the natural (Roy Rappaport, 1999: 167). Joseph Campbell (1988: 208) put it most succinctly when he said "The divine is what we say it is." But to admit this is to admit the sacred is arbitrary, an artifact, and not ordained (Joseph Campbell, 1959: 55).

Although sacred views set the highest values or ends of a social group - the ends that instrumentality serves - the sacred is at its heart instrumental. It binds people to a specific order in the face of many possible orders, providing behavioral guidance.

Cosmological axioms are legitimated by the sacred but not logically derived from it (Roy Rappaport, 1999: 263-73). Axioms describe the general principles that govern the workings of the material world and the basis for lower order rules. For example, the cosmological axioms of the Enlightenment worldview include beliefs that universal laws govern Nature and human society, and these laws can be discovered and used to rearrange the world to better suit human desires without unmanageable consequences (David Ehrenfeld, 1979: 5, 16-7).

The legitimacy of axioms does require holding them to be unchanging, and they are in some sense subject to testing because they purport to describe how the world works. Their level of generality, however, means that only a major unexplainable crisis or the accumulation of substantial evidence over a long period of time is likely to bring them into question. Only in the late 20th century did widespread doubt emerge concerning the 17th century belief in the human ability to manage Nature successfully. Small adjustments are usually easily incorporated, and changes in axioms do not call the sacred into question.

That axioms change does not mean bigger changes are easily embraced even in the face of evidence. Einstein's rejection of aspects of quantum mechanics (god does not play dice with the universe) is a good example, as was the Catholic Church's slow acceptance of Copernican astrophysics. In complex societies openness depends on factors such as differing interests or differential exposure to new knowledge.

Specific views and rules of conduct are derived from axioms and transform the general into the specific (Roy Rappaport, 1999: 314). This level of worldview comprises understandings, values, and rules that regulate daily behavior, such as defining who is family and the obligations owed them, what kind of car is desirable or okay to drive, what is good to eat and what is okay to eat, when it is appropriate to dam a river or convert a forest into a tree farm or cornfield, how many children it is good or right to have, and what policies or candidates should be supported. Some rules are embodied in law, but legal enactment does not always reflect importance.

People and groups vary in their openness to change at this level, although the rules, values, meanings, and attitudes that regulate daily behavior are subject to constant pressure from a changing social and natural environment and from deliberate efforts by governments, nongovernmental organizations (NGOs), and businesses to change things. The legitimacy of views governing daily behavior is not dependent on holding them to be unchanging. But humans are creatures of habit and invest emotionally in their views. Behavioral plasticity declines after adolescence. The ability of conservation NGOs to change behavior often depends on changing views and trying to do so in the face of opposing efforts to maintain them or move them in a different direction. Success depends on many factors, but among them is the ability to place mobilization messages in the context of extant axioms and the sacred. The fit may not always be easy, needed messages may not always be derivable from extant axioms, but often a hook can be found. When US conservationists seek regulatory solutions they often confront resistance from interests who invoke faith in "free markets." Most free-marketers have not read their sacred and axiomatic texts, however, and so it is possible for conservationists to enlist none other than Adam Smith (1976 [1759]: 83–7), who argued that regulation was important in achieving justice. Capitalism, he believed, could only be successful within the constraints of a strong moral community that placed limits on self-aggrandizing behavior.

The lowest level of specificity in a worldview comprised, inter alia, understandings and rules defining *important events in the social and natural world and the appropriate response* (Roy Rappaport, 1999: 266–7, 273). Elections call for voting or denigration; the announcement of a planning process for a public forest falls on deaf ears or calls forth lobbying. The dynamics of legitimization, change, and mobilization that govern the previous category of views also govern these.

There are three other aspects of worldviews (or frameworks or cultural orders) that conservationists will find important in mobilization. First, worldviews include understandings of how reality works, but as Roy Rappaport (1999: 21) notes, although the world is "subject to causal laws, not all . . . are known." Because people must explain the world, they fill in what they don't know based on values and axioms. Knowledge and beliefs become conflated, and new information about what causes lightning, for example, may not be persuasive to a group who have long thought it reflected the gods' anger. Facts by themselves are rarely persuasive.

Second, for some groups the sacred is not confined to basic and untestable propositions about the proper order of the universe at large, but includes all levels of views contained in certain texts or pronounced by certain authorities. Daily rules are sacred and unchanging. This is a core characteristic of fundamentalism, making it difficult to reach such groups.

Third, the erosion of aspects of any worldview is not just a function of innovation or the existence of alternatives. Some individuals and groups in a society always find it to their advantage to bend or break some rules while trying to otherwise maintain the order which provides predictability. A central purpose of a worldview is to maximize adaptive behavior by providing rules that proscribe short-term and narrowly self-serving actions in favor of longer term and group interests (Roy Rappaport, 1974: 62–3). Rational calculation cannot achieve this because it seldom transcends the tyranny of small decisions. Nonetheless, as Morris Freilich (1972: 291–310) argues self-interest often wins out in this battle, which can make some individuals and groups appear successful while the society as a whole becomes maladaptive and declines.

Myth and story

Some readers of this book may enjoy a good scientific paper more than narrative film by a great director. But not most people. Many conservationists, especially scientists, don't understand this. Communication that results in mobilization almost always utilizes story to reach people emotionally and cognitively. We are storytelling animals. We make sense of and navigate the world through story. It is overwhelmingly through stories that all of the aspects of worldviews we have been discussing are communicated. Stories tell people what is good and bad, right and wrong, how the world works, what the purpose of life is, and what ought to be done in particular situations (John Marcus, 1960: 223; Percy Cohen, 1969). Stories offer answers to the hard questions – about death, the need for acceptance and for association with a greater purpose (Henry Murray, 1960: 305–6; G. Kirk, 1970: 79; Joseph Campbell, 1988: 37–67).

There is a dual hierarchy to stories. First, they mirror the hierarchy of views they express – the sacred, the cosmological, the everyday, and events that require action. Second, stories reflect the hierarchy of social and cultural organization: civilizations, societies, groups within societies or sometimes ones that span societies, families, and individuals. The term "myth" usually refers to stories about the sacred and the cosmological. These stories provide the framework for stories about day-to-day affairs (Kirk, 1970: 86–131). Myths express and shape the identity of groups, helping to foster social unity by describing common origins and purposes and legitimating them (Joanna Overing, 1997: 7; Susan-Mary Grant, 1997: 88–98; Richard Hughes, 2003). They are not simply about the world and self, but "they are the things we see with." (Graham Bennett, 1980: 167). They typically involve narratives of origin, redemption, and suffering; rebirth and renewal; kinship and collective descent; and the bonds to place (George Schopflin, 1997: 28–34). Myths link the concrete and local to the universal: birth, growth, death, the stars, male, and female (Joseph Campbell, 1959: 461–2). These grander stories give meaning to more personal ones, and provide a "library" of scripts for groups, leaders, and individuals (Jerome Bruner, 1960: 281; Ira Chernus, 2004). All of the great worldviews have at their roots an epic story, not scientific treatises. Even the Enlightenment humanism, which has generated libraries full of treatises, has stories at its heart – about the triumph of knowledge over medieval darkness, the march of historical progress (Marxist and otherwise), and these in turn contain stories of particulars: the voyage of the Beagle, the conquest of polio. Leaders claim power and purpose and justify policies through reference to myths and lesser stories. Candidates sway voters with stories.

Stories are not history in the social science sense of history, but they are often told as if they are history and are accepted by many as history. This is so because they often express the truth of experience and a widely shared framework for thought. As such they influence social science and natural science hypotheses. Stories provide a means by which the elements of experience can be fitted into a meaningful pattern which is also aesthetically pleasing. Joseph Campbell (1959: 61–77) believed that the most appealing stories tapped into what psychologists call primary process or childlike thinking. Thus stories that are danced, painted, and conveyed through music, song, and visual images may touch most deeply. A story is experienced as good when the pattern is pleasing, and it reduces the world's complexities into a more manageable representation. Although stories are not strictly testable in the scientific sense (Northrup Frye, 1960: 116), they do need to work, to explain things accurately enough to elicit adaptive behaviors. In this sense some stories work better than others.

Shaping human needs, emotions, cognition and consciousness: maturation, ritual, and social networks

Maturation

Humans are not very diverse genetically, but our behaviors vary widely. Jack takes in stray dogs, weeps at the sight of a logged forest, and regularly joins demonstrations against the destruction of wild places. His sister Jill grew up pulling the wings off flies, went into research involving vivisection, and invests in factory farming enterprises.

During an individual's development the world leaves indelible marks on how needs and emotions are experienced and expressed, on the content of beliefs, and on many other aspects of personality that affect behavior throughout life. This occurs as surely as differences in rain and soil and sun shape the growth seeds of the same species into different trees. Even before birth the genetic endowment of an individual finds expression via interaction with its surroundings.

Indeed, the importance of the developmental process on later life is a major reason why there is such bitter political conflict over the control and content of parenting, education, the media, and other institutions. It is why churches, governments, and interest groups begin indoctrination at an early age. It is why armies and navies prefer young recruits. Conservationists have not paid enough attention to influencing personality development, although this is starting to change given prodding (Gail Melson, 2001; Richard Louv, 2005; Gene Myers, 2007). The investment has a long-term payoff, however, and this book is focused on the near term: how can conservationists mobilize groups of people given extant personalities, cultures, and social relations?

It is nonetheless important to highlight some aspects of the development process that are particularly important in shaping personality and cultural attributes affecting conservation mobilization. An understanding of developmental dynamics is important in deciding which audiences to target for mobilization and how to approach the audience. I will look first at aspects of developmental dynamics, the entities having a role in development, and the developmental outcomes important for conservation.

Like other animals, we grow according to a genetically encoded plan. Different aspects of personality undergo development at different times, in sequence (Erik Erikson, 1968: 92–5). The development of the self and ego occurs before the capacity for analytical thinking, for example. Later development depends on earlier stages, and if something goes amiss early on it can have profound effects on characteristics that develop later (Erik Erikson, 1968: 208–31; James Masterson, 1985: 20–9). Although people do change in adulthood, barring major crises the basic aspects of personality do not change much after adolescence (Erik Erikson, 1968: 95; Martin Teicher, 2002b).

Two primary processes shape the experience and expression of needs, emotions, and cognitive attributes during development: acculturation and socialization. Acculturation is the transmission to the growing child of the rules, meanings, values, attitudes, descriptions of reality, and problem-solving techniques that are meant to guide behavior. Socialization is the transmission of learned behaviors, largely through example, and is closely tied to learning how to channel associated emotions and need states. Learned guidelines for behavior and learned behavior are often not in accord with each other. The rules and values say "do not lie" or "do not be domineering," but observed adult behavior may reward lying and dominance.

Many institutions are engaged in acculturation and socialization. Parents and other family members or custodians usually play a major role in human development through the quality of their caregiving. The presence, absence, and quality of these emotional ties have lasting consequences. Religious institutions, schools, organizations like scouts, and the media all play profound roles, usually from a very early age; they provide both deliberate indoctrination and a wider world of social relationships and roles. In the United States, television has become the primary baby-sitter. US fast-food restaurants now operate over 10,000 playgrounds, distribute millions of toys, and spend more than $3 billion on ads, most aimed at kids; they also have direct ties with film studios and toy makers (Eric Schlosser, 2001: 47–8). Governments affect developing personality by means of everything from nutrition delivery programs and early learning (or propaganda) programs to economic and military policies. Unemployment in the United States contributes to domestic violence (US Congress Joint Economic Committee, 1976, 1984), war creates orphans and damages the personalities of many participants, affecting their ability to parent. In later stages of development, political organizations such as parties and NGOs play a larger role; many have youth divisions, aimed at recruitment and consolidating ideological and organizational commitment. Changes in social institutions and power relationships affect attitudes about the body and associated emotions (Ronald Aminzade and Doug McAdam, 2001: 26).

The power differential between children and the adults they interact with means that when the child's needs, desires, and natural impulses to seek their satiation and fulfillment clash- with what family, group, and society are prepared to give, the latter prevail. Children must adapt to these circumstances; if chronic, they have lasting effects on personality that may include alienation from themselves and others, support for institutions hostile to their own interests and hostile to Nature, and much unhappiness, the causes of which are unconscious (Norman Brown, 1959; Sigmund Freud, 1961; Abraham Maslow, 1968; Wilhelm Reich, 1949, 1972 [1934]; Marshall Sahlins, 1972; Ernest Becker, 1973; Dorothy Dinnerstein, 1976; Marvin Harris, 1977; Paul Shepard, 1990; Morris Berman, 2000; Alice Miller, 2001). Because children are dependent on people who may not provide what they need, less basic needs (e.g., a caregiver's loving response) are sacrificed to more basic needs (e.g., maintaining the mere presence of a caregiver).

Developmental outcomes.   How conservationists ought to influence children's developmental experiences is a matter for another book. For now it is useful to identify those particular developmental factors that are useful in identifying individuals and groups most likely to be sympathetic to the natural world – admittedly only one basis for engagement in conservation politics. The qualities of care that are favorable for conservation include meeting a child's material and emotional needs consistently, childhood-long interaction with Nature and other creatures, affirmation of the capacity for emotionally grounded physical pleasure, and support during identity formation for healing and for the embrace of occupations, ideology, and political activity that have positive ecological consequences.

When a child's needs are largely met throughout the various stages of childhood and adolescence it grows accepting of those needs, their expression and related emotions, and their source – the body. Empathy with other life begins with acceptance of and comfort with one's body (Harold Searles, 1960: 120–37; Erik Erikson, 1968: 92–7). Material security – clean water, adequate medical care and nutrition, low infant mortality, the absence of war, and so on – is especially important in earlier developmental stages. In addition to at-homeness in the body, material security underlies an openness to and connection with the larger world.

It is also closely associated with higher status for women and lower birth rates (Pippa Norris and Ronald Inglehart, 2004: 84–108).

Extensive interaction with the nonhuman environment during the entire developmental process is central to a solid sense of self and healthy self. A healthy self is connected to, and recognizes a kinship (not oneness) with the nonhuman world; it is a recognition that humans and the rest of the world are made of the same stuff and share the same fate (Harold Searles, 1960: 55–99, 101–2). Genuine interaction with animals, including pets, is often an important part of this, generating empathy, moral recognition of nonhumans, and enhanced awareness (Peter Kahn, 1999: 18; Shierry Nicholsen, 2002: 79; Ulrich Gebhard et al., 2003: 97).

Affirmation of libidinal impulses, which are quite powerful and associated with bonding, belonging, longing, love and creativity as well as sexuality, is related to feelings that the physical world is something worth defending. Because sexuality is a main source of joy and the primary means of reconnecting with our earliest (and necessarily lost) oneness with the world, Dorothy Dinnerstein (1976: 144) argues that its availability is critical to psychic health and the impulse to protect the integrity of the world.

All of these elements in the child's experience give rise to a self connection with the world, trust in one's impulses, a self with empathy for the world, and a self that takes pleasure in its needs and emotions. They are the foundation for the sense of fullness that Henry Thoreau (1964 [1854]: 335) had in mind when he wrote that "(a) man is rich in proportion to the number of things he can afford to let alone." It is when fullness is lacking and there is inner poverty that people seek satisfaction in the accumulation of things and power and thereby destroy the world.

Adolescence is of special interest to conservationists' near-term political calculations because it is in this developmental phase that an occupation is chosen, ideology settled, political organizations joined, and the main outlines of one's place in the world set (Erik Erikson, 1968: 128–35, 156–88, 300, 309). The many roles that constitute identity are integrated intrapsychically and behaviorally into a more or less workable whole. By the end of this period individuals normally find a place in the dominant social, economic, and technological order – an order destructive of the natural world. Because a psychological openness marks this period – the process makes accessible previous developmental stages providing an opportunity to overcome earlier failures of nurturing – it is possible to help catalyze the impulse to heal and to create or strengthen connections with the natural world. If this can be tied to choices about livelihood that help transform existing economic roles and relations, the effects can be potent.

Raising children that as adults are connected to their needs, emotions, and Nature is not the same as raising children to be conservationists. Action requires more and that is what Part 2 of this book is all about.

It is also good for conservationists to be aware of those factors affecting development that portend poorly for conservation by creating disconnected and nonempathetic personalities. The same factors are involved, but the quality of care is different.

Thus, failure to meet the emotional and material needs of the developing child tend to create personalities in which impulses are not trusted because they have been chronically unsatisfied or punished (Erik Erikson, 1968: 96, 102). This experience results in seeking high levels of certainty about, and control of a "hostile" world (Michael Milburn and Sheree Conrad, 1996; Pippa Norris and Ronald Inglehart, 2004), a core attribute of fundamentalism. A feature fundamentalisms share is norms that limit women to reproductive and drudge economic roles which keep birth rates high (Pippa Norris and Ronald Inglehart, 2004: 5, 22–4, 107, 223). Associated attributes include the diminishment of skepticism and curiosity which are essential aspects of creative and critical thought, and resistance to personality growth and change. Cognitive rigidity makes rejection or denial of information that conflicts with early acquired beliefs common (Leon Festinger et al., 1956; Leon Festinger, 1957; Allen Johnson and Douglass Price-Williams, 1996: 82; Eddie Harmon-Jones and Judson Mills, 1999: 6–7). Change that cannot be accommodated in the existing worldview is experienced as threatening, generating defensive behavior.

A world that denies food or safety, or punishes autonomy or physical contact and the expression of related emotions, makes the experience of the body and Nature a source of pain rather than genuine happiness; happiness is instead found in an afterlife or on another plane. Death is rejected as real. When death is not taken as real, life is devalued (Ernest Becker, 1973), as the sad history of inquisitions, crusades, and jihads suggests. The lives of other creatures are also devalued and Nature is seen as corrupt or fallen.

Those whose needs go chronically unmet have little basis for empathy, insight, or connection with the natural world, i.e., no comfort in their own body (Harold Searles, 1960: 101; Wilhelm Reich, 1972: 62). When needs are chronically frustrated, ignored or punished, common responses include hurt, anger, sometimes rage, denial that the urge even exists (making the injury invisible and hard to address), splitting off of the original impulses and their projection on to others, including Nature (Claude Bernard, 1957 [1865]: 103; Roderick Nash, 1982: 23–43; Neil Evernden, 1985; Paul Shepard, 1990: 75–92). Deprivation leads to arrested development, which, depending on when it occurs in the growth process, has different consequences, most of which are associated with negative attitudes toward the natural world. This book is not the place for a detailed analysis; I simply note the most important consequences for conservation.

Dorothy Dinnerstein (1976: 109), for example, attributes much of humanity's destructiveness toward Nature as the literal, not metaphorical, consequence of "amoral infantile greed" run amuck and magnified by technology. Michael Milburn and Sheree Conrad (1996; also John Jost et al., 2003) have documented the close relationship between physical abuse in childhood and the defensive identification with authority and control figures and ideology throughout life. When healthy sexual impulses are subject to consistent shaming it leads to varying levels of discomfort with or rejection of sexuality thereby undercutting empathy for other creatures and the projection onto Nature of shameful and dangerous impulses that must be checked (Philip Grevin, 1977; Roderick Nash, 1982: 36–9), or a crusading supermoralism that seeks to control and subdue others and Nature (Erik Erikson, 1968: 119). The unavailability of bodily (including sexual) joy and spontaneity – a primary means of reconnecting with our deepest selves – generates compensatory behavior: the imposition of various control schemes on the world to demonstrate prowess and importance (Dorothy Dinnerstein, 1976: 144; Paul Shepard, 1990: 108).

Although children within the same family are treated differently and different personalities result, it is also true that families and groups share certain patterns of treatment that produce similar attributes. The characteristics that militate for and against conservation are group as well as individual attributes. Thus class position, nationality and ethnic background, gender, and larger social forces strongly affect development and hence motivation. Although children are predisposed to empathy toward animals (Kahn and Kellert, Kahn 1999: 188–8; Myers and Saunders 2002: 158–170: 164, 187–8), civilization-wide religious or philosophical traditions affect what adherents feel toward other creatures, such as the degree of compassion they deserve (Paul Waldau and Kimberly Patton, 2006). National cultures with a stronger sense of community tend to produce people who take obligations to the natural world more seriously than those stressing individualism (Christopher Rootes, 2004: 634). These differences also affect how mobilization occurs. Child-rearing practices that affect everything from a sense of efficacy and the capacity for empathy to how interests are defined and what is permissible in their defense vary by class, education, ethnicity, and the individual (Erik Erikson, 1968; Robert Coles, 1986; William Chafe, 2005). Most importantly, opportunities to experience the natural world and the quality of that experience – which nurture empathy and set the benchmark for what is considered natural (Peter Kahn, 2002) – vary by social category. The many factors affecting material security depend on the position of the child's family in society (e.g., class and ethnic group), the position of one's society in the international order (poor or subaltern country, rich or dominant country), and whether war or peace predominate (Pippa Norris and Ronald Inglehart, 2004).

The purpose of highlighting some consequences of developmental plasticity important for conservation is to encourage their recognition, and to make clear that these attributes are deeply rooted and not easily changed once set. Attributes that predispose people to be hostile to Nature must be worked around or the group written off, while favorable attributes should be built upon. Personality attributes also make some people or groups more or less difficult to mobilize.

Ritual

Ritual is an important interactive process that plays a major role in acculturation and socialization – especially once children become more verbal and involved in the larger world. Ritual remains an important regulatory mechanism throughout life, providing continuity. Ritual's centrality to mobilization is examined in Chapters 14 and 18. Here I note how it works to bind people to an institutional order.

Rituals are invariant patterns of codified behavior performed by members of a community at certain times and places. Ritual permeates all societies, including modern ones in which the meaning of many rituals is contested. US Independence Day (July 4) originally celebrated successful armed insurrection against illegitimate authority. Today US elites seek to evoke affirmation of US power for counterrevolutionary purposes while others celebrate its original meaning.

Ritual includes rites of passage and rites of exchange; they mark the passage of seasons, commemorate past events, the starting and ending reigns or wars (Catherine Bell, 1997: 94). A mass, a handshake, an oath of office and inaugural celebration, and a holiday are rituals. Rituals may be celebrated with an individual emphasis, as when US Thanksgiving is celebrated at home; or they may involve large gatherings such as in a church or mass celebrations. The same rituals may occur in several localities simultaneously, thereby linking participants to national and even international events, structures, and order. Ritual infuses elections, birth and death, demonstrations, and even very boring meetings. A trial has the practical effect of determining a dispute between parties, but the robes, alterlike benches from which the judge presides, and the highly stylized interactions between parties are the stuff of ritual. Ritual aspects are not explained by the need to reach a decision, but by the need to legitimize it via participants' acts of fealty (Sally Moore, 1977: 152–3). The symbols used in ritual include words, icons such as flags or sacred texts, and encoded actions include dance, song, and chant.

Ritual is directly regulatory; it functions to obtain public commitment to an order, and to its purposes, values, meanings, rules, and attitudes. Ritual establishes conventions and obligates participants to abide by them. Some rituals are about acceptance or reaffirmation of the sacred – they bind one to a set of unquestioned assumptions about ultimate meaning and purpose. They bind one simultaneously to the holy – that which must be defended and kept integral at all costs (Roy Rappaport, 1999: 23–68). Such rituals connect people to the beginning of things, to creation, to the eternal (which is the "timeless now," not an unending future).

Most ritual is secular, but serves the same function: to invest various aspects of a particular order with legitimacy through public acts of acceptance. Ritual may combine the sacred and secular, as when voting invests the electoral process (and whoever wins) with legitimacy, but may also be seen to affirm the "chosen people's" mandate to carry out god's will (David Kertzer, 1988: 37; Richard Hughes, 2003: 33; 106, 154–73). Even when ritual seems to be concerned with the petty or apparently lacks emotional investment, it provides continuity and predictability (Sally Moore and Barbara Myerhoff, 1977: 13–4).

Part of any order includes transformations between roles and statuses or states of affairs. Ritual marks these, making that which is invisible or ambiguous, visible and clear (Sally Moore and Barbara Myerhoff, 1977: 14, 22; Robert DaMatta, 1977: 259; Roy Rappaport, 1999: 69–106). It imposes distinctions and then mediates them. By virtue of ritual children become men or women, two people are joined and a family is created, classes and states turn war into peace or peace into war. Ritual clarifies roles and relationships. In doing so, it makes visible the principles that govern or regulate the transformation and affirms them (Terence Turner, 1977: 61, 68). Ritual is no less important when a transformation is embodied in formal law. Ritual contributes to the law's meaning, purpose and legitimacy. Ritual also renders the uncertain, uncontrollable or unpredictable, certain, controllable, and predictable and makes the random meaningful (Sally Moore, 1977: 24; Sally Moore and Barbara Myerhoff, 1977: 60–1).

Ritual binds people to the broader social order typical of modern societies or earlier empires and to particular groups within the larger society, based on regional, class, ethnic, or value distinctions. Ritual helps create, build, consolidate, and legitimate organizations, including political parties, interest groups, and social movement (David Kertzer, 1988: 14, 17, 23, 65). It defines an individual's relationship to group, society, and the universe and mediates them. Group loyalty is usually both complementary and competitive with loyalty to a society in varying mixes, but it may be subversive of the larger order at times.

Ritual – which usually takes the form of a public, collective and physical action when its purpose is to express acceptance of social obligations – binds people more powerfully than a private promise (Roy Rappaport, 1999: 119–24). Public performance is more powerful than public statements, especially in the face of personal ambivalence. Action taken collectively before the community impresses itself profoundly on the actor. Ritual invests "meaning with *public* morality" (emphasis added) (Roy Rappaport, 1999: 27).

When a ritual is intensely physical the results are more profound because it generates deep pleasure or other intense emotional states, as with group singing, dancing in unison or even marching over extended time periods (Eugene D'Aquila and Charles Laughlin, 1975, 1979; William McNeill, 1995). Even absent intense physicality ritual takes people out of their routines and into a different sort of time, reinforcing group solidarity (Roy Rappaport, 1999: 209–35; Francesca Polletta, 2004: 173). Ritual may also bring the extraordinary into the daily. In either case it links the two, integrating experience and making adherence to the worldview more likely.

Ritual is often dramatic and drama is more powerful emotionally than mere verbal statements (David Kertzer, 1988: 30). Publication of the US Declaration of Independence did not much arouse people. When it was publicly read aloud, followed by burning the British king in effigy, enthusiasm was sparked.

The coded behaviors that make ritual ritual – its sameness regardless of who performs it, or where and when it is performed – and the familiar symbols used in ritual, reinforce in participants the incontrovertibility of the truth of their understanding of things (Sally Moore, 1977: 21).

Ritual's primary aim is not to diminish personal doubts or cynicism about the order whose acceptance it codifies, although it may do so. Rather, its role is to contain their potential corrosive effects through public displays of acceptance. What is subversive of an order or worldview is publicly expressed doubt, repudiation of dominant ritual, or investment in competing ritual (James Scott, 1990: 204–5; Roy Rappaport, 1999: 132). The potential subversive effects of private beliefs are also checked by the need for belonging and the need to be accepted by a group in order to obtain material benefits (access to work, for example).

Ritual's ability to bind is paradoxical because although it is clearly artifice, it simultaneously discourages inquiry into that fact and that the cultural order is arbitrary and created rather than natural and preordained (David Kertzer, 1988: 88; Roy Rappaport, 1999: 139–68). It achieves its authority not only by means of the intense physical states often associated with it, or by its reassuring qualities and the fact that its absence generates anxiety, but also by reason of the quid pro quo just noted – one must demonstrate acceptance of the order or group to be accepted by it and receive its material benefits (Sally Moore and Barbara Myerhoff, 1977: 13–24).

Absent a powerful ritual life the conservation movement will lack serious appeal.

Social networks

Human dependency on family and society for our growth is part of our intensely social nature. Humans adapt not just as individuals to air, water, nutrients, temperature, and trophic cycles, but to families, groups, and societies. Humans earn their living as groups. Starting in childhood and throughout the rest of our lives social networks are as ubiquitous and important as the air. Governments, political parties, and educators know the influence of peers on obedience and resistance to authority. Marketers use social networks to sell products. They call it viral marketing – a term apt in ways marketers probably don't intend.

Social networks consist of people who interact with each other on a regular basis or otherwise maintain a relationship. Networks typically consist of family, friends, coworkers, and others known on the basis of shared institutional affiliations such as school or church, community (neighbors, people we meet at the store), and shared value orientations that are the basis for some organizational affiliations such as interest groups or hobby clubs (Christopher Ansell, 2003; Pamela Oliver and Daniel Myers, 2003).

Social networks are politically important because they are a primary means of transmitting information, generating and reinforcing common emotions, and because *people make choices about political action as cohorts*, not individuals (Mario Diani, 2003a: 8). People join and leave movements or NGOs together (Pamela Oliver and Daniel Myers, 2003: 178, 184). Commitment and activism increase and wane by network cohort. Reporters who are networked strongly influence each other, covering the same stories and attributing the same degree of importance to them. Networks of organizational leaders within an industry, a movement, or across a major interest (e.g., big business or labor) can lead to concerted political action in the absence of formal coordination (Margaret Keck and Kathryn Sikkink, 1998). People bound by strong ties, because they share many cultural attributes, are discrete targets for mobilization. Reaching broader audiences requires targeting many strong networks or targeting those few people who have strong relationships with others in many networks.

To utilize network connections effectively the distinction between strong and weak must be noted. Strong ties are typified by emotional closeness, frequent interaction, and a high degree of mutual though not symmetrical influence. Close connections are highly correlated with similarity of interests, values, and other cultural attributes (Florence Passy, 2003: 23). Weak links are less emotionally charged, relationships exert less influence, and there is much greater diversity among those weakly connected. Much information within a network is only circulated among those who share strong ties, but is likely to have greater influence because of the strength of the ties. Weak links, although less influential, are critical to information spreading widely and quickly, not only within a network but also between them. The latter is true because people maintain weak links in a variety of networks. The importance of weak links extends beyond their capacity for wide and rapid diffusion of some information. Weak links are essential to integrating social networks and thereby perform an important role in communication within society as a whole. If only 20% of weak links are severed a society "collapses" into smaller groups (John Bohannon, 2006: 914–6). The similar loss of strong links has no similar social effect. Networks mediate between institutions and individuals, affecting behavior and decisions (Florence Passy, 2003: 23).

Social networks, which do span societies, also reflect important characteristics of individual societies. In more conformist societies, for instance, influence is most successfully exerted on a network by influencing those looked to by participants as leaders (Jeffrey Broadbent, 2003: 205–6).

Group and organizational dynamics

During my first school years I often played in woods not far from my home. The second growth conifer forest was huge and mysterious, both a refuge and vaguely ominous. On the hottest days it was dark and cool. In the winter the fiercest tempest barely penetrated; the hurtling rains from pacific storms were reduced to a steady drip from limb to limb to litter. Only about 12 hectares, there is little in the way of large fauna, but animal sign was everywhere – burrows, nests, trails, droppings, and faint echoes of stealthy movements.

In my eighth spring stakes with colored markers appeared, rapidly followed by chainsaws and bulldozers. The forest, in the language of economists, planners, and other confused creatures, was being improved. How could this happen? I was angry and outraged. I was losing a refuge but other creatures were losing their homes. I could swear a streak even at that age, and attendance at a Catholic parish presided over by a particularly mean-spirited priest gave me some command of the language of eternal damnation. It helped to get attention from adults, but that attention and calling forth higher powers had no effect. Effective monkey wrenching was beyond my vision and resources. Nature did have one whack at this development – a freak typhoon on Columbus Day 1962 laid flat or snatched the roofs from many of the newly completed houses. But the ticky-tacky artists had soon pasted things back together and the forest's fate was sealed for my lifetime.

The political lessons continued. In high school I helped start an independent newspaper that administrators tried to suppress. "We're here to learn," they said. Reading about democracy was okay, apparently, but not engaging in it. By the next year, this being the mid-1960s, I was involved in civil rights support work and organizing antiwar and antidraft actions. Through all of this a common thread ran: to be politically effective it was necessary to work with others. In politics "one man is no man" (Bertolt Brecht, 1977 [1926]). Pooling time, skills, money, and other resources in an organized way is necessary.

The internal dynamics of organization profoundly shape political behavior – beckoning some, pushing away others, and sucking up huge amounts of energy in its own right. Energy (resources) are required to sustain organization, in internal struggles over policy and the personal, and to achieve its goals. I suspect it was internal organizational politics rather than combat with opponents that led philosopher-playwright Jean-Paul Sartre (1947: 61) to say that hell is other people.

The necessity of collective action requires that the motivations of those within the organization accommodate each other and that personal motivations and organizational imperatives accommodate each other. Politics is about the internal dynamics of organizations as well as cooperation and competition among organizations. Thus politics is about the internal dynamics of organizations as well as cooperation and competition among organizations. The most salient aspects of these dynamics include:

- Unorganized groups are the ground on which organizations are built, but although action by unorganized groups (or individuals) can have major political effects, for example, a riot or an assassination, the unorganized are disadvantaged compared with the organized in the ability to realize political goals. The unorganized act like a wildfire – outcomes are chaotic. Organizations, though lacking spontaneity, possess intelligence – the potential to reflect, evaluate, and act with focus.

- People seeking to join an organization confront limited choices and must therefore make greater or lesser compromises or forego organized political involvement. This is true even when they seek to change the organization more to their liking. Those who start their own organization are also limited by available political and geographical niches.

- The competing motivations within an organization that shape its behavior are many. Founders often resist sharing control or altering the original mission. Older and newer participants often differ significantly based on length of experience and lessons learned. Activism takes an emotional toll. Weariness affects behavior. New recruits may have more energy and hope, but often don't learn from those with more experience and needlessly reinvent the wheel. Because recruitment tends to occur in cycles, organizations have cohorts at different stages of energy and weariness and with different histories. Even those joining at the same time come to an organization with different histories and depth of political understanding, different skills and therefore abilities to influence the organization as they would like. Positions within the organization – leaders, paid staff, volunteers, activist members, other members – provide different levels of influence and shape individual behavior via the opportunities each position offers.

- To become and remain successful, organizations must create and sustain a sense of community that binds people. They must be more than an entity with political goals. Regardless of whether NGOs pursue community, participants in all roles bring a myriad of often conflicting motivations that have little to do with the NGO mission, but which shape behavior within it, aiding or undercutting effectiveness.

- Resources can't just be used in the pursuit of programmatic goals; they must be invested in maintaining the organization and usually in attaining additional resources.

- Political effectiveness requires constant innovation, which can run up against habit and timidity. Inertia, groupthink, and the mutual reinforcement of wishful thinking in place of learning are real tendencies in all groups. Organizational cultures and democratic structures can ameliorate some of this and make self-renewal and reinvention easier. Too much creativity and a lack of discipline can lead to erratic action and lack of focus.

- Success may result in demobilization among those who think the war, not just a battle, has been won. Success will also make an organization a target for containment or repression by opponents and sometimes the object of derision or worse by competitors. NGOs led by the ambitious and narcissistic are especially prone to such behavior.

- Larger organizations are in many respects less cost-effective, but with more resources in total they can achieve goals smaller organizations cannot. Size shapes organizational behavior in many ways. Holding together larger numbers of people requires broader and vaguer purposes to contain internal division and maintain group identity. Larger organizations require more hierarchical and centralized organization just as an orchestra requires a conductor and that can attract leaders more interested in their own power than the organization's mission. Members often have less control over leaders in larger NGOs. In an effort to maintain and grow resources larger organizations become dependent on major donors or government, which can moderate their mission regardless of what members or even staff want. Perpetuating the organization replaces the mission.

Many of the chapters in Part 2 address internal organizational dynamics in detail. The purpose of this section is to note the salient ways in which organization shapes behavior and possesses its own dynamic, not reducible to individual motivation.

Organizational cooperation and conflict, social structure, and structural evolution

Behavior is not just shaped by the motivations of individuals or by groups of individuals with more or less common motivations acting collectively. Political behavior is shaped by organizational conflict and cooperation such as the struggle of NGOs to realize their differing interests through campaigns or insider deals, including efforts to bolster their strength through coalitions and alliances. Somewhat like individuals who must accommodate each other within a group, groups constantly react and adapt to each other's actions and predicted actions as they seek to counter each other. Virtually all conservation bills, in drafting and during the legislative process, are altered to gain enough support to overcome opposition or to prevent opposition mobilization (Ronald Libby, 1999; Frank Wheat, 1999; Roger Kaye, 2006; Judith Layzer, 2006). Conservation campaigns must adjust to opponents' attacks, their efforts to mobilize their own forces and undercut conservation support, or to complete insider deals. Overcoming opposition usually requires forging coalitions, which, in turn, requires adjusting some goals or methods.

These interactions take place within a sociocultural system that shapes behavior by defining what is possible. Structure is inherited from the past, shaped somewhat by present actors, and changes according to a punctuated equilibrium model. Existing societies are the product of long histories of nonlinear but increasing division of labor and accompanying economic, political, and cultural inequality. The division of labor – into those who decide, follow, produce, manage, and consume – defines groups that relate to each other in particular ways within a society; it also defines how societies relate to each other and how groups in different societies relate to each other, such as Canadian mining companies and Asian workers. The groups and relationships defined by the division of labor and the accompanying infrastructure (e.g., factories, ports, and technologies) constitute social structure. Structure includes social processes and how society's members understand the explanations and justifications for existing relationships, what is possible and changeable, the desirability of change and its risks, and the consequences of breaking the rules. Structure also includes groups, which are value based, that is, that arise out of the struggles among groups defined by the division of labor and that take on a life of their own, such as conservation. Structure, then, includes the overarching pattern of interaction determined by the social organization, physical infrastructure, and culture that constitute the extant division of labor. As we saw in the previous two chapters, a group's position in this structure permits some behaviors and forecloses others, provides differential control of power, wealth, and other political resources, and shapes interests.

Structure is what the individual confronts when, fed up with oil companies, war over oil, and oil-generated pollution, they decide to renounce their car, but can't get to work or shop without one because of the high costs imposed, for example, two hours on a bus. Structure is what conservation NGOs confront when trying to pass legislation that runs contrary to the interests of those who control powerful institutions, that requires basic change in the organization of production or social relationships, or that requires different infrastructure. In an effort to halt roadbuilding and resulting land conversion conservationists not only come up against those who by reason of their position in society possess significantly greater political resources such as wealth and command of major institutions, but existing infrastructure as well. Oil, auto, and related industries benefit politically from existing transportation, residential, and commercial infrastructure, which they helped to bring about and which would be very expensive to alter. Millions are employed in building and maintaining such infrastructure or depend on it, making them wary of change. The US political structure makes major change more difficult than maintaining the status quo, adding to the structural obstacles. Structure even limits the powerful. A publicly held company that chose to internalize the cost of refraining from environmental harm when competitors did not probably faces lower profits and exposure to being taken over by another business. Furthermore, its management could find itself subject to major shareholder complaints of violating the legal obligation to prudently maximize profits. The law reflects and also imposes structure. It's no different for political leaders. Brazilian President Lula, an avowed supporter of protection for the Amazon, reversed himself in 2005, and restored logging licenses in the face of violence and threats of violence from loggers (Larry Rohter, 2005: A3). US President Thomas Jefferson ignored John Adam's advice that indulging in pomp and circumstance was necessary to maintain presidential influence, and it hurt his effectiveness (David Kertzer, 1988: 163).

Social structures are dynamic. They change over time as a result of interactions between groups, interactions among societies, and ecological interactions involving resource use, pollution, and the like. Change is usually incremental. Continual interaction among individuals and groups in the pursuit of goals – the waxing and waning of alliance and coalitions, the playing out of personality and organizational imperatives and constraints, and the occurrence of unpredictable events such as the Exxon Valdez oil spill, scandal among the leadership, the application and playing out of new technologies which weaken some and strengthen others, strategic blunders – generates small changes. Classes and regions wax and wane. Not all incremental changes are accommodated, as when groups seeking a place at the table mobilize but are repressed, or the application of new technologies is resisted by those who profit from older ones, or external forces make business as usual untenable but businesses (or states) persist. These accumulate overtime, adding stress to existing relationships. Periodically the concatenation of accumulated stress can no longer be managed and a period of relatively rapid structural change ensues. "Events" make visible and usher in changes that have been long underway such as the rise of new economic sectors, the dominance of new technological regimes, the decline of societies relative to others, or societal reconstitution in a more complex form. The industrial revolution, the US civil war, the Algerian revolution, and the penetration of market relations into every part of the globes are examples of structural transformation at various scales. At a very large-scale structural change has evolutionary attributes – societies have tended to greater complexity and greater division of labor, to more hierarchy, concentration of power and sophisticated social control and to become larger geographically and demographically. More complex societies have displaced and absorbed less complex societies. Structural change at the social and global scale is path-dependent in many respects; the past forecloses some options (as when soils, forests, and energy sources are exhausted) and makes others possible (as when population size, technological expertise, or control of surviving institutions enables domination of others or exploitation of previously untouched ecosystems). European and US development depended on the exploitation of what came to be called the Third World. Today's developing countries have no similar regions of the world to look to for cheap labor and raw materials.

Structural change alters the forces affecting human behavior at every level of aggregation, from the individual to those of the largest players in world systems, redefining limits and possibilities. Although some leaders of countries see the storm clouds of climate change, rising energy demand to fuel growth does not easily permit energy use to be curtailed. The goals of the Convention on Biodiversity, a treaty signed by most states – with sincerity and earnest intent by some – are being overwhelmed by growing human population and consumption. Structural change does not just limit – it also provides opportunities. Periods of rapid change are turning points – like the switches in a rail yard which send trains down different tracks. When conservationists (or others) correctly identify turning points, understand the levers, and can get to them, it is possible to consciously influence the human direction. It is easier to influence the small scale – turning the influx of amenity migrants into a constituency to protect species (Ray Rasker and Ben Alexander, 1997) – than it is to influence the next energy source for a society or civilization.

Organizational cooperation and conflict

Most of politics involves organizational conflict and cooperation. NGOs trying to influence decision makers must bring political resources to bear on them. Success depends in large part on bringing more critical resources to bear on decision makers with regard to the issue than opponents do. Organizations have little choice but to devote energy to mobilizing more resources in order to gain or sustain relatively higher levels than opponents and make credible their promises and threats. They can achieve relatively high mobilization in three ways. First, by undercutting resources available to opponents. This includes acting quickly, before they can mobilize; or acting quietly so they don't mobilize. Second, new resources can be mobilized to enhance the capacity to deliver support for the cause. And third, by forming coalitions and alliances with other organizations.

Because all of Part 2 is focused on mobilization and organizational cooperation and conflict, only the briefest summary is provided here. Conservation NGOs, even as they seek to out-mobilize opponents, must adapt to opponents, strength and ability to block them. NGOs may need to alter goals, settle on a longer-term campaign, or seek help they would otherwise not seek. Strategic and tactical choices will be adjusted with the intent of doing what's necessary to overcome opponents. This includes anticipating opponents' actions, gathering intelligence on them, and deciding whether and what counter measures to take.

Mobilizing new resources requires NGOs to invest existing resources in that effort. The costs may rise and fall based on efforts of opponents and competitors to mobilize the same resources. Costs include changes the NGO must make to attract new resources. When mobilization entails coalition- and alliance-building or targeting other politically organized targets there is almost always some quid pro quo that requires adjustments in NGO behavior. Coalition partners and new members that join en bloc have their own ideas about goals and how to do things and it takes energy to address them.

Thus, whether NGOs are relating as dance partners or boxers, they must adapt.

Both mobilization and programmatic work put enormous pressure on NGOs to innovate. Innovation is important tactically to keep opponents off balance and gain the attention of new potential adherents. But tactical innovation is problematic because it involves stepping outside established repertoires of behavior, which can trigger repressive responses. The media may denigrate or ignore, opponents may resort to violence or other forms of attack designed to intimidate and scare sympathizers and allies away. The actions of opponents and allies in turn constrain and present new options.

NGOs must operate within a political universe of overall limits. Powerful cultural symbols are in finite supply and conservationists must compete with opponents over their meaning, use, and ownership. Strategy and staffing are important in such contests. Money is limited as are people likely to become activists. As conservation NGOs have become more dependent on paid staff and hence on foundations, attracting financial resources has become a major preoccupation. The search for competent and successful fundraisers and campaigners is almost constant.

Because no conservation victory is ever safe in the face of human appetites, mobilization must be sustained. To sustain mobilization requires building communities around conservation so that conservation issues and being part of the conservation community become central to people's lives rather than an afterthought. Perhaps more than anything else, the creation of the sort of community that sustained the battle for civil rights in the United States and the end of apartheid in South Africa has been neglected by conservationists. When NGOs do address community it will redefine their organizational behavior and organizations in fundamental ways.

Social structure

The US Endangered Species Act is one of the most powerful conservation laws in the world, made so by its emphatic requirements to recover endangered species and citizen suit provisions. But in the context of an economy structured around endless and accelerating growth, that is, one based on endless conversion of habitat, it is reduced to a defensive tool. If politics is the art of the possible, social structure largely determines what is possible. Erik Erikson (1968: 293, 309) observed that human behavior was the result of a person's developmental history and the possibilities of their particular historical time and place.

Societies comprise groups that are differently positioned by reason of their function within society. Functions are defined politically (rulers, ruled, bureaucrats), economically (big and little capitalists, labor, underclass), by gender, age, and culturally (opinion leaders, dominant or minority ethnic group members, and religious leaders and followers). The organization of groups thus defined, along with infrastructure (the tools used to produce and deliver goods and services, and the energy that powers the tools), laws (e.g., defining ownership, systems of national accounting, fiduciary duties), custom (nonlegal rules and expectations, including how and when to bend and break laws and rules), and how a society's members understand, and accept or reject existing relationships, constitute social structure.

The natural world shapes the contours of social structure. The presence or absence of fertile soils, the availability of water, its volume and ease of access and seasonality, weather patterns and climate, the types of energy and minerals available, topography, and vegetation, all shape settlement patterns, transportation, trade, communications, types of construction, and political and economic possibilities. The Columbia River system was once so rich in fish that it supported some of the few nonagriculturally based complex societies. The relationship is not one way. Only particular social forms and technologies can make use of certain resources. Construction of huge dams, such as those on the Columbia that have destroyed salmon runs in order to provide cheap hydropower, or drilling for, processing and utilizing petroleum on a mass scale only occurs in industrial societies.

Social structure is often conflated with the nation-state (a society) or a state's subunit. Human history, however, is much about conquest and migration. Thus "a society" or state is usually comprised of many peoples, who possess not just differing cultural characteristics but interact as social and sometimes as political units, for example, Scotland and Wales within the United Kingdom. By the same token societies and states or empires have always been part of larger systems; systems comprised of trading partners of varying degrees of inequality, of colonizing polities and the colonized, and so on. Whole societies generally do not interact – particular groups such as rulers, militaries, and economic entities do – although whole societies may be affected by the actions of others, as when great wealth or virulent diseases are transferred. The point is that important relationships among groups today frequently transcend borders and are often global in character. The current (2008) energy and food crises make plain such relationships, but so do institutions such as the World Trade Organization and Davos meetings.

The organization and relationships among groups, although in some flux, are relatively stable much of the time; structures persist although conflicting interests clash and relationships are altered. The ruling and electoral coalitions that included most extractive industries recently lost control of the US congress – a shift in power that is generally good for conservation. This shift in power somewhat weakens those elite factions out of government; it does not alter the fact that large-scale profit-seeking businesses as a whole make the economic decisions and exert enormous influence on, or control over political decisions affecting their autonomy and the pursuit of growth as the dominant social good. The stability of structure rests upon two primary factors. Those who occupy positions of great wealth and power in the social structure seek to protect their position collectively and generally will act with great ruthlessness in doing so. Because of the resources they control they are uniquely positioned to reward others for going along and to impose great costs on those who challenge them. Secondly, there are hosts of cultural-psychological reasons. Humans are creatures of habit. Although it can be advantageous to go the wrong way down a one-way street to reach a goal, most see advantages to systemic predictability and reliance; big change is uncertain. Many aspects of social structure people would find most objectionable are invisible. The very wealthy and the very poor are usually hidden and thus emotionally distant: they don't arouse intense envy or anger, or outrage at injustice. Major change usually occurs when the powerful are unexpectedly weakened and others are positioned to act.

Understanding structure is important for a number of reasons. It is central to formulating good campaign strategies and overarching strategies.

- Understanding structure reveals who holds real decision-making power.

- Structure provides an understanding of the discretion that individual decision makers have. Individuals do make a difference (Nixon's paranoia, Bush's ineptitude and lack of connection with reality, Gorbachev's desire to be accepted by western leaders [Timothy Allen et al., 2003: 156–7]), but structural constraints limit their discretion and describe the situations in which leaders' personalities are likely to get them into trouble with other powerful actors. In developed democratic states governments have little discretion about fostering economic growth and enforcing order, for example (John Dryzek et al., 2003: 1–3).

- Analysis of social structure makes clear the relationships among groups within and across societies, thereby enabling conservationists to better identify those groups best able to influence decision makers.

- It reveals the range of interests decision makers or opponents may have and that are susceptible to pressure. Often bringing pressure on an entity's interest unrelated to conservation (e.g., video rentals) may work to gain action on the conservation interest (pulping primary forests to manufacture paper used for mail-order catalogs).

- It helps reveal an accurate picture of potential allies' and opponents' resources, strengths and weaknesses, and what options are open or foreclosed to them by reason of their position.

- It helps conservationists know whether they are trying to break through a "load-bearing wall" that is likely to be fiercely defended, or are trying to breach an obstruction less valued by opponents or decision makers. An international oil company that can drill elsewhere may give way to an effort to protect an area when a smaller firm with no or few other options will not.

- It can help avoid setting campaign goals that are predicated on breaking down such a wall and are likely to fail, resulting in disappointment and demobilization of supporters. Or, if such goals are necessary – and ultimately conservation success depends on breaking through such walls – it alerts campaigners to the need for extraordinary resources and the need to prepare for a long campaign.

- It contributes to useful categorizations, for example, allows for identification of similarly situated elites under similar pressures such as those whose economies are close to breaking out of the cycle of debt and that may be most tempted to sell "natural resources" to put them over the top (Thomas Hall, 2000: 12; Albert Bergesen and Tim Bartley, 2000: 308). When decision makers are similarly situated then similar strategies may be employed.

- Structure reveals the degree to which human economic activities are dis-connected from ecological feedback loops resulting in decisions that are ecologically stupid. Thus, losses in currency markets may lead a holding company to direct the liquidation of forests to raise cash; the bursting of a finance bubble can cause metal prices to rise making destructive mining profitable. Because the ecological consequences will not affect profits or the immediate drivers of profit seeking, other means of pressure are required.

- It suggests in what circumstances elites will need to employ propaganda, such as selling the logging of ancient forests as a means to fire prevention and "healthy forests." This can give conservationists more time to fashion effective counter moves that expose obfuscation.

Gaining knowledge of specific structural relationships is not always easy because efforts are made to conceal them. The Bohemian Grove does not issue guest lists. Because many audiences are unaware of, for example, how concentrated economic decision making or the media is in the United States, mobilization that depends on exposing structural relationships can come across as conspiratorial and therefore suspect. Regardless, both insider (reliance on members of elites) and outsider (mass mobilization, protest) strategies depend on a good understanding of who is connected to whom and the nature of the influence or power they have over another and in what policy areas this influence is exercised. The German Greens, when in power, could influence the social democrats and through them the European Union (EU), in a policy area such as climate change, but not on limiting overall economic growth (Neil Carter, 2006: 122–7). Protest tends to be ignored or condemned in Britain but obtains a hearing in some other EU countries (John Dryzek et al., 2003: 43–53).

Recognition of structural limits does not mean they should constrain conservation demands. Indeed, effective conservation depends on changing what is possible. Pushing the envelope tactically, strategically, and programmatically can bring limited structural changes. Greatly increasing conservation's relative level of mobilization and sustaining it would constitute a structural change. Greatly increasing it did bring about structural adjustments in the late 1960s and 1970s in some countries. But the countermobilization of industry checked conservation. By contrast, the US civil rights movement, although its mobilization was countered, has been better able to sustain a higher level of mobilization than its most strident opponents and has made change permanent. Blacks are still discriminated against but Jim Crow is dead.

Structural change and evolution

Structures change and evolve, altering human behavior at every level of aggregation from family to actors in world systems and redefining limits and possibilities. Structural change is usually incremental, punctuated with periods of relatively rapid change. Rapid change is usually the result of the cumulative effects of smaller changes – the tipping point phenomenon – or caused by conflict resulting from resistance to change, the resolution of which results in new societal relationships and organization. Not all change is (de)evolutionary, that is, change that involves significant transformations in complexity or social organization into new forms.

The reason for this book, which is focused on near-term politics and therefore takes existing structure for granted, is twofold. Because structural change is continuous, it affects the near term, if only in slight ways. More importantly, many societies and the global system are in a period of more rapid structural change. Levels of social complexity and economic centralization and interconnection are reaching limits; the energy sources that have underwritten explosive growth for a century are peaking; and the destruction of species and ecosystems is degrading ecological services on a global scale. This transformation creates an evolving context for near-term political campaigns that must be reckoned with because changing relationships create opportunities and new roadblocks.

Secondly, action in pursuit of near-term goals by conservationists and others affects the course of structural change to the degree such actions alter power among groups and major choices are made among alternative policies. New opportunities are presented as a result and to some degree can be created – if they are recognized.

Seeing the near term in the context of larger structural issues is not easy. The results of direct efforts at major structural change, such as revolutions that aim to "skip historical stages," are reminders of this difficulty and that the unintended consequences always outweigh in importance the intended consequences. This path dependence is true at lesser scales as well. That the concatenation of the Endangered Species Act of 1973, decades of overcutting of Pacific Northwest forests without concern for their ecology, and the 1992 presidential elections would lead to the Northwest Forest Plan and sharply diminished cutting to protect the northern spotted owl, would not have been easy to predict. Unfortunately, because conservationists were not well organized globally, a smaller cut in the United States meant more cutting elsewhere in the world. It's not clear whether conservationists anticipated this, though it was easier to foresee with no decline in demand for lumber.

Similarly, the nonlinearities of natural and human social systems would have made it difficult to anticipate that the 1999 Seattle demonstrations against the World Trade Organization would have contributed significantly to the breakdown in negotiations, even though they were already faltering (Naomi Klein, 2004: 219–29). It was only after NGOs arrived in Seattle that they realized their strength and took forceful action to shut the meeting down.

Despite the difficulties of anticipating effects on the course of structural change, trying to understand and plan for structural change is important. Most NGOs do so, though neither consciously nor systematically. Observers note that generalists appear to possess better anticipatory insights than experts in a field because the latter know too much and get caught up in the details, losing sight of the pattern (Philip Tetlock, 2005: 67–188). Others argue that artists are best at seeing trends because they focus on patterns (Marshall McLuhan, 1964: 62–73; Murray Edelman, 1995: 37–52, 69), and sensitivity to changing patterns is critical to taking adaptive action (Anthony Wallace, 1970: 200–1). But which artists to watch and how to interpret them; or does one look for an emergent pattern of perception in artists' work as a whole? Discerning the emergent also depends on the capacity to see nonstereotypically (Forgas et al., 2003: 18), to be open to seeing new categories and arrangements.

There are regularities in structural change at a variety of scales that if attended to will provide conservationists with useful insights and leverage. Insight does not equal leverage and conservationists are not very powerful (a tiny tugboat trying to alter a huge ship's course). But the weak have a long track record of beating the strong with guerrilla strategies. Although conservationists do not seek state power or see violent combat as a means to their goals, the strategies of guerrilla movements are all about dealing with much more powerful foes and offer lessons.

Cycles, waves, and opportunities.   A number of political economists, historical sociologists, and others have identified regularities in structural shifts (Jack Goldstein, 1988; Brian Berry, 1991; Andre Frank, 1993; David Fischer, 1996; Berry et al., 1998; William Thompson, 2000; Koopmans, 2004; Kreisi, 2004; Mitchell Allen, 2005; Sing Chew, 2005; William Thompson, 2005).

One regularity identified cycle consists of a 55-year cycle of price inflation–deflation, which also marks replacement of major infrastructure and technology (termed a K-wave after economist Nicholas Kondratieff) and two shorter growth cycles (named after Nobel laureate Simon Kuznets) within it. Picture the two-humped *Camelus bactrianus* superimposed on the one-humped *Camelus dromedarious*. The first growth cycle of 20–25 years approximates the rising price phase of the K-wave and the second is synchronic with the K-wave decline (25–30 years). When price troughs and growth troughs coincide as they do at the beginning and end of a K-wave, economic conditions are typically termed a depression or recession. Brian Berry's (1991) analysis of US cycles implicitly predicted the timing of the current recession, but it is important to note that the cycle length varies. Like oceanic tides the cycles are not synchronous across all economies; and how they unfold varies with political interventions and regime type (Rudd Koopmans, 2004: 36), the size of the economy, and levels of integration into the global economy. K-waves are also subject to longer-range perturbations and contingencies, for example, this recession is witnessing inflation due to oil shortages caused by war and the rapid growth of China's state controlled economy.

- In the downturn phase of the price–investment–technology cycle (in the US: 1981–2009?), profits come under pressure from market saturation and volatility, increasing the intensity of technological research and development to reduce labor costs and increase efficiencies. The technologies and associated infrastructure that appear most lucrative will attract major investments and when they become widely adopted they will fuel the next wave's upswing in prices and growth. *Once a cluster of innovations is adopted it cannot be derailed.*

  Investment decisions critical to the next wave are made by business people trying to protect their industries and innovators trying to create new ones, by investment bankers and fund managers seeking higher returns, and by governments responding to pressure and campaign contributions. Venture capitalists may be the single most important gate-keeping group (Brian Berry, 1991: 183), and with limited influence conservationists might do best to focus on them and governments. Some venture capitalists fund conservation and they are more inclined to look at a broader range of factors in making decisions, including their own passions. Choice of technology and infrastructure will not halt anthropogenic extinctions, as David Ehrenfeld (1979) long ago pointed out. But some technologies are better than others in their impact on the natural world, especially if shared with rapidly developing economies.

- In the United States, which differs from parliamentary systems with proportional representation, recessions typically mark shifts from conservative to liberal leadership (not to be conflated with which political party holds power). High inflation combined with slow or no growth typically ushers a shift from liberal to conservative rule, for example, the ascendancy of Reagan and his coalition dominated by resource extraction, energy, highly polluting industries, and a good portion of the financial sector (Kevin Phillips, 2006). Although both major parties are committed to growth, liberal elite factions tend to be more sympathetic to conservation. To the degree such shifts regularly occur based on structural factors, conservationists can focus their energies on factions likely to hold power. This may bear little fruit given the hostility of some elite factions, and in even with sympathetic factions *only concerted efforts by conservationists in the electoral process will result in the connections and IOUs that can be parlayed into policy gains.*

- Upswings in growth and price cycles increase grass-roots political action and calls for cultural change that contributes to reshaping the political agenda (Brian Berry, 1991: 144–66; Berry et al., 1998). Grass-roots movements are invigorated by prior years of conservative rule and attendant rising inequality, the consequences of deregulation and unrestrained accumulation, the failure to address social problems, increasingly by ecologically related crises, and by the perception that liberal rules offers improved opportunities for change. An increase in grass-roots politics is not the same as success, however. Some have noted that although successes are indeed clustered, they do not coincide with growth and price upswings but tend to follow crises in which elites need support from weaker groups, which increases the latter's bargaining power (Jack Goldstone, 1980). In the United States, for example, the 1970s conservation victories came when a besieged president needed organized support. The powers of the US civil rights' movement and women's movement were much enhanced by World War II. The labor of both was needed and neither group was willing to return to second-class status. These movements helped create the political space for other movements such as emergence of conservation as mass movement.

- Clustering of policy change is not always due to structural factors. Its episodic nature may be due to media herd effects, court decisions, or the unanticipated effects of crosscutting legislation (Frank Baumgartner, 2006; Helen Ingram and Leah Fraser, 2006).

- As growth and price upswings gather momentum, the competition for resources needed for growth can become acute and lead to increases in conflicts over resources (Joshua Goldstein, 1988; William Thompson, 2000). Populations are usually not mobilized for war by direct appeals to material aggrandizement, but by calls to mission, patriotism, and the like. But because resources lie at the root of many conflicts, including global conflicts (e.g., the Japanese Greater East Asian Co-prosperity Sphere versus the American Open Door) and this becomes apparent over time, an opportunity is provided to mobilize people against growth destructive of biodiversity because it is also destructive of human societies. It is a message that must be carefully crafted because many don't want to hear it. Oil, agricultural hinterlands, and water are likely to figure prominently as sources of resource conflicts in the coming years.

- Global wars, according to Joshua Goldstein (1988: 364–75), Brian Berry (1991: 157–65), and William Thompson (2000: 88), are associated with every other K-wave peak. These wars tend to involve efforts by rising powers to displace declining hegemonic powers that have become overextended (e.g., the German and Japanese challenge to the United Kingdom and its allies). Such wars are perhaps not inevitable given existing international institutions, and it is possible that proxy wars may substitute. But shifts in hegemony are inevitable. Conservationists may not be able to affect such events or even meaningfully prepare for them (witness the difficulties in dealing with ongoing smaller conflicts in east Africa). But they are of enormous importance. A Chinese or EU–Chinese hegemony will entail different conservation consequences than the current US-dominated world. The particularities of war (e.g., the severe oil pollution in Southwest Asia) and the resources over which they are fought have specific effects on protected areas and biodiversity. War often marks a decline in law enforcement in affected parts of the world. The interruption in food supplies can cause people to invade wildlands.

Deep drivers.   If the temporal clustering of change provides useful insight into how structural change shapes behavior and opportunities, it is limited insight. Neither cycles nor waves explain their changing content and what has moved humans from a globe dominated by hunter–gatherer bands, then agricultural villages, then ancient and classical states, empires and trading systems, and now modern states and a global world economic system. Human social evolution is driven by individual and group motivation as shaped by experience, but these have given to several structural factors that evolve according to their own dynamic, shaping human motivation and imposing their own imperatives regardless of motivation. If conservationists are to grasp and address the extinction crisis they must understand how these factors shape behavior toward the natural world. At heart is a macrodynamic governed by the relationship between population and economic intensification and its offshoots: hierarchical, expansionist, and technological imperatives that have become dynamic forces in themselves.

Population – intensification. Although humans have been causing extinctions at an increasing rate since they left Africa some 40,000 years ago, the scale and rate of extinctions have increased geometrically since the Neolithic age. At the beginning of this period – some 15 millennia ago in southwest Asia and later in other regions – it became impossible for "excess" people in a band to simply move next door. Next door was occupied. A drying and warming trend was also reducing available food. More food had to be extracted from the land base available to each group. Marvin Harris (1977) refers to this process as intensification.

Deliberate human selection of wild grasses followed by their domestication led to increased food production and eventually to agriculture. In some cases animal domestication was the path taken. Increased food production allowed population to grow further, and a new dynamic started governing behavior: population growth and pressure > intensification of production > population growth and pressure > intensification, and so on (Marvin Harris, 1977). It was a very different way of life than what went before and had major consequences. People had to settle in villages in order to reap the return on their investment in crops and animals. Diets became less varied and protein-poor, stature and life expectancy declined as infectious and other diseases became more prevalent (Marshall Sahlins, 1972; Richard Steckel and Jerome Rose, 2002). People worked many more hours to sustain themselves (Marvin Harris, 1975). These factors and a sedentary life that permitted accumulation, created pressure for increased consumption; consumption joined population increase as a driver of intensification.

The transition to agriculture involved fundamental experiential and cultural change: from living in Nature to being outside of it and trying to control it. This created domestic and wild, inside the village and outside (Paul Shepard, 1990). This was the human fall from grace recounted in so many stories born in that era; a fall that left humans frightened of the world, anxious about death, and preoccupied with suffering, immortality, and salvation (Paul Shepard, 1990; Morris Berman, 2000). The need to explain the hostility of Nature (such as drought) and estrangement from it left animism behind and gave birth first to tribal agriculture's Earth-mother goddesses, then to the "mature" worship of distant and arbitrary sky-gods associated with the first states and still familiar today. To this last must be added a secular version: Nature as machine and subaltern, still ours to manipulate (David

Ehrenfeld, 1979; Neil Evernden, 1985). Only recently has there emerged in dominant religious and philosophical doctrine an effort at rapprochement with Nature.

The intensification dynamic based on population pressure continues down to the present in much of the world, although it interacts with many other factors and is therefore less directly visible. In much of the world increased consumption has become a major if not the major proximate driver for intensification. It is not confined to elites although it serves their interests as organizers of production. In addition to consumption that satisfies needs, status or profit, consumption plays a role in temporarily assuaging mortality anxiety as we discussed in Chapter 2 (Sheldon Solomon et al., 2004 Kasser et al., 2004).

Hierarchy. As societies intensify production and populations grow, the division of labor increases and becomes more complex, giving rise to a new dynamic. Anthropologist Richard Lee (1972: 343) observed that "(w)hen the size of the local group grows beyond the scale where everyone knows everyone else well, new modes of behavior and new forms of social organization must crystallize in order to regularize the added complexity." When human groups reach 350–400 a new specialization emerges: the generalized decision maker or headman. Egalitarianism, enforceable in band society, can no longer be enforced in larger groups (Christopher Boehm, 1999). The effects of this are compounded because those who come forward as decision makers are almost always self-aggrandizing. Morris Berman (2000: 73–5, 106) argues that every society produces power seekers; they are the product of loveless or humiliating childhoods, and power rather than cure is their salve. It brings attention and promises control of a hostile world.

Once centers of power are created, they reproduce within society and accrete more power as tumors trigger angiogenesis. As societies grow the village headman becomes a chief or bigman, then a priest-king, emperor, and today premiere, first secretary, CEO, chairman of the board of directors, chairman of the state council, prime minister, or president (Johnson and Earle, 2001). Ruling groups replace individual rulers and they increasingly employ layers of retainers – administrators, bodyguards, managers, and enforcers – who benefit disproportionately from their position and become tied to rulers.

This change – away from decision making by all adults – would have been momentous enough in itself. But it is the group's diminished control over the decision maker that marks the significant transformation. Although ruling groups make decisions for society, their decisions are filtered through their own interests. Those interests are often at odds, not just with those over whom they rule or exploit economically, but with the maintenance of the social order itself. They frequently overreach, for example, to enhance their personal wealth or power, by ratcheting-up extraction from other groups, going to war, or exhausting scarce resources, causing rebellion, bankruptcy, and ecological ruin (Marvin Harris, 1977; Diamond, 2005).

Of particular concern for conservationists is that hierarchy breaks the ecological feedback loops that constrain the degradation of Nature and decreases long-term flexibility by foreclosing options. When the powerful are insulated from the direct consequences of their actions by the structural allocation of the costs to others, they do not act, if they act at all, until the consequences become so serious that social disturbances arise or their power is otherwise affected.

The consequences of hierarchy are exacerbated by centralization. Complexity, size, and the pursuit of wealth and power lend themselves to centralization, which requires imposing broad and uniform categories on the world to make it manageable. Political scientist James Scott (1998: 3–5) likens these categories to a large-scale map which lacks the detail needed for accurate navigation. The implementation of decisions in centralized structures is also likely to be cumbersome (Roy Rappaport, 1976: 58–60). Complexity apart from centralization can be crippling; together they make control mechanisms unwieldy and finally inadequate (Joseph Tainter, 1988). The psychological predisposition of most rulers, along with the ubiquitous human predisposition to dislike criticism or aspects of reality that don't conform to desire, means information from other groups that might serve as warnings of trouble are often ignored or repressed, which aggravates matters and makes the social system more brittle.

The cultural consequences of hierarchy are equally problematic for the natural world. Although people have been resisting and rebelling since the emergence of hierarchy they have not succeeded in ending it, only creating newer and often stronger hierarchies, with greater capacities for intensification. One problem, Paul Shepard (1990: 19–46) argued, arises from the experience of hierarchy and subordination; those in complex societies have no other model for organizing the world. When control of the sacred is transferred from the whole group based on its collective experience, to the elites, its primary role becomes adapting people to the hierarchical social order, rather than incorporating ecological experience into rules of conduct. Ecological rules may remain, but without adjustment based on adequate ecological feedback they become outdated.

Hierarchy confronts conservationists with a vexing question. Can the juggernaut of biological destruction be halted with its own tools? If hierarchy is an essential element of all societies with more than a few hundred (let alone hundreds of millions) people, and if those at the top are bent on maintaining their position, can society's control mechanisms be used for conservation beyond a limited degree? Some think not (John Dryzek et al., 2003), but others are more hopeful (Marvin Harris, 1977; Diamond, 2004). Regardless, hierarchy is like gravity in the natural world – it bends and shapes reality.

Technology. If intensification drives technological innovation, the emergence of hierarchy puts it on steroids. By controlling production and distribution, elites extract surpluses that are invested in innovation such as improving seed productivity, constructing irrigation works, metallurgy, and so on. The rapid development of technology is made possible by a hierarchical division of labor and technological development that in turn reinforce the trend to a greater division of labor. Many technologies depend on the cooperation of large numbers of people – today in the hundreds of thousands – from top bureaucrats and engineers to posthole diggers. Technology extends a society's capability to manipulate the world and mine soils, reshape rivers, consume huge tracts of forest, and harness or destroy other species.

Although elites have not explicitly thought about the extended reach technology provides in terms of calories, technology is about the capture of calories – calories necessary to sustain human activities and feed the machines. (Marvin Harris, 1975; Leslie White, 1987 [1975]). If a mountain lion expended more calories capturing and eating its prey than the prey offered it would starve. Humans have learned to use technology to gain energy subsidies. Making and controlling fire, exploiting the labor of other creatures and other humans (they produce more than they consume and the difference is taken), building dams, waterwheels, irrigation works, tools for construction and record-keeping, engines, the instruments of policing and war, and the extraction of petroleum are all means of capturing energy; and they all take energy to develop and use. It's calories we eat and calories that make and fuel our technology, part of which is turned to getting more calories and part to sustaining a growing population and higher consumption. It is energy that allows humans to gain access to minerals, water, and land and to reshape them. It is no coincidence that "energy" and "power" are used interchangeably. Energy capture keeps people warm, makes food easier to digest by cooking, and enables humans to impose themselves on ecosystems in a way no other species can.

Technology shapes human behavior because its adoption requires particular forms of social organization, including unequal social relationships, and because it must be fed. Factories need coal, cars need gas, and aircraft require refined metals. Technology enables humans to gain more resources, exhausting them and creating more waste than living systems can absorb without degradation. Soils salt up from irrigation or lose fertility from too frequent cropping and whole regions are denuded of forests and other vegetation. Easter Island's history is a grim reminder of what even technically simple societies can do. What industrial societies can do is even more impressive: the US Dust Bowl, the dead zone in the Gulf of Mexico, the smogs of Shanghai and Mexico City, the PCBs, dioxins, and other chemicals ubiquitous to drinking water supplies, and the unseen devastation of the oceans from bottom trawling and industrial fishing.

Technological innovation, based on the capacity for tool making and altering social organization, is driven by intensification and then by hierarchy, and comes into its own dynamic. Incremental technological improvements are always being made by those who directly use a technology, but much innovation is driven by competition between elites in different societies aiming to achieve advantage over one another, competition between elite factions or firms, and by efforts to solve problems generated by intensification and earlier technological innovation, e.g. fish-killing dams. Those societies, for example, organized around private property and accumulation of wealth, foster constant technological innovation. Although it is true that resulting efficiencies can reduce the per capita human footprint, concomitant population growth and higher levels of consumption overwhelm efficiencies, as has occurred with automobiles.

Maintenance of long-term flexibility is the best evolutionary strategy for organisms or societies (Roy Rappaport, 1976: 43), and here the contradiction inherent in technological innovation is glaring. Technological development appears to increase a society's flexibility to respond to its own growth, other societies, and a changing natural world, but it decreases it over time by degrading the ecological systems on which a society depends, thereby closing future options. The weakening and collapse of societies is often closely associated with ecological ruin – a ruin more easily brought about by the technological magnification of human propensities to manipulate.

Beyond the more primitive toolkits, technology breaks the real-time ecological feedback loops that inform a society when it is harming the ecosystems on which it depends (Roy Rappaport, 1974, 1976). Technologies not only create more serious problems than they solve (David Ehrenfeld, 1979) but also they enable societies to "overpower" natural processes and ignore the effects such as soil erosion and pollution build-up, delaying the consequences until the situation becomes difficult to reverse. Competition seriously aggravates this break. Farmers knowingly push production at the expense of the soil's integrity in order to stay in business. Entire societies do the same, ignoring longer-term negative consequences in order to prevent subjugation by others, to subjugate others, or to maintain control of resources that could be used by another.

Conflict and conquest. Intensification increases intersocietal cooperation and conflict, which, like hierarchy and technology, takes on a life of its own; not independent of intensification, but interdependent with it. The division of labor, always intersocietal to some degree, becomes more extensive and short- and long-distance trade increases. Trade may be on relatively equal terms or reflect differences in power and leverage between the parties resulting in the net transfer of wealth/calories and varying degrees of politico-military domination. Trading relationships are important because they define, as much as political control, social structure, and the largest unit of social evolution (Wallerstein, 1974). Even the most expansive political empires trade beyond their boundaries. Trade permits a division of labor over a larger geographic area and is likely to encompass a wider range of ecosystems; this allows the consequences of locally ecologically destructive practices to be ignored because production or extraction can be shifted to intact areas within the trading zone. Trade connections provide channels for diffusion of innovation and expertise, increasing (the always temporary) control over Nature and in many circumstances increasing the rate of innovation.

Intersocietal conflict on an ever larger scale results from intensification. One means of obtaining more resources is to take them, if not because one needs them immediately, then to deny them to an opponent. Fertile land, forests, minerals, additional sources of cheap labor, and other sources of energy can be obtained through conquest. Conquered peoples may be pushed aside or slaughtered and their resources taken, or they may be left in place to work for the support of their conquerors. Successful conquest often hinges on which society is more powerful, that is, who controls more energy in the form of troops and technology. Societies that have harnessed relatively greater energy subsidies have generally dominated, conquered, absorbed, or destroyed societies that have controlled less energy or used it less effectively. The Mongols, who conquered much of Asia from horseback, only maintained their conquests when they installed themselves on the thrones of conquered institutions and abandoned their historic way of life. Collecting, hunting, and gathering societies – societies with the lowest subsidies and the most ecologically benign of all forms of human social organization – have disappeared except in "peripheral" areas of the globe. Energy, intensification, and conflict remain closely linked in the world today and not just in Southwest Asia (Johns, 2002; Kevin Phillips, 2006).

Intensification, hierarchy, technological innovation, and intersocietal cooperation and conflict constitute intertwined macroforces driving structural change and hence human behavior. Efforts to change institutional and individual behavior must take them into account because they constrain options. Even the most powerful actors are constrained and can't simply act contrary to these pressures. When they do they often find themselves out of power. Thus, even those aware of these pressures and where they are leading – such as the consequences of global warming – are unable to act on them because existing pressures enforce a tyranny of the short term (David Fischer, 1996: 253). The next election needs to be won. The expectations of Wall Street analysts for net returns must be met by the next quarterly report. More oil is needed in the next few years and Iraq is sitting on a big pile of it. Expediency often trumps the long term. Fire fighting takes priority over fire prevention. Market worship enshrines the micro and short term. Thus even politicos and strategists who know better rarely act little better. Add to this the backward-looking mentality of so many decision makers – it is not just the generals who prepare for the last war – and it is clear conservationists confront profound obstacles.

Addressing these forces directly is the subject of another book. But a near-term political focus must take them into account if foolish strategies are to be avoided and opportunities identified and seized.

Summary and a nod to nature

Karl Marx and Frederick Engels (1975 [1844]: 110) said that history fights no battles, only real, living individual humans do. It's a useful reminder to not get caught up in abstractions or to give more explanatory power to structures and institutions than is their due. Thinking about institutions is more prone to misleading abstractions than is thinking about individuals. Notwithstanding, conservationists should realize that thousands of individuals do not spontaneously show up on a battlefield on the same day to engage in mutual slaughter on their own. They are ordered there under penalty of punishment, exhorted by appeals to patriotism, and some seek glory or reward. Those who order them there, usually from a place of safety, have political and economic ambitions. A battlefield does not exist without organized armies, command structures and governments, tax collection systems, interests, conflicting interests, armaments makers, transportation systems, food production systems, and so on.

With huge tasks before it conservation's political effectiveness hinges on a thorough understanding of the many factors that affect human behavior that in turn affects Nature. Individual motivation, structural factors, and everything in between is important to near-term politics. In each situation some factors are more important than others, but none can be ignored. Mobilizing people based on their existing feelings and ideologies doesn't mean biologically anchored propensities, contradictions implanted via acculturation, organizational imperatives or the evolution of large structures are strategically unimportant. The notion that "if we only get the facts out people will do the right thing" is hopelessly simplistic. It is one of the apparent ironies of life in complex social systems that on the one hand humans create such systems – piecemeal by accretion and episodic rapid transformations – and on the other hand seem to be at their mercy, inheriting what was constructed before (H. Young, 1998). Humans also inherit an Earth increasingly and tragically diminished.

The Earth and its systems, though much altered by human activity, are a powerful force in shaping individual and institutional behavior. Humans are adapted to and dependent on the Earth's air, water, soils, plants, and other elements and there is no way around that as the fragility of a space capsule suggests. Despite our hubris we are a biological species among others.

Climate exerts a powerful influence (Eugene Linden, 2006). Glaciations have prodded, aided, and hindered human migration over millennia. Relative climate stability over the last 10,000 years has contributed to the rapid growth of complex societies and extensive human domination of the landscape. Events like the little ice age profoundly affected agriculture, trade, and warfare. Very long-term processes and cycles, such as those associated with the Earth's elliptical orbit, axis shifts, and wobble, play a major role in climate and weather.

Villages and cities have arisen based on their proximity to water, forests and grasslands, and the Earth's morphology and topography. Annual flooding and prevailing winds have enriched some regions' soils, enabling them to withstand human abuse, and withheld or blown away soil in others. The oceans' currents have shaped long-distance trade routes.

Volcanoes have destroyed cities and sometimes civilizations, as have earthquakes and tsunamis. Volcanic eruptions have changed the weather for years or decades, affecting agricultural production, fertility, balances of power, and individual life chances. They have deposited the stuff that became over centuries fertile soils many feet deep across hundreds of square miles.

Natural processes don't just act on humans and other species. Species are part of natural systems and affect the systems of which they are a part at every scale (Tim Clarke et al., 1999; James Estes et al., 2007). Human behavioral flexibility and consequent ecosystem disruption makes for very complex interactivity. Diseases, for example, are often dependent on human population densities and movements and have wrecked havoc on entire civilizations, influencing the outcomes of conflicts (Europeans and Americans since the 16th century), destabilizing powerful regimes (Rome), and altering the balance of power between labor and capital (the Black Death in Europe). Disease has drained the vitality of some groups, and given their competitors cause for joy. The effects of earthquakes and storms depend on the built environment. As one engineer put it, earthquakes don't kill people, buildings do. Regardless of whether hurricanes are growing stronger due to anthropogenic climate change, their destructiveness is increased by the modification of barrier islands and foolish settlement and building practices. Logging often contributes to flooding.

Cyclical occurrences like sunspots, eclipses, and comets become the very stuff of culture, guiding behavior. The movements of the stars informed one of the earliest efforts to make understandable and predictable the behavior of emerging complex societies. Astrology is still with us. Millions consult stock forecasts and US President Reagan's wife relied on astrologers in arranging his schedule. So it is not just values and human purposes that distinguish human groups, but their notions of what is real.

There are probably few people around who think thunder means Zeus is angry. Yet there is always much that is unknown about the world. Because of the need to map and explain reality, the gaps in *knowledge* are filled in with *beliefs* consistent with human purposes, and the two are conflated (Roy Rappaport, 1999: 296, 304). It's a convenient method and often works well enough. But beliefs can get in the way of knowledge, which is often important to good policy. The Vatican's effort to suppress Galileo's findings was doomed by their importance to commercial navigation. But knowledge does not always have commerce on its side; often it's the reverse. Specialization means that people may have good knowledge in one area of their lives but embrace fantastical beliefs in other areas (Noble, 1997). All societies possess systemized knowledge (Ecological Applications, 2000), but it is not even, and people compartmentalize, approaching some aspects of the world critically and others on the basis of faith or convenience. People frequently exempt their impulses from review: scapegoating when they know better, or, although understanding how pheromones work and how early emotional relationships shape later ones, they believe that finding their true love was ordained.

With Nature dear to their hearts, and the future of biodiversity on the line, conservationists cannot afford to indulge the temptation to confuse knowledge and beliefs, especially when it comes to why people act and what is likely to motivate action good for the natural world.

Part 2
Conservation as if life depended on it

We've simply got to get the hogs out of the creek. ... (T)his is not a chore to undertake in your best trousers, politely pleading: "Here hog, here hog ... pretty please." To get the hogs out of the creek, you have to put your shoulders to them and shove.

Jim Hightower, *Get the Hogs Out of the Creek* (1995–1996: 32)

Actors

When Europeans began the conquest of the Americas there were about 100,000 grizzlies in what are now the western contiguous United States. By 1999 only about 1000–1500 bears remained, though this number has increased a bit since then (Chris Servheen et al., 1999). The passenger pigeon is gone, as is most of the tallgrass prairie and long-leaf pine forests. Ocean-going factory ships scour the oceans like bulldozers clearing land for a subdivision. It is difficult to quantify the loss of small species that live in fecund soils or in dense wet forests, but the overall rate of extinction is accelerating and the evidence of hemorrhage is everywhere. Each species extinguished takes unknown others to the grave because of interdependency (Lian Koh et al., 2004: 1632–4). If a repeat of the last great extinction of 65 million years ago is to be avoided there is not much time; recovery becomes more difficult as the cascade gathers momentum. Conservationists must become more politically effective sooner rather than later.

There is good news. Conservationists can take steps that have a record of improving political effectiveness whether it involves the long term such as building new institutions and shaping the next generation, or winning campaigns in the next months and years. Across the range of conservation activities – monitoring ecosystems and species and reporting findings, buying land or working with landowners to restore it, litigating and lobbying, mass education, electioneering, networking regionally, nationally, and globally, and direct action (Margaret Keck and Kathryn Sikkink, 1998: 217–8; Michael Gunter, 2004: 52–61) – opportunities exist to incorporate the lessons of other movements and make better use of conservation's own lessons. Opportunities exist regardless of organizational form – interest groups, political parties, social movement organizations (SMOs), quasi-public or governmental agencies.

Interest groups have typically been identified with everything from lobbying and litigation to land management and restoration work, political parties recruit and try to elect leaders, and SMOs organize grassroots protest, raise awareness, and bring pressure through disruption aimed at achieving broad social change. Practitioners and social scientists alike see these forms as becoming less distinct from each other. Parties now directly engage the grass roots and utilize SMOs for election mobilization (or create phony grassroots groups), interest groups engage in protest or act a public bodies administering programs or setting policy, and SMOs engage in lobbying (David Meyer and Sidney Tarrow, 1998: 20). Social movements in some countries still develop into revolutionary movements, although conservation SMOs generally have not.

In the United States and a few other countries interest groups have been the central players in conservation, with SMOs playing a larger role starting with the 1960s when conservation became a mass movement. Party forms have played a bigger role in Europe where systems of proportional representation offer greater electoral openings, but parties have better served environmental interests than conservation interests (Christopher Rootes, 2004: 663; Neil Carter, 2007: 48). SMOs are active in many European countries but very differently engaged – some receive government support, others challenge government. In Asia and Africa international nongovernmental organizations (NGOs) have allied with, supported, or helped create organizations that blend many attributes of interest groups and SMOs. International NGOs sometimes act as quasi-governmental agencies, funding and administering programs. United Nations (UN) agencies and treaty secretariats are also important actors, although much more so in some countries than others.

Throughout the book I refer to the conservation movement and include in that all organizational forms and types of conservation activity. Because of the blending of activities I also use the term "NGO" to cover SMOs, interest groups, science groups, and factions of parties unless there is a strategic or tactical reason to make a more precise distinction.

Action

Despite the enormous pressure on the conservation movement to achieve significant change in the near term, it has not broadly embraced what political scientists call transgressive action. It is the type of action most likely to result in rapid deep-going change (Doug McAdam et al., 2001: 7–8, 12) and refers to collective action outside normal channels, "represent(ing) a sustained challenge to power holders" by means of "public displays" (Charles Tilly, 1999: 257). Because those in power tend to ignore demands for significant change, innovative and disruptive tactics that create surprise and uncertainty are necessary to gain serious elite notice (David Meyer and Sidney Tarrow, 1998: 25–6). Achieving notice, however, frequently brings repression and in authoritarian regimes the repression can be harsh.

In response to state repression those engaged in transgressive action often start to conform protests to "scripts" that trigger milder state responses (David Meyer and Sidney Tarrow, 1998: 6, 21) and the results of this have generated much debate. In one view scripted protest easily becomes purely symbolic, imposing no real cost on the powers that be and hence cannot generate much in the way of concessions (David Meyer and Sidney Tarrow, 1998: 21, 22). It is also uninspiring to movement members (diminishing solidarity and recruitment) and can politically marginalize NGOs adopting scripts. When some NGOs script protest it becomes easier for authorities to isolate those who continue with transgressive action and concentrate repression on them. On the other hand such routinization can greatly broaden support because most people want predictability rather than risk when they protest. A peaceful but massive demonstration can have far more influence than a few hundred people breaking windows and burning cars (David Meyer and Sidney Tarrow, 1998: 24). John McCarthy and Clark McPhail (1998: 108–9) argue that although scripting of protest results in less disruption it opens new disruptive possibilities because protestors gain access to more arenas. Mary Katzenstein (1998: 196) notes that interruption should not be confused with disruption and argues that truly disruptive (and effective) politics can occur in the course of protest or in the course of ordinary politics such as lobbying, elections, or litigation. Similarly accommodation can be associated with protest, not just ordinary politics. Greenpeace's actions were considered radical in the early days of environmental protest, but its demands were not (Christopher Rootes, 2004: 615). Both Suzanne Staggenborg (1997: 437) and David Meyer and Sidney Tarrow (1998: 12–3) believe that less radical methods do not mean less radical goals. But do less radical methods mean less likelihood of radical (i.e., go to the root of) change?

Notwithstanding the importance of transgressive action in raising fundamental issues and pushing for fundamental change, more mainstream and less directly political activities such as lobbying land restoration continue to be vitally important. This book's focus is on how to make more resources available for all forms of meaningful conservation action and thereby improve the likelihood of success.

Three factors in achieving goals

Achieving conservation goals hinges on three factors. First, conservationists need a clear and uncompromising vision. How that vision is attained will certainly require negotiation and compromise, but there can be no compromise that sacrifices a healthy Earth or wild Nature to extinction. Clarity of vision and firmness of purpose are indispensable resources.

Second, conservationists must have a hammer – the wherewithal to reward, punish, and otherwise influence decision makers. Even sympathetic decision makers need support. US Interior Secretary Bruce Babbitt (2004) said to conservationists on taking office: "Don't expect me to do the right thing. Make me do the right thing." What he meant, of course, was that he was going to get plenty of pressure from opponents of conservation and needed conservationist support to hold his ground. Support in the form of campaigns, votes, and financial contributions, help establishing legitimacy, and provision of political cover in the face of opposition all contribute (DellaPorta and Tarrow, 2005: 6). Escaped slave and abolitionist Frederick Douglass (1985 [1857]: 204) understood this point, especially regarding unsympathetic decision makers. He observed that "(p)ower concedes nothing without a demand. It never did and it never will."

Achieving goals requires "endless pressure endlessly applied" said activist Brock Evans. He might have added that "enough pressure" was also required. The inability to bring enough pressure is a major conservation failing. This can only be remedied by mobilizing significant new conservation resources such as much larger numbers of people willing to act on behalf of biodiversity and ecological integrity as if their lives depended on it. Although achieving conservation success will always depend on the collective power of coalitions and alliances, and although it will often depend on past achievements that bring the state, with its resources, in on the side of conservation, at heart the hammer is the strength of the conservation movement itself.

It's not clear that conservation can arouse large numbers of people to the level of passionate commitment that many social justice movements have done, such as the African National Congress in its battle to topple apartheid. If conservationists cannot do that it is likely they will fail. If such a passionate mass movement is possible it will not be achieved unless conservationists can build a strong community for which conservation is integral. Conservation must become more central to people's daily lives, their social relationships, rituals, and myths. Short of that it will remain an afterthought.

Third, conservationists must be persistent and tireless. The opponents of conservation will never take seriously those who fight for the wild if they believe conservationists will grow weary and fade away. The hammer must be sustained over time. Conservation victories are not only seldom secure, but achieving them often takes a very long time.

I turn next to the importance of vision (Chapter 5) in creating a strong movement and community, then to forging both movement and community (Part 2A), and finally to maintaining them (Part 2B).

5 The role of vision

The important thing/is to pull yourself up by your own hair/to turn yourself inside out/and see the whole world with fresh eyes.

Peter Weiss, *Marat/Sade* (1965: 26–7)

Carlton Albert, Zuni head councilman, commented on the tribe's successful effort to stop a huge open-pit coal mine, by saying that "(i)f there is a lesson to be learned, it is never to give up and stay focused on what you want to accomplish." (Robert Cox, 2006: 243). Without a clear vision there can be no focus. Only with clarity about how the world ought to be – a place of healthy ecosystems with all of their native species – can goals be defined and action guided. Without a clear vision there is no context for understanding the present let alone the path forward.

Clear vision is an organizational imperative. Without it a NGO is as impotent as a vision without organization. Leaders, staff, activists, members, and supporters all need this clarity although they will differ in the depth of their understanding. They will also participate differently in developing, defining, propagating, and bringing the vision to ground. Too many NGOs lack clarity of vision, treat vision as something separate from daily work, or have lost vision through entropy. Vision performs a guidance role only when living, only when revisited and adjusted based on experience, only with new infusions of passion, and only when made part of daily work. A living vision is the sacred made concrete.

Movements comprise many NGOs and by their nature are not coherent wholes (Charles Tilly, 1999: 256). NGOs may have a clear vision, but can a movement as dynamic and diverse as conservation have a clear vision? Can so many organizations in different social, political, and financial circumstances, with differing histories and working on different pieces of the puzzle, share a common vision without it being imposed – the unanimity of the graveyard as it is sometimes called? Diversity of views and conflict over vision are inevitable and healthy if they do not degenerate into medieval theological disputation. Shared aspects of vision are always negotiated, the result of ongoing interaction and deliberate efforts at seeking common ground. The process itself contributes to trust and a more reliable division of labor. It is true that movements consisting of NGOs that share many elements of a vision, although each emphasizes different goals, strategies, tactics, and competes for resources, will generally be more effective. This is so if only because they spend less time bickering. But they are also more likely to coordinate their actions (if only quietly and behind the scenes), and it is more difficult for opponents to divide them. Because conservation's vision is a primary beacon calling people to the movement, shared elements of vision permit the propagation of the same overarching message.

Because new insights constantly percolate and because the ambiguity of language allows differences to be obscured and then later emerge, a shared vision is always in flux. There are times when movement ferment makes approaching a common vision more difficult than at other times. In the past decade and a half some sweeping themes have emerged in conservation. But more recently some powerful NGOs have moved away from these themes, calling for the subordination of conservation to economic development. It is unfortunate given that the conservation movement appears poised to reassert itself.

Unifying themes

Across the globe – in Australia, the Russian Far East, China, India, southern Africa, and the Americas – NGOs have come to recognize the need for large-scale conservation solutions (David Johns, 2003; Graham Bennett and Kalemani Mulongoy, 2006). These solutions are centered around the creation of national, regional, and continental systems of connected protected areas. Large, strictly protected areas are deemed vital to support healthy populations of all native species over the long haul, including highly interactive species such as top predators which often require large areas (Noss, 1993; Reed Noss and Alan Cooperrider, 1994; Michael Soulé and Reed Noss, 1998; Brian Miller et al., 1998–1999; John Terborgh, 1999; Michael Soulé and John Terborgh, 1999; Carlos Caroll et al., 2001). Extensive protected areas are also needed to ensure that enough of all ecosystem types are included so processes including succession, fire, and nutrient cycles, can operate unencumbered. Because even the largest protected areas are unlikely to be able to achieve these goals in isolation – and because biodiversity has evolved on the basis of lands and oceans interconnected continentally and globally as well as regionally and locally – connectivity between protected areas is seen as essential. Connectivity provides for dispersal, genetic flows, seasonal migration, and response to change.

The turn to large-scale conservation resulted from similar conclusions by conservation advocates and scientists: existing approaches were not appreciably slowing the loss of species and ecosystems. Although many species, and to some extent ecological processes, can be accommodated in small patches (e.g., the Oregon silverspot butterfly, J. McIver et al., 1991) or in areas heavily disturbed by humans, habitat loss, degradation, and fragmentation make it impossible for other species to persist. Even slight disturbances can be highly problematic (Kent Redford and Brian Richter, 1999). Proponents of continental and oceanic systems of large and connected protected areas recognize they are not the solution to every threat; even systems of large protected areas are not immune from anthropogenic changes such as wind- or water-born pollution, acid rain, and climate change.

One thread in the emergence of large-scale conservation was the creation in 1985 of the Society for Conservation Biology by biologists who sought to create a mission-oriented discipline like medicine, concerned with healing species and ecosystems and keeping them healthy. Although not an advocacy organization, many of its most distinguished members have played pivotal roles in the marriage of science and advocacy. They have engaged the policy process (Craig Thomas, 2003) and played important roles in grass roots and global advocacy NGOs such as the Wildlands Project and Wildlife Conservation Society. Its members have generated a body of knowledge vital to defining what must be done to recover species, the land, and oceans.

Another thread involved the Wildlife Conservation Society's Paseo Pantera Project. Initiated in the late 1980s it sought to create a connected system of reserves in Mesoamerica. Archie Carr III, Mario Boza (a founder of the Costa Rican national park system), and Jim Barborak pioneered this effort. Unfortunately Reagan's war on peasants, corruption, population growth, and the "sustainable development" agenda of international bureaucracies have slowed or deflected elements of this effort.

In the early 1990s the Wildlands Project, founded by leading North American biologists and activists, set out to create a continental system of connected protected areas. In conjunction with partners, the Wildlands Project generated several regional conservation plans for North America (Wildlands Project et al., 2003; Carlos Carroll et al., 2004; Wildlands Project, 2000, 2003, 2004, 2006) based on the biology of key species and knowledge of ecosystem processes. In addition to leading campaigns to make these plans a reality, the Wildlands Project catalyzed the creation of NGO networks and groups such as the Yellowstone to Yukon Conservation Initiative which work on behalf of large-scale conservation. The Baja to Bering marine protected areas initiative and ocean zoning proposals were inspired in part by these large-scale terrestrial efforts as was The Wilderness Society – Australia's continental Wild Country program. Scientists associated with these efforts have facilitated the development of the biology and ecology by enlisting colleagues, publishing, and organizing path-breaking conferences such as the one that generated *Continental Conservation* (Michael Soulé and John Terborgh, 1999; see also Graham Bennett, 1995).

Many larger North American and global NGOs were wary of these developments as were many traditional wilderness advocates. The proposals for large networks asked for "too much" and were seen as radical and subject to easy attack. Wilderness advocates feared that a focus on biologically rich, or recovery of rich but degraded lands, would distract from protecting what was left of intact areas – especially because these intact areas were at high elevation and not very rich biologically. Wilderness advocates, however, came to recognize that even pristine places were losing their wild qualities because they could not sustain animals, plants, and ecological processes. In the meantime advocates and scientists around the globe were independently coming to the same conclusions as North Americans while also being influenced by them. By the end of the century serious debate about the need for large-scale conservation was past, although what exactly it meant in practice continued to be debated.

In North America grass-roots groups and chapters of national groups have formally and informally coalesced to bring about regional systems of protected areas and in doing so have demonstrated how a common vision can help effectively organize conservation work. The wildlands network plans are like the picture on the front of the jigsaw puzzle box. Putting in place the pieces of the puzzle – restoring extirpated species, removing exotics, closing roads, and changing incentives for private landowners, expanding existing or creating new protected areas, restoring landscape, or aquatic connectivity – can be undertaken by NGOs doing what they each do best: this one lobbying, another working with landowners or buying land, still another working with agencies (see Dave Foreman, 2004 for a fuller discussion). Similar coordinated efforts are underway in Europe, the Russian Far East, parts of Asia, and elsewhere. The creation of the Northern Hawaiian Islands Marine National Monument was achieved by several groups operating with a common vision.

A step backward

If large-scale conservation is a globally appealing vision to many, many NGOs, there are also voices espousing a very different vision: conservation in the context of development. In this view the only realistic path to conservation is to acknowledge the inevitability of growing human populations and endless material growth. Wildlife and ecological processes must find whatever place they can in the midst of these forces. The precursors to this view are embodied in the Bruntland Commission (1987) report, which identifies human population and consumption growth as the root causes of biodiversity loss and simultaneously calls for a 5- to 10-fold increase in economic output to meet future human needs. This view was more recently expressed at the 2003 World Parks Congress where some leaders of the World Conservation Union (IUCN) called for protected areas to pay their way by means of mining and other economic activities within their boundaries (John Robinson, 2004; Locke and Dearden, 2004). They argued this was necessary to alleviate world poverty and obtain support for the protected areas. Bittu Saghal, an activist who leads efforts to protect tigers in India, responded that poverty had not been alleviated by exploitation of the vast unprotected majority of the Earth and it was foolish to think sacrificing the last remnants of the wild would therefore achieve this goal. It would simply destroy biodiversity.

Another IUCN proposal reflecting this view is reminiscent of US President Reagan's effort to keep the official measure of inflation low by excluding from the index the cost of buying a house. Two newly proposed categories of protected areas – fascinatingly named Category V and Category VI – are in fact extractive reserves, not protected areas. They are defined by development criteria, not by prescriptions to protect Nature (Locke and Dearden, 2004). By including them as protected areas, 10% of the Earth's lands count as protected, rather than the current 4%. This means many countries will be able to meet their commitments to protect Nature without really protecting it. This sounds a lot like fraud. Development to provide a better life for impoverished humans is a worthwhile goal but if there are any benefits to Nature they will be incidental.

Still more recently Peter Kareiva and Michelle Marvier (2007) argue that because conservationists have failed to build adequate political support for biodiversity and the protected areas needed to prevent its destruction, they should abandon trying to create protected areas where biodiversity is richest. Instead, they say, conservationists need to focus on protecting areas that are most important for people (e.g., those areas providing important ecological services) because support can be built for protecting these areas on the basis of self-interest. Conservationists will just have to settle for saving whatever plants or animals can be accommodated within these areas.

It is not difficult to understand why industry would advance such schemes cloaked in human welfare arguments, but why would conservationists? Why would failure to build support for a goal result in abandonment of the goal rather than a new strategy for building support? Is it a failure of imagination or of political will? A lack of awareness of alternative political tools or a misplaced notion *real politick*?

Before this view found expression among the influential conservationists just noted, it guided many conservation efforts. John Oates wrote about the results of these efforts based on his 40 years of research and conservation field experience in Cameroon, Ghana, India, Liberia, Nigeria, and Sierra Leone. He observed that

> ... (T)here are serious flaws in the theory that wildlife can best be conserved through promoting human economic development. It is a powerful myth

that has made all those involved in its formulation feel good.... (I)t seemed
to provide the best of several worlds: both people and wildlife would
benefit.... (I)n reality... the approach has had disastrous consequences
for many wildlife populations.... (I)t has led all concerned to assign low
priority to basic protection efforts that, because they must involve some
enforcement, are considered to be 'antipeople' and therefore opposed
to human development efforts. Yet the conservation and development
approach has done little for human development in most places where it
has been applied. The material benefits... to ordinary rural people... have
often been slight relative to those that have flowed to political leaders
and bureaucrats in the countries where the projects have been put into
practice, and to the consultant experts and conservation administrators
(based mostly in North America and Europe) who have planned the projects.

John Oates (1999: xv)

The path forward

This book is predicated on the notion that conservationists should not consider
deviating from their goals of protecting biodiversity and from what the best
science suggests are the most efficacious means of doing so, including large
and connected protected areas, until and unless they have exhausted all political
alternatives. They have not. Strategic mobilization of coalition partners and allies,
mass mobilization of carefully targeted groups, and mobilization of more activists
have not been undertaken with a good understanding of all the tools available
and with the lessons of other movements at hand. Moreover, as societies continue
to destroy their ecological bases – and the bases of all life – Nature will speak
more unmistakably, providing new opportunities. Crisis is not necessarily the
mother of insight – people often dig in and embrace denial more fervently. But if
conservationists are prepared to interpret Nature's voice compellingly, this is one
more basis on which people can be mobilized.

Part 2A
Forging the hammer

Find out just what any people will quietly submit to and you have found out the exact measure of injustice and wrong which will be imposed upon them, and these will continue till they are resisted with either words or blows, or with both.

Frederick Douglass, *The Significance of Emancipation in the West Indies* (1985 [1857]: 204)

Bringing government, business, and other institutions in line with conservation goals depends on the skills, capacity, and other resources of those working to achieve it (and that of those resisting it). Organization integrates and provides structure, form, and purpose to resources. Organization transforms caring into coordinated political action. It is organization with its many eyes and opposable thumbs that can be in many places simultaneously and has a memory capable of keeping good political accounts. The creation, maintenance, growth, and intelligent use of organization is at the heart of being able to influence policy and the choice of leaders. Organization is the hammer. It has the capacity to reward and punish over time or to hold out the promise of rewards and punishments.

If this talk of hammers, rewards, and punishments makes politics sound like it's all a bit morally primitive, it is. It's not that politics lacks decorum, politeness, nuance, or people with principle. All of these are attributes of politics. But it is also true that politics at root is about power, and principle, unless backed by power, usually doesn't fare well.

Organizations must not only invest in influencing decision makers but also in maintaining and strengthening themselves. Because most organizations cannot achieve their goals alone and require coalition partners and alliances, they must also invest in these. Because resources are always scarce, apportioning them between program and organizational maintenance and growth (acquiring new resources) is always an important question. Some grass-roots groups neglect attending to the organization believing that resources come as a result of substantive achievement; sometimes they do. Other groups become so focused on organization building or maintenance that this becomes their mission. Such groups achieve nothing while absorbing valuable resources.

How can conservation nongovernmental organizations (NGOs) increase their capacity and fashion a hammer effective enough to accomplish their goals while adhering to their mission? Programmatic achievement, as just noted, typically attracts participants and other resources. Those already part of an NGO usually become more strongly identified with it and more fervently committed with achievement (Amatai Etzioni, 1968: 403; Doug McAdam et al., 2001: 319–20). In some cases, however, success gives supporters and members a sense that the work is done and the time, energy, and money previously committed are no longer needed (John Dryzek et al., 2003: 158). The effect of programmatic success on mobilization is part of the analysis that follows, but because it is contingent my primary emphasis will be on the other major factors associated with mobilization. In Chapters 6–16 I outline the steps and tools needed to do this based on a synthesis of practitioner literature, social movement literature, and experience. I do not present a formula because there isn't one. Creativity, insight, intelligence, and boldness are needed.

The following chapters provide a framework for addressing the steps in mobilization.

- Identifying goals based on the vision of a healthy Earth (Chapter 7).
- Identifying the main strategic questions that must be answered to achieve mobilization (Chapter 8).
- Determining the criteria for identifying the targets of mobilization (Chapter 9).
- Understanding what moves to action the targets of mobilization (Chapter 10).
- Identifying the best messengers and the means of reaching mobilization targets (Chapter 11).
- Developing messaging that integrates effective emotional and cognitive elements (Chapter 12).
- Developing stories that convey the message and resonate with mobilization targets (Chapter 13).
- Using action to mobilize groups (Chapter 14).
- Using the right tactics from the mobilization repertoire (Chapter 15).
- Determining what's working and what's not (Chapter 16).

Before addressing these questions a quick review of mobilization is in order.

6 The centrality of mobilization to politics

During a discussion on 13 May 1935 concerning the relative strength of European armies, French Foreign Minister Pierre Laval asked Soviet leader Joseph Stalin to improve relations with the Pope because the Vatican could be an important ally against Germany's growing power. "Oho!" said Stalin. "The Pope! How many divisions has he got?"

Churchill (1948: 135)

US President Woodrow Wilson, knowing that a Senate majority opposed ratification of the League of Nations treaty nonetheless believed he could bring the Senate around because two-thirds of the US population supported the League. Wilson tried to rally the public to pressure the Senate, but the Senate did not budge. Many conservationists have adopted a similar strategy, sharing Wilson's flawed belief that public opinion translates into votes in Congress. It rarely works that way.

What translates into votes, as John Muir discovered in his battle with Gifford Pinchot over Hetch Hetchy, is organized political pressure. Muir won the public argument on the merits if newspaper editorials were an accurate indicator, but he lost the vote in Congress; Pinchot was able to mobilize and brought to bear more pressure than Muir.

Mobilization is the process by which people become and remain politically active and organized. Each person has only so much time, money, and other resources at their disposal. These can be spent in private pursuits such as earning a living or play; or some of this time, money, and skill can be devoted to collective action. The degree to which resources are devoted to collective action is the degree to which a person or group is mobilized. Sociologist Amatai Etzioni (1968: 405) describes mobilization as the transformation of a poorly combustible material into a volatile one. An organization can obtain new resources by mobilizing them from outside of the organization (e.g., recruiting new members or donors), or by increasing the level of mobilization of those already within the organization (e.g., gaining additional commitments of time or skill not previously available).

In most societies the level of mobilization is low. The majority are not routinely engaged in politics, and when they do engage it is often in a minimal way (e.g., voting). Few actively campaign, contribute money, protest, or take other actions. When a group can increase their level of mobilization relative to others it can make a significant difference on political outcomes. Groups do not start from the same place, of course. An enormous disparity exists among individuals and groups by reason of their social position and their wealth, skills, access to the state, and other resources that are attributes of social position. But those with fewer resources can still prevail on an issue if they bring greater resources (and use them more intelligently) to bear on a particular issue. More powerful opponents are often otherwise occupied.

The political landscape is constantly shifting as groups mobilize themselves while trying to prevent the mobilization of opponents or by staying ahead of opponents' countermobilization. When a group can sustain higher levels of mobilization relative to others, it can generate not just policy changes (Giugni, 1999: xix) but changes in the structure of society. Change in the levels of mobilization is one of the primary transformative engines of society (Amatai Etzioni, 1968: 393).

In the 1960s and 1970s US conservationists mobilized millions of people in support of their goals, including many people new to the political scene, people previously involved in other areas of politics such as antiwar or civil rights work, and even some who had previously been opposed to conservation goals. Conservation in the United States was transformed from a mostly insider affair into a mass-based political movement. The rise of this new political force constituted a change in the balance of power among those groups concerned with forest, water, population, and many other issues (Benjamin Cashore and Michael Howlett, 2006: 144). Many factors affect mobilization, but out-mobilizing the opponents of conservation was the proximate reason conservationists prevailed in so many legislative and other battles. Bringing "more" resources to bear means not just more people or more voters or more ads, but skill in using them, taking advantage of structural opportunities presented by a growing economy, a heightened sense of possibility created by other social movements of the time, and an elite under siege because of an unpopular war.

The environmental and conservation groundswell caught many powerful conservation opponents by surprise. It's not that they didn't try to block wilderness or pollution control legislation, but that they were often out-pressured and out-maneuvered. Conservationists were successful in framing the issues and maintaining the initiative. In many cases opponents were focused on other issues, did not fully appreciate the implications of laws such as the Endangered Species Act, or joined the battle late.

What caused the great swell in conservation mobilization? The deliberate effort to mobilize was indispensable, as was the ability of activists to recognize and act on the political opportunities the period presented. Social change is often the harbinger of the rise of social movements (Doug McAdam et al., 2001: 42–3).

Specific to the rise of the conservation movement was a growing awareness and direct experience of the downside of "progress." The middle decades of the 20th century evidenced horrendous pollution, loss of ecological services, and rapid destruction and fragmentation of habitat in the face of road building, sprawl, industrial grazing, agriculture, and logging. These trends made many groups in society and policy makers receptive to action being advocated by conservationists and environmentalists. Policy makers, like surfers attracted to a huge wave, sought to ride it. The great environmentalist President Richard Nixon even appointed a commission chaired by Nelson Rockefeller to examine the threats posed by rapid population growth. The panel recommended stabilizing US population at its current levels (about 200 million) to protect the country's quality of life (Commission on Population and the American Future, 1972). Naive do-gooders and businesses seeking to lower wages and break unions had other ideas, however.

By the mid-1970s social change was enabling countermobilization that weakened conservation relative to its opponents. Nixon was the last Republican President in the 20th century to embody an echo of Teddy Roosevelt's conservation ethic. By the time of his resignation conservation opponents had regrouped and utilized their considerable power to water down some earlier legislation and further weaken it during agency rule-making. Although some of the industries engaged in resource extraction were in economic decline, such groups exerted disproportionate influence by their desperate actions, indelibly shaping the future (Barrington Moore, 1966: 505). Other opponents remained quite strong, and once focused on the challenges posed by conservation they acted to "preserve (their)... access, influence, and power." (Robert Durant et al., 2004: 495). Led by extractive industries and other business sectors that pined for the good old days of the 1870s–1890s, an unholy alliance was founded of profit-*uber-alles* extremists, culturally conservative busybodies reacting to *Roe v Wade*, opponents of affirmative action, government regulation, and public services. Since 1980 this coalition has dominated US politics, electing presidents, congresses, and governors hostile to conservation.

Although this coalition is largely unable to win on the merits of the issues, they had the resources to successfully undercut the credibility of conservation nongovernmental organizations (NGOs), to reframe the issues with important constituencies, and sow confusion and uncertainty when all else failed. This coalition led by powerful sectors of the elite has had remarkable success in branding conservationists – and anyone else who believes social problems require public solutions, that is, government action – as elitists. And they had significant success in defining a market dominated by a few hundred corporations out of millions of businesses, as democratic (John Micklethwait and Adrian Wooldridge, 2004). Recognizing that most people do not want ancient forests logged, under guise of responding to the threat of fire (most people do not yet understand how important fire is), they gained passage of the Healthy Forests Restoration Act of 2003 which allows logging of big, healthy trees in areas distant from human habitation (Jacqueline Vaughn and Hannah Cortner, 2005). By raising doubts about climate change they helped undercut efforts to mobilize in support of policies to address it.

In response to the rise of this coalition the conservation movement continued to mobilize people and its members and budgets have increased down to the present, ebbing and flowing along the way (Ronald Shaiko, 1999; Gary Bryner, 2001: 36–7). The movement has been able to block repeal or attempts to weaken important laws and often prevails in the courts. In some cases conservationists dropped the ball, as Michael Gunter (2004: 117) argues regarding US failure to ratify the Convention on Biodiversity. There was no serious conservationist effort to mobilize support for ratification. But at root the movement is limited by not being able to recapture the advantages it enjoyed in the late 1960s and early 1970s. Almost invariably the opposition has out-mobilized conservationists by buying experts, media time, even the media itself, elections, and much else. Battling conservation has become another cost of doing business. The cost of battle raises opponents' overhead, but they are able to pass these costs on to consumers who buy their products and services such as oil, metal, lumber, mortgages, and credit cards. Opponents command much wealth (not just money) and their location in the economic structure (as employers, large-scale investors, and economic decision makers generally) puts conservation at a huge disadvantage.

Growth in a movement's resources or level of mobilization does not directly translate into political success (James Jasper et al., 1997 [1993]: 405; Dieter Rucht, 1999: 221–2). Although mobilization is the bedrock of political effectiveness, the resources it provides may be used well or poorly. Strategy, timing, social structure, tactics, leadership, vision, the unity of ruling elites are all important factors in determining the outcome of struggles to shape the direction of society.

Opportunities for conservationists have ebbed and flowed like membership, but lack of opportunity does not adequately explain the inability of the growth of conservation mobilization to keep up with the need. Much of the mobilization strategy has encouraged recruitment of check writers or letter writers in support of a professional staff who does the work. This does not create a base of support that votes on conservation issues or takes other action that would make a difference. Too little has been invested in grass-roots activism. Politically weak constituencies have often been targeted for mobilization rather than strong ones. There have been related failures in understanding mobilization.

Conservationists have generally not been prepared for the consequences of their success which inevitably trigger countermobilization and repression (James Jasper et al., 1997: 397). Conservation's opponents have not obtained their wealth and power by being politically timid; they respond forcefully to perceived threats.

Conservation mobilization has neither focused adequately on using social networks in recruitment nor understood how these networks can result in unanticipated spikes and dips in recruitment and disengagement (Buchanan, 2002: 42–3, 107; Roger Gould, 2003: 239). Many NGOs still speak to groups they target for mobilization as if they were speaking to themselves and without understanding them. Inadequate use is made of brokers who are essential in reaching groups many conservation NGOs have few or no links to (Laurence Wallack and Lori Dorfman, 2001: 398). These factors will be addressed in the following chapters, along with changing opportunities presented by election cycles – or the absence of elections – and other factors such as the vigorousness of the media, the legality of mass protest, and the availability of fora for outreach. The factors contributing to mobilization are not fungible. And the same factor that in one society fuels a wave of protest may in another generate fear and uncertainty because repression is the norm (Roger Gould, 2003: 239).

Many conservationists are democrats (emphasis on the small "d") and resist mobilization's top-down nature (Amatai Etzioni, 1968: 389, 402, 406). Indeed conservationists generally do better in democracies. There are no benign dictators, ruling their people with humanitarian and ecological wisdom; dictators almost always have their hand in Nature's till, selling off whatever assets they can. Mobilization, then, begins with those who have a vision, organize themselves, and reach out to others to join in the cause. Those who start the process tend to be better educated or possess insights and skills not widely shared. They may be ordinary people confronting extraordinary situations. Like Aldo Leopold or Peter Illyn, they may have had an epiphany that changed their lives by putting them in touch with what was most sacred to them. Whatever the factors that move innovators – and often a significant dose of narcissism is involved – mobilization involves those who are in some sense leaders trying to catalyze behavior from others that accords with the desire of the former. The same is true of organizational relationships. It is usually the same organizations who again and again take the initiative, acting as catalysts and hubs in coalition and alliance building, although in some cases the process is more symmetrical (David Dozier et al., 2001: 235).

Building on the past

Conservation mobilization has changed the world. It is a mass political force within many countries and at the international level. Even though the opponents of conservation can regularly outmobilize its proponents, the growth in the absolute level of conservation mobilization represents a net gain for the defense of Nature. When conservation was a "gentlemen's" or outdoorsman's movement the number of battles that could be engaged was far fewer than it is now, with thousands of grass-roots groups and many major globally oriented NGOs. No longer can private agreements among upper-class friends commit the conservation cause to timid or bad deals. It's not that bad deals aren't attempted and made, but few can escape being exposed for what they are.

And conservationists do win battles. Because they are stronger in a particular fight, more agile, and sometimes opponents are unable to reframe the debate because their positions are indefensible no matter what the spin and how much they have to spend. Agencies such as the US Forest Service, once the captive of industry, are now divided. These new circumstances result, at times, in the willingness of opponents to compromise or to find less contentious ways to make a profit. Even greenwashing attests to the power of conservation. But conservation opponents are much less likely to be caught napping or diverted as in the early 1970s. Indeed, they are quite vigilant, with trade groups, professional spies, and astro-turf NGOs monitoring, infiltrating, and disrupting conservation organizations (John Stauber and Sheldon Rampton, 1995; David Helvarg, 2004).

So it's not that conservationists don't know much about mobilization. They must become much more effective at it. Becoming so is not mysterious.

7   From vision to goals

He who organizes must be organized.

John Rosen, Spanish Civil War Veteran and Chicago Organizer (Note posted on a disorganized desk, 1968)

Mobilizing others requires that those doing the mobilizing have their act together, including clear goals, a sound strategy, and the right people in place. Direction is defined by goals and taking action to attain them. Vision is too general to guide action although every conservationist ought to be able to answer the question "Why are we doing this?" and state how the action in question furthers the vision and the goals established to reach the vision. Goals make specific the purposes for which mobilization is undertaken by defining achievements on the path to the vision – usually achievements that can be attained in one to five years and toward which progress can be measured. Goals are not the same as milestones or benchmarks, which are set largely to inculcate a sense of accomplishment. Goals emerge primarily from identifying the next steps toward the vision. They must also be informed by a good grasp of the political landscape.

There are two broad types of goals: programmatic and organizational. Programmatic goals describe substantive achievements, for example, ~400,000 hectares of new wilderness within three years. Organizational goals describe achievements that strengthen the organization to better enable it to achieve programmatic goals, for example, gain new staff with a particular expertise or add activist members. Mobilization is usually undertaken in support of a campaign to achieve programmatic goals, but it may also be undertaken to increase the capacity of the organization in general so it can achieve more in the future. General mobilization of new resources as well as maintenance of existing resources must ultimately serve program goals, which ought always to predominate. If an organization becomes its own end it becomes an obstacle to achieving its vision.

Organization building is important. Obtaining adequate funding, staff, skills, connections to media and brokers, a good strategy and strategic thinkers, strong and creative leadership, good internal communications, and a strong support base requires attention and the investment of resources just as program goals do.

Some sergeants are needed to recruit new troops and train them just as surely as sergeants are needed to lead troops into the political arena. Allocating resources is an important strategic question.

In goal setting the relationship between programmatic and organizational goals – how they support each other – should be kept in mind. Programmatic work, for example, almost invariably creates opportunities for mobilization, often around the organization as well as the program goal. At the same time there is always tension between program work and related mobilization, and organization building work and associated mobilization, because resources are always scarce. Ignoring this tension creates confusion.

Clarity in goal setting also requires that the distinction between means and ends be considered. The protestors who shut down the 1999 Seattle World Trade Organization (WTO) meeting sought to do the same at a subsequent WTO meeting in Washington, DC. They forgot, says activist John Sellars (2004: 184) that their goal was not to shut down a meeting but to change the WTO's direction or greatly diminish its power. Shutting down the Seattle meeting, along with other factors, did derail its work for a time. But the disruption was a means. By treating that means as a goal protestors not only distracted themselves but also set themselves up for defeat. The DC police learned from Seattle and were ready to ensure the meeting was kept open. Activists realized their error after the "failure" in DC to disrupt the meeting or slow WTO work.

Major events can require changing or adjusting goals. Although it does not alter vision, a major change in political leadership such as the election of the extractive industry backed Bush administration in the United States meant that plans to strengthen the Endangered Species Act (ESA) and introduce an Endangered Ecosystem Act had to be shelved. Not all proactive goals disappeared (some work shifted to the state level), but new, defensive goals came to the fore. A disaster like the Exxon Valdez oil spill can create new opportunities and a shift to more ambitious mobilization goals and program goals. When the stock market tanked in 2001, many foundations curtailed their funding causing dependent nongovernmental organizations (NGOs) to adjust their goals.

Some events that ought to cause a reconsideration of goals can be anticipated, but often they cannot. Human societies are less complex than whole ecosystems but they are nonetheless nonlinear and quite complex (John Peet, 1992: 78–81; Stuart Kauffman, 1993: 173–235), making forecasting difficult. It is still possible to think in terms of alternative scenarios: if this outcome, then that goal, if that outcome, and so on. Some events, like US presidential elections which suck up money that might otherwise go to NGOs, can be planned for. But goal-affecting changes in the political landscape caused by the simultaneous occurrence of events that by themselves would have little effect cannot easily be anticipated. For example, the combination of a few or many events, such as drought, rising energy prices tight credit, diversion of crops to biofuels or a crop disease outbreak that leads to food shortages, can push large numbers of people into wild or minimally damaged lands, destroying and fragmenting habitat.

Many changes that lend themselves to goal reconsideration are less apparent. It took time for conservationists to understand how effective industry attacks on them were. If more resources had been devoted sooner to countering attacks, less credibility loss might have resulted, making subsequent campaigns easier. Only belatedly did many conservationists recognize opportunities presented by the emergence of religious conservatives into US conservation politics and that mobilization goals should include them.

Changes in the landscape more frequently call for adjusting objectives – the various elements and steps that are part of achieving goals – or elements of strategy. But when goal changes are called for and NGOs are inflexible they lose effectiveness. Similarly, organizations become ineffective when they are managed by anecdote, changing goals in response to ripples instead of changes in the current. They move about a lot but get nowhere.

8 From goals to strategy: answering strategic questions

There never is enough money to organize anyone. If you put it on the basis of money you're not going to succeed.

Cesar Chavez (Richard Jensen and John Hammerback, 2002: 66)

A strategy is the plan for getting from here to there; from the world as it is (extinction crisis) to how it needs to be (biologically healthy). Among other things it includes plans for mobilizing resources, including which ones, how to obtain them, when they are needed, and what is to be done with them. Strategies address much more, such as when and under what conditions to engage in changing policy. This chapter is concerned with those elements of strategy that address mobilization. Mobilization is most often linked to a programmatic campaign but is undertaken to augment a nongovernmental organization's (NGO) strength and influence generally. Strategies will vary by campaign and by whether mobilization is in the service of a campaign or mobilization itself is the aim. Campaign-oriented mobilization is usually focused on constituencies that can help in the near term, while organization building mobilization may look to a wider range of audiences over a longer term. In both cases similar questions must be addressed. Goals obviously define the *there* of strategy – where is the NGO going? But they also direct attention to what is relevant in the *here*.

Here

The elements of the existing political landscape most important to developing a sound mobilization strategy are discussed below.

- An understanding of the strengths and weaknesses of the NGO developing the strategy. What is its track record of success, failure, and ability to learn? Given proposed goals and the resources needed to achieve them, what resources are in good or short supply, including money; expertise in communications, media, organizing, administration, and strategic direction; number of activists, members, and supporters; experience and commitment of leaders and other staff, board and activists; knowledge of target constituencies; access to decision makers and brokers; capacity to integrate and utilize those mobilized? No action would ever be taken if conservationists waited for resources to be available in desired quantities. But there's a difference between putting the basics in a pack and starting on a journey, not knowing for certain when you'll arrive, and getting on an airplane that lacks the fuel to make it to the next airport.

- An understanding of other conservation NGOs' programs, strategies and direction, resources and capabilities, leadership, and intramovement relationships. Which organizations are vital and which are stumbling or senescent? What work is not being done or being handled poorly – what are their geographical, political, and functional gaps? Some groups are too timid or lack resources to be effective. Many regions lack coverage of particular niches such as private-lands work. Often there is no NGO considering whole systems or connectivity between regions. An important part of any assessment is an examination of which NGOs could provide a good complement to the NGO undertaking the assessment, that is, can another NGO enhance its strengths or backfill its weaknesses by closer cooperation or even merger?

- An understanding of the opportunities and threats that are relevant to goals over the next few years. Some opportunities and threats can be identified with a high level of confidence while others are relatively speculative. Elections are predictable in many countries, but not always the candidates. Economic cycles lend themselves to predictions about the growth of resource extraction. Identification of action-forcing events is very important. These are events or crises that make the status quo less tenable and enhance the probability of policy change. Examples include (Robert Repetto, 2006; Helen Ingram and Leah Fraser, 2006; Charles Davis, 2006) an undesirable "natural" event such as a destructive flood or massive landslide caused by logging; a legal victory requiring an agency to change course; new scientific findings that trigger law enforcement (e.g., listing under the US Endangered Species Act); a change in political leadership, a leader's need for conservation support, a leader who has "got religion" over conservation issues; passage of legislation that weakens close relationships between industry and government or opens the agency to conservation pressure; dramatic weakening in the power of opponents on conservation matters due to internal divisions, scandals, political losses on other fronts or major engagements on other fronts; international economic or political events or trends that affect the relative power of states, regimes, or others (e.g., unpopular wars or economic crises may weaken some leaders and strengthen others, alter relations between states and multinational corporations or between states and International Monetary Fund (IMF) or World Bank in ways that help or hinder conservation). In some instances these events can be brought about or magnified by NGO actions.

- An analysis of which groups are likely to be supportive, opposed, or neutral to specific conservation goals. Identifying groups and resources that they can bring to the political process does not in itself determine which groups should be targeted for mobilization, ignored, or made the focus of efforts to weaken them. It provides initial information on who the likely players are, their capacity, interests, and related attributes. Specifics count. Ranchers may not want an area developed for industry any more than conservationists, but they may also resist tourist development associated with park or wilderness status; common ground may be found in some other designation that won't attract people, as in the case of protecting Steens Mountain (Tara Gunter, 2001). Ranchers will strongly resist removing exotic species like cows, however. Birders may make good allies or new members when the goal is protection of bird species and habitat, but they may not be desirous of having grizzlies present where they wander. Analysis of the political landscape will keep surprises to a minimum, whether it involves allies that seem to come out of the woodwork or especially nasty opponents.

- An understanding of whether or to what degree achieving goals will require structural change. If structural change is needed, and barring the opportunities afforded by a major systemic crisis, a strategy must address how to obtain the support of critical elements of the elite, exploit divisions among the elite, and mobilize high levels of mass protest. When an elite is united it is almost never defeated – except by itself. Insider–outsider strategies tend to be most effective in all circumstances, but they are essential in the face of structural obstacles, that is, basic power relations and social organization (John Dryzek et al., 2003: 155, 161; Marco Giugni, 2004: 220–1, 226). Only with mobilization of a faction of the elite, or when significant division undermines elite unity, is it possible to make inroads into the interests they so fiercely defend: their power and their commitment to economic growth.

- Elite factions may come to side with conservationists based on their personal values, the recognition that some of their interests are contrary to those of other elite factions (e.g., those particularly destructive of the natural world), or a recognition that certain changes are inevitable and it's best to act sooner rather than later. Elite voices bring legitimacy to demands by conservationists. When members of the elite are mobilized they can use their power and provide direct lines of communication to others who wield power, making it more likely conservation concerns are on the agenda and the range of acceptable solutions include ones that really constitute solutions. Elite allies can provide early warning about state or corporate strategies and repression.

- Mass mobilization for protest and direct action are essential when goals face structural obstacles. Sustained mass mobilization constitutes a countervailing *force* to dominant interests by making the status quo a nonoption and thereby enhancing the bargaining power of elite allies or other pursuing insider efforts to change policy. Environmental and conservation success in the United States in 1960s and 1970s was made possible by support from some elite factions, by low levels of governmental and social legitimacy, and by the mobilization of thousands of new grass-roots groups. These circumstances allowed conservationists to gain concessions because those in power needed to respond positively to some claims (John Dryzek et al., 2003: 137; Benjamin Cashore and Michael Howlett, 2006: 144).

- Division within elites was a critical factor in bringing an end to the South African apartheid regime and the US-backed oligarchy and its death squads in El Salvador (Elisabeth Wood, 2000). Mass mobilization, protest, and resistance (including armed resistance) made the status quo untenable by driving up the costs of repression and cutting into profits. The costs of compromise appeared much less to some elite factions who split with hard liners. In some very repressive societies it is not possible to gain access to domestic elites and mass protest can be dangerous. Yet structural change at some level is essential, for example, creating enough openness to allow conservationists to operate in relative safety. In these circumstances an outside strategy will rely on contact with foreign NGOs or officials who can in turn utilize their domestic elite connections to bring pressure from the outside (Clifford Bob, 2005: 4–12). It doesn't always work – insular regimes such as Myanmar's are especially resistant.

Obviously goals that require structural change are more difficult to achieve; some seem downright impossible. Fighting a losing battle is usually not a good idea, but not always. Mobilization around a compelling policy goal, even if it cannot be achieved, can produce higher levels of mobilization and contribute to organizational or movement momentum. It may also provide a means of framing issues, exposing opponents' venality and generating identification with conservation values and interests. In some instances it is necessary to fight because what is at stake is so central to the purpose of conservation. The loss of an iconic forest or species without a fight creates demoralization; to live with oneself requires giving battle. To avoid a sense of futility interim goals may help maintain excitement and purpose. Nonpolitical benefits that come from being part of a conservation NGO – a sense of belonging, mutual caring, and joy in the natural world – are especially important in sustaining people in difficult times.

- Identifying the decision makers who have the power to make goals a reality. Goals and objectives, by describing the policy and behavioral changes to be achieved, define the decision makers who must be influenced. If wilderness designation in the United States is the goal, then key congressional committees and leaders and the president must be brought around; so must affected governors and often county commissioners. If the goal is wilderness in Mexico then the decision makers will be ranchers, ejidos, or other landowners. When conservationists wanted to stop the slaughter of dolphins in the tropical south Pacific Ocean they mounted a campaign against US tuna companies as well as the US Congress and US agencies; all had important decision-making power (Judith Layzer, 2006: 313–45).

- Decision makers need to be identified as specifically as possible because it helps to answer the critical questions on which a sound strategy is based. Knowing who they are as individuals is usually as important as understanding the institutions they lead or play a role in. Strengths and vulnerabilities are personal and not just institutional. The networks in which decision makers are embedded are personal in addition to being defined by social position.

- It is not always easy to determine who the decision makers are. Finger pointing is a fine political art. Legislatures point to agencies they can't control; business blames government; governments say that markets must be allowed to operate to ensure prosperity or say their hands are tied by treaty obligations. Certainly the latest round of globalization has strengthened the hand of the more powerful states and corporations in those parts of the world where states are weak. The powerful have much to gain from avoiding accountability even as they demand it from their debtors. In such circumstances the advice of the Watergate informer "Deep Throat" remains timely: follow the money.

To there

With an understanding of the political landscape relevant to its goals in its possession NGOs can shift their focus on how they will reach their goals – the there. To do this a number of additional strategic questions can be answered based on the identification of decision makers. These questions include:

- Who must conservationists mobilize to gain the outcome they want from decision makers? When in the mid-1990s US House Speaker Gingrich sought to sabotage the Endangered Species Act he was unresponsive to pressure from conservation groups. Only when the Evangelical Environmental Network was mobilized – a constituency Gingrich's coalition was dependent on – was the effort blocked (Bruce Barcott, 2001). When animal activists wanted to alter factory farming of chickens they went after McDonald's, the biggest single buyer of chicken in the United States; as a food retailer they were much more susceptible to end-consumer pressure (Robert Cox, 2006: 260–2) as were tuna packer-retailers (Judith Layzer, 2006: 313–45). Businesses and other nonstate actors are often much more vulnerable to pressure than government, depending on the country involved and other circumstances (Doug McAdam, 1997 [1983]: 398–405). In some parts of the world, of course, business enterprises are more powerful than governments and have their own repressive forces or can command those of the state. And influencing virtual monopolies such as the oil industry through economic pressure is much more difficult than influencing lumber retailers.

- In the case of private landowners it is usually less a matter of pressure applied by conservationists than the persuasiveness of the messenger or economic incentives. Incentives may be governmental (taxes, subsidies) or market-based (higher prices for predator-friendly beef). Mutual influence and peer pressure in a closer knit communities plays a big role; if one farmer or rancher goes organic or predator-friendly and does well others may follow.

- Many groups will have influence with a decision maker, and although redundancy is good, conservationists must be economical. Not all groups are created equal. Some are easier to mobilize than others. Some are much more powerful than others generally, but specifics are important – who has the most influence with decision makers identified. The Zuni campaign to halt a major coal mine focused on several audiences who could delay permitting of the project and raise its costs – nonnative American religious leaders, civic leaders, and regulators themselves (Robert Cox, 2006: 267–8). Litigation also threatened lengthy delays. When courts offer political leverage, very specialized resources must be mobilized –top legal talent. Some groups that influence decision makers cannot be reached directly by conservationists; brokers or other intermediaries must be relied upon. Consumers were mobilized to pressure McDonald's in order to pressure factory farmers. Zuni mobilization of mainstream religious leaders helped sway the New Mexico congressional delegation to call on the Secretary of the Interior to do a legally adequate environmental analysis.

- Relationships among opponents matter as well. Opportunities usually exist to divide them thereby lowering their relative level of mobilization and influence. Beating up the biggest bully on the block may cause others to take note as insurance companies do when a personal injury attorney wins big – they tend to settle rather than go to trial.

- There are different ways of working with those able to influence decision makers. Asking a group or individual to act once or twice is common. "When you're playing cards with the premiere next weekend encourage him to move on a new national monument or make a good an appointment to a regulatory body." Coalitions (partnering with other conservation NGOs) and alliances (working with nonconservation NGOs) are also common. Working together over time is usually necessary to achieve a goal – conservationists usually cannot win without friends. But they need to be the right friends. Cooperative relations may extend beyond a particular campaign to the longer term and have been successful in the labor movement. Partnering with organized groups almost always brings more to the table than mobilizing the unorganized. Partnerships require the investment of resources, however, and don't make sense unless the investment has a good likelihood of being worthwhile. Coalition building is often difficult and slow going, involving negotiations over goals and protocols for cooperation.

- The other primary means of influencing decision makers is to make conservation NGOs stronger by gaining new adherents and other resources: more members from politically important demographics, the capacity to significantly increase media outreach, higher levels of commitment or activism from existing members, and so on. Mass mobilization – the outsider component of insider–outsider strategies – is an important lever. Elections are determined by a few points in some districts. Mass protest can disrupt, challenge elite legitimacy, and give cover to sympathetic political leaders. There is no single mass, of course – no one public – but specific groups of people that have influence with specific decision makers.

- An important factor in selecting mobilization targets is the effect of their mobilization on the conservation NGO doing the mobilizing. New members and different types of members can profoundly affect an NGO's political center of gravity. Groups mobilized as coalition partners or allies have their own agenda; if the tent gets too big too quickly it can become unwieldy or goals become diluted. In some circumstances charging ahead alone makes sense; a rightly timed bold initiative can redefine the political landscape, turning the radical into the mainstream.

- Whatever the means of influencing decision makers – mass mobilization, coalition and alliances, or some combination – building relationships that can withstand the corrosive pressures of politics is critical. Personal interaction and forthright communication assist in creating good relationships. Coalitions and alliances are prone to stumble under deliberate attack by opponents who exploit differences and mistrust.

In Chapter 9 the selection of targets and prioritizing targets will be examined in detail.

- What is wanted from those being mobilized? How will they exert their influence on decision makers? Protest, use their connections with the powerful, apply specialized skills (legal, media, administrative, direct action), provide funds or other support to professional staff (lobbyists, land managers, scientists)? If members of other conservation NGOs are being targeted are they being asked to do something different? If an NGO is targeting its own supporters for increased mobilization is it more of the same kind or of different kinds? How will those newly mobilized be fit into the organization, coalition, or alliance so that the desired actions are most likely to be forthcoming? Can the leaders of coalition partners or allies deliver their organizations or members? Are the targets of mobilization being sought for a single action, repeated similar actions, a time-limited or open-ended commitment?

  Because timing is important in affecting decision makers, it's important to ensure constituencies are ready at the critical junctures. How long will they take to mobilize? Capacity and predisposition are also important: it makes little sense to organize trial lawyers to sit in trees – they probably won't. Whether targeted audiences will play the desired role also depends on whether they are mobilized on the basis of shared values, a common goal, or quid pro quo.

- What is known and needs to be known about target audiences to move them to action? Action is what's important. Most people do not act on conservation issues because it is neither central to their lives nor do they feel intensely about it (Deborah Guber, 2003: 3, 38, 105–9, 177). They can be brought around to action but only if it is understood what motivates them to act. Whether organized group or unorganized cohort, the target audience's interests, social position, views, values, emotions, and stories must be understood. Who do they listen to? How are they best approached? Who else will they bring along or neutralize? Are opponents trying to mobilize them as well, or to keep them neutral? Chapter 10 explores in detail what needs to be known about mobilization targets and how to go about finding it out.

  More can always be known but politics doesn't usually permit very complete knowledge; events move too fast. Humans are well adapted to acting with limited knowledge when they are not too self-conscious (Gerd Gigerenzer, 2007: 4–5, 17–9). Part of good judgment is knowing when to rely on judgment and when to utilize more rigorous analysis – the deliberate and methodical review of problems and potential solutions with their pros and cons.

- What is the plan for mobilizing a target audience? When the key attributes of target audiences are understood a sound plan can be developed for mobilizing them. Minimally a plan should identify the messenger, channels of contact, messages, the stories that will carry the messages, and the tactics that will be employed. These elements are discussed in detail in Chapters 11–15. A good plan will also describe what actions are expected of those mobilized and identify the obstacles to mobilizing specific audiences. How mobilization will be sustained, discussed in detail in Part 2B, is worth addressing from the beginning because keeping people mobilized is often necessary and is less costly than remobilization. This is true, even though sustained mobilization requires building a strong community and sense of identity.

- What resources are needed to undertake mobilization and what resources are available? Edward Bernays (1955: 19–23), whose clients were commercial interests and government, stressed the importance of lining up resources before launch, being utterly realistic about what is available and how that limits a campaign, and having organizational matters in place (staff recruited, tasks defined, etc.). There are never enough resources for NGO goals, as Chavez (Richard Jensen and John Hammerback, 2002: 66) observed; social movement NGOs must strive to reach goals they lack the resources to attain with the notion they will gain them along the way. It is a view that is necessary, but too little attention is given to how the needed skills, staff, and other resources will be obtained. Many NGOs exhaust talented staff because of this failure. Lack of detailed thinking about this problem is compounded by the necessity of allocating resources between influencing decision makers and maintaining and garnering more resources, that is, mobilization. The two tasks are complementary but require different talents. Money may be shifted from program to mobilization, but staff may not be as easily shifted. Having the right talents and other resources available takes deliberate decisions about priorities. Identification of products needed for mobilization, assigning clear responsibilities, creating an organizational home for the work, and establishing time lines so elements build on each other all assume the allocation of resources. The pressure to fight fires displaces longer term work if there is no commitment to the former. Conservation has more than its share of emergencies that must be attended to, but addressing the extinction crisis is more akin to a marathon than a sprint.

- How can countermobilization and similar actions by opponents be minimized? If opponents raise their level of mobilization in response to conservation mobilization and the relative levels remain the same, then not much has changed. Quiet or insider strategies are two means of remaining under opponents' radar, as is acting when they are distracted by other issues or weakened by other fights. Distractions can be arranged. Getting a head start on a mobilization campaign can create enough momentum that opponents can't catch up. Acting quickly or boldly can create opportunities for mobilization not available to opponents. The most powerful opponents have enormous resources that need only be redirected against conservation to overwhelm conservationists, all else being equal. For conservationists to prevail they must use resources more intelligently.

Opponents have been particularly successful at undermining conservation mobilization by discrediting conservationists' motives and integrity with key constituencies and by reframing the issues (Ronald Libby, 1999). Robert Cox (2006: 382, 396–7) describes a two-stage process opponents use. The first is to legitimate their framing of the issues with expert opinion while spouting "symbolic reassurances" and nebulous remedies. In the second stage opponents try to disassociate themselves from the problem and pass the buck; failing that they focus on discrediting critics, blaming the victims or others, and accusing the media of exaggeration. In the case of the Exxon Valdez spill the corporation tried to shift blame to the ship's captain, then it tried to minimize the damage done, and finally tried to demonstrate that it had acted responsibly and quickly to contain the damage. Sewing doubt about the reality of conservation problems has also proven effective (Robert Cox, 2006: 344–53); in the face of doubt it is difficult to raise concern above the threshold necessary for action.

Physical and economic repression and threats can dissuade potential supporters and drive away existing supporters. Opponents utilize differing means and it is important to understand what to expect from which opponents: intimidating lawsuits, character smears, threats to livelihood, or physical violence.

- How will success be monitored and evaluated? Programmatic goals can take a long time to achieve and good interim measurements can be hard to devise. Getting an agency to formally consider a good policy option is a measure of progress but may not be indicative of its likelihood of adoption. The same is true with mobilization given its nonlinear qualities. There is a threshold for action and it's not easy to know how close to it a campaign is getting until it happens. Approaches to effectiveness are discussed in Chapter 16.

Other strategic issues

The sum of NGO strategies does not equal a grand movement strategy. There are advantages to good coordination such as the ability to play good cop/bad cop, avoiding duplication of effort, being better able to concentrate resources where needed, and so on. Larger organizations have the advantage of being able to shift resources to areas of opportunity but may lack the organizational agility to do so. Conservation is likely to remain a domain of many NGOs. A grand strategy is not attainable without commitment to a common vision and that is just the first step. Integrating the big picture with the local depends on the ability to move up and down geographical scales and levels of social aggregation, blending diversity with the right degree of uniformity, and much else that is hard to attain. It also requires that interaction among NGO leaders be dense. At the very least NGOs need to share information with each other about their goals and strategies and not take offense at the political necessity of publicly distancing themselves from one another. Such distancing should avoid feeding efforts by opponents to isolate and pick off movement groups labeled as radical. They are vital to pushing the envelope; the abolition of slavery, breaking down barriers to political participation by the poor, labor, and women to name a few, were not achieved by adhering to the rules set by their opponents – such groups were radical and needed to be. From the perspective of smaller grass-roots groups, some of whom are considered radical, it is frustrating that some NGOs spend so much and do so little while flying first class. Moving resources to more effective NGOs is a positive step but needs to be undertaken without the venom that lends itself to public rifts.

A strategy should be written because writing exposes gaps in plan elements and reasoning, and other inadequacies, opening the path to greater clarity and coherence. Writing also helps tease out assumptions which can otherwise go unstated and unevaluated. A written strategy provides a record for later assessment and evaluation. It also provides a stronger basis for developing operational and work plans – the daily, weekly, and monthly lists of nuts and bolts and who does what by when. Documentation of strategy is particularly important in larger organizations and for more complex operations and campaigns. Conservationists are in a long struggle that will consist of many campaigns, and each campaign will have many battles. Good maps are needed. Written strategies can also fall into the wrong hands.

A methodical approach to strategy leads to better strategies, but the formulaic quality of much strategic planning is unhelpful. Proceeding methodically by addressing a series of relevant questions is different than adherence to bloated templates that keeps creativity, intuition, and judgment out of the mix. Good method includes involving the right people in developing strategy – those who have a good sense of NGO and movement history, knowledge of organizing and organization, knowledge of other movements' successes and failures, those with campaign experience, those who are not fond of overanalysis, and those who come to the process with fresh and unencumbered eyes.

Effective strategies are organic; they match circumstances and goals. Even non-mechanistic and well-grounded strategies are rendered obsolete by changing circumstances. Effective action requires a view of the world and a plan that is both stable enough to provide a framework for decisions over time and includes a means to adjust to changes in the political landscape. The responsibility for achieving this balance falls to leaders, their connection to reality, and their agility. They cannot afford to be so vested in a plan that emergent opportunities or threats are obscured. Sometimes change unfolds slowly and there is time for more consideration. At other times events develop quickly and change is drastic, making considered review impossible.

The arrest of Rosa Parks for refusing to give up her bus seat to a white person and the response of the civil rights movement is a good example of adaptation to contingency. Parks's arrest was neither the first of its sort but part of a long-standing pattern meant to humiliate and intimidate, nor was her decision to stay seated spontaneous. The assessment by the civil rights leadership in Montgomery and elsewhere that the time was right for decisive action was based on their recognition of a number of trends (Doug McAdam et al., 2001: 38–40). These included Black migration into cities north and south following World War II, which supported stronger local organizations that were networked nationally, increased numbers of people in these cities along with enhanced economic strength; cold war criticism by the Soviets of US race relations that resonated in the third world causing some US political leaders to support an end to segregation; the recognition by many of these same leaders of growing Black political strength and the calculation that more votes would be won than lost by backing desegregation.

The call for a bus boycott by civil rights leaders was a shift in strategy decided over the course of public discussion in the days following the arrest. They could not have predicted 90% of Black bus riders would join in. Over 13 long months the boycott was sustained until it succeeded, but it required constant tactical innovation to get people to work during this period, dealing with mounting harassment, and keeping people's spirits up. Unanticipated national media coverage – neither press nor electronic media had up until that time deemed the civil rights struggle very newsworthy – was effectively utilized, bringing unprecedented pressure on officials.

More recently the US conservation movement confronted a major change in the political landscape with the installation of George W. Bush as US President. Bush's appointment to the White House by the US Supreme Court called for a major readjustment in strategy. Responses varied, but the movement as a whole did not adjust quickly apart from shifting more resources to protecting past gains – an obvious move. Litigation was a major part of this effort. Some NGOs reoriented campaigns to the state level, working with wildlife, natural resources, and highway departments (on connectivity) and were able to work around the unholy marriage of economic conservatives and medieval fundamentalists that settled into Washington, DC. No one, however, anticipated the thoroughgoing assault on science and on reality that was to follow; Galileo would have shuddered in recognition. Neither insiders nor outsiders responded quickly, in part because they were not prepared to give battle to claims that seemed so patently unreasonable. Indeed, uncertainty remains as to whether the effort to manipulate science was based on lies calculated to serve the economic interests of administration backers (and many in the administration), on denial, or on 'true believerhood'; administration actions were certainty systematic and deliberate (Chris Mooney, 2005). Only in the last years of the Bush regime, with a Congress somewhat less accommodating, were conservationists able to gain some traction in reclaiming lost ground on claims that policy should be informed by science.

Conservationists have done better in adapting to other circumstances such as the migration of retirees and professionals to rural areas creating both development pressure and new constituencies for protecting "natural amenities." They grasped the opportunity of a prime minister retiring at the peak of his power and looking for a legacy with the result of several new national parks with boundaries established on the basis of biological criteria. Conservationists have made use of media herd effects, but have remained less effective in portraying complex issues as compelling stories.

Sometimes changing circumstances open up new, proactive opportunities, as when important opponents are weakened. At other times conservationists are thrown on the defensive. Defensive work is not just important because it blunts some negative effects of the status quo but because it offers an opportunity to mobilize around the necessity of systemic transformation (Immanuel Wallerstein, 2004: 272–3). Defensive work thus lays the foundation for transforming rather than fixing or reforming systems based on commodification; they cannot be reformed out of commodification.

Conservationists could learn some things from whitewater boatmen – attentive to both waves and current, balancing analysis with informed intuition. A thoughtful strategy should not be abandoned lightly, but as with running a river, improvisation is essential. That strategies become outdated does not mean that the need for a good strategy is a relic.

9   Who will do the heavy lifting: targets of mobilization

Who should conservationists seek to mobilize? Not enough systematic attention is directed to this critically important question given that conservationists have much to achieve with very limited resources. Hitting one's target is a good thing, but so is making sure the target is worth hitting. In the last chapter we saw that goals identify the relevant decision makers. The next question is who can influence the decision makers? Conservationists can influence some of them, but to generate the influence needed many other groups must be mobilized – either into conservation nongovernmental organizations (NGOs) or as partners and allies. In addition to assessing which groups have influence there are many factors to consider in deciding which ones to target. Which groups are able to exert the most influence? Is the group likely to act? On what basis – shared values, its own interest in the goal, quid pro quo? If the basis is quid pro quo does the mobilizing NGO have the chips to trade? Are conservationists able to directly mobilize a group? If not, what intermediaries are needed? Is timing such that priority should be given to less influential groups that can be mobilized quickly, including an NGO's own membership and those of other conservation NGOs? When is it necessary and appropriate to go directly to another organized group's membership rather than to, or in addition to, its leaders? When should groups be mobilized into an NGO as members, supporters, or activists, and when should they be mobilized as coalition partners or allies? All of these questions are really variations on the larger question: What is the best use of limited resources to gain goals?

Two approaches

In the United States (as an example) NGOs mainly rely on two variations of an insider–outsider strategy, rather than methodically addressing the questions just noted. Grass-roots groups usually take a concentric circle approach to mobilization and national groups and larger regional groups mostly rely on a lobbyist–public strategy. Both have shortcomings in that neither approach has resulted in the desired levels of mobilization. A few NGOs, of course, engage only on protest and direct action.

In the concentric circle approach NGOs begin by mobilizing groups that are politically and socially closest to them, and proceeding over time to mobilize groups that share less with them but that are better connected politically. By recruiting more supporters and building alliances with other organized groups NGOs aim to enhance their clout with decision makers. Such groups may also litigate and undertake direct action. The first circle mobilized consists mostly of those in the social networks to which an organization's board and staff belong, hence the importance of the founding group for the future. The next circle consists of those demographics and organized interests that share some goals, interests, or values with conservationists; for example, hikers, climbers, some hunters and fisherman, and ecotourism operators. The next circle consists of organizations and constituencies that are generally not focused on conservation but are sympathetic or care about an issue that is important to conservation, such as those concerned with good land use, quality of life groups, urban open-space groups, many educators, and watershed groups. The next circle consists of groups that are sometimes labeled "nontraditional allies," including religious groups, Native American groups and governments, and civic leaders. This category may also include unorganized demographic or value-based groups such as newcomers to rural communities, urban educated females, or undifferentiated mass audiences. Mass outreach is limited by available resources so is often focused on opinion leaders who are thought to influence unorganized groups. Increasingly some grass-roots groups are trying to recruit those with access to wealth, access to or skills with media, and those with similar connections.

Groups may fall into more than one circle. Native Americans may be allies but also decision makers if they control a land base. Some elected officials and agency staff are allies as well as decision makers.

Beyond these circles the landscape becomes amorphous except for those who oppose conservation goals. In place of assumptions about the groups between the outer circle and opponents, a more thorough analysis of who influences decision makers will reveal potential targets for mobilization, including some presumed to be hostile or uninterested.

The limitations of the concentric approach are several. NGOs frequently fail to get very far beyond the inner circles and so fail to gain the resources needed to exercise greater political influence. This occurs because mobilization work typically proceeds along lines of established interaction; most who join a movement organization know someone who already belongs (Ann Mische, 2003: 285). Those networks consist of family, friends, fellow employees, neighbors, those involved in the same recreational activities, or having similar values (Lewis Rambo, 1993: 36, 108; Doug McAdam et al., 2001: 41; Mario Diani, 2003a: 17, 178), or those known from the institutional venues in which the networks are embedded such as schools, places of employment, and civic organizations. Even when these institutions can be oriented toward movement goals (Doug McAdam et al., 2001: 44, 116) as many churches were in the case of the US civil rights movement, the networks are significantly limited by the original connections and rarely surmount these limits unless a concerted effort is made. Conservationists are like other people in that they like "talking to themselves," but this does not explain the failure to reach other networks. Instead, NGOs seem to lack the knowledge about how to systematically connect with new networks and use tools like brokers to do so.

Some NGOs do better than others because their leaders recognize the need for reaching the outer circles and focus on it, or because the organization's founders include people from a range of networks including influential ones (Pamela Oliver and Daniel Meyers, 2003: 192; Aldon Morris and Suzanne Staggenborg, 2004: 180–6). NGOs with a diverse founding circle will have access to diverse networks; those with a monolithic founding circle will face more obstacles in reaching other networks.

Because the default route of mobilization is along existing network connections, reaching out to new networks requires deliberate work. Among conservationists who recognize this, not all have followed through. Several years ago a thoughtfully crafted effort to mobilize a number of groups in the outer circle was undertaken in a rural region on the US west coast, including business, labor and civic leaders, high-school teachers, and churches. Initial contacts were made with individuals from many of these groups. The objective was to cultivate these sympathetic contacts so they could become effective messengers to their cohorts. The process never went beyond initial meetings with these contacts, many of whom were quite enthusiastic about talking to friends about something for which they cared deeply. Mobilization stumbled when those responsible failed to continue to engage these contacts.

More recently another NGO working in the US Pacific Northwest made a concerted effort to engage rural communities and did carry through. They gained acceptance and some support in rural communities for their goals of protecting wild places and for recolonization of the region by grizzlies and other native species. This approach has enjoyed success elsewhere in the United States and Canadian west and in Australia.

The two primary means of reaching new networks – mass media and brokers (Doug McAdam et al., 2001: 334, 157; Florence Passy, 2003: 28, 42) – are discussed extensively, along with other channels of communication, in Chapter 11.

Conservationists have generally done a good job of networking among themselves and this has enhanced their effectiveness. Lessons of successes and failures are quickly transmitted as are calls for help. Many NGOs have been quick to take advantage of electronic communications, geographic mobility, international meetings, and other means of creating and sustaining interaction up to the global scale (David Meyer and Sidney Tarrow, 1998: 12, 18; Doug McAdam et al.,

2001: 337; Robert Durant et al., 2004: 492–3). In the mid-1990s the NGO Sacred Earth Network equipped dozens of emergent groups across the former Soviet Union with computers and E-mail, cheaply linking them to the world. Instant communication with other NGOs, access to the scientific literature, and freedom from an unreliable post were among the benefits. Charles Tilly (2004: 98, 104) rightly cautions that electronic communication cannot replace personal relationships, and networks are not coalitions. But electronic communication greatly facilitates cooperation.

The other major approach to mobilization in the United States is typical of larger NGOs and centers on professional staff supported by the financial resources and numbers of members to lobby decision makers (some NGOs also litigate). Lobbyists claim the weight of their organizational membership who may respond to requests to write letters, make calls, support a boycott, make special contributions, or join a genteel protest. Lobbyists also claim the support of the vast majority of Americans based on polling data. A few organizations engage directly in electioneering. The professional lobbyist–public support approach has had major successes, especially in conjunction with grass-roots campaigns conducted by smaller NGOs. Electoral outcomes have been affected. Wilderness battles have been won (Frank Wheat, 1999; Roger Kaye, 2006). Businesses have been forced to behave themselves (Kevin Danaher and Jason Mark, 2003: 111–36; Judith Layzer, 2006: 313–45). Victories are most common when those elite factions dependent in part on conservation support are the decision makers. But decision makers for and against conservation know that support for conservation is shallow among most people. Few people cast their vote or take other action based on conservation concerns (Deborah Guber, 2003: 50–1).

In contrast to the United States, Green parties have fared well in some European countries (those with proportional representation), but their programs are only marginally concerned with conservation (Neil Carter, 2007 48, 123–4). Parties must be multi-issue to thrive and this has meant distancing themselves from NGOs (Christopher Rootes, 2004: 623). European NGOs vary widely. They tend to be moderate and bureaucratic in countries that fund them or that formally make them part of the policy process (John Dryzek et al., 2003: 7, 23–53, 99–102). Protest and radicalization are more likely in countries that exclude them, although if protest is not accepted in the political culture or it is criminalized (Britain), both are limited. In Australia the Wilderness Society has successfully combined grass-roots direct action with credible and successful lobbying based on mobilization of members and effective alliances. The most strident protests in Europe have been over nuclear power, while in the United States and Australia they have been over wilderness (Christopher Rootes, 2004: 620).

In many parts of Africa, Asia, Latin America, and the former Soviet Union conservation NGOs are historically more recent and often depend more on international conservation NGOs that support them financially and with scientific and political advice. International nongovernmental organizations (INGOs) also lobby governments directly and work through international regimes such as Convention on International Trade in Endangered Species (CITES), United Nations Environment Programme (UNEP), the World Conservation Union, and others. Protest is usually not an aspect of this work, although litigation may be. In some countries highly placed international conservationists may still be able to "lunch with the prime minister." Protest is more commonly associated with indigenous groups whose conservation concerns are interwoven with matters of land claims, unwanted development, and poverty (Christopher Rootes, 2004: 620); protest is often scripted for media attention and to avoid triggering the worst of repression. Grass-roots groups that protest also lobby and avail themselves of available treaty regimes in these regions.

The approaches to influencing decision makers are quite varied globally but show similarities based on regime type (Koopmans, 2004: 29, 36), which itself is the result of varying historical factors. Nowhere have these approaches achieved the overarching conservation goal of ending anthropogenic extinction.

Prioritizing targets: the organized and unorganized

Who can best influence the decision makers relevant to a conservation goal? Determining this requires looking at a number of factors in addition to the influence the group has with decision makers. Is the group likely to act as desired? Do conservationists have the connections and other resources to compel the group to act? If timing is critical can action be catalyzed within temporal constraints? Are interim actions required to gain additional time? Groups that have the influence to gain time (protestors, attorneys, someone who can make a well-placed call) may be different than those who can influence the final decision making on an issue. Is it more effective to emphasize undercutting opponents' level of mobilization rather than increasing conservation mobilization? In answering these questions much hinges on the organization of the target groups.

Conservationists often target groups on the basis of their sympathy for conservation rather than their capacity to influence decision makers. NGOs usually give inadequate consideration to the relative costs and benefits mobilizing the unorganized versus those already organized. Mobilization is almost invariably aimed at groups. Even when an individual is the focus – a governor, a chief executive officer, a legislator, a media star, or someone with unique abilities or capacities – it is their institutional affiliation or network and the resources associated with it that are being engaged. It's not that individuals are unimportant – for example, a group is best mobilized by reliance on one leader rather than another – but that their capacity rests on their social position in various hierarchies.

Those groups that can influence decision makers, then, may be organized or unorganized. There are big differences among organized groups, but the greater political gulf is between groups that are organized and those that are not. Karl Marx (1979 [1869]: 187–8, 1976 [1896]: 211–2) highlighted this distinction when he described the differences between a "class-in-itself" and a "class-for-itself." The distinction is now widely used by social scientists in the study of social groups generally. A group-for-itself is one that is conscious of some salient shared attributes or circumstances and has organized around them to further common interests or values. The consciousness of common interests or values that underpins organization is a major political resource; it is a threshold resource that permits organization and the concerted acquisition of additional resources and the pursuit of goals. A group-in-itself refers to a group – for example, a demographic or ethnic group, or a category of consumer – that has not organized itself around common characteristics either because it is not conscious of the commonalities or does not regard them as salient. Unorganized groups are labeled as groups mostly by outsiders such as marketers, politicos, or academics, but not exclusively so. I use the term "group" to denote both unorganized and organized collectives unless otherwise modified.

Mobilizing organized groups brings significant resources to bear beyond what mobilizing the unorganized usually brings because the latter must first be organized to be effective, even if only minimally organized, for example, to vote. Organized groups such as businesses, unions, recreationists, churches, or grass-roots NGOs have experience with collective action, have leaders and internal communications that make them more formidable assets, and are more economical to approach. They may also have political experience and thus be more likely to act politically and to do so intelligently. Leaders, of course, cannot always deliver their members without further work, but at least they control organizational resources. Mobilizing individuals who have political experience brings some of the same attributes.

The unorganized are nonetheless important in many instances. They bring previously untapped energy and access to new social networks. In electoral systems mobilizing unorganized voters can play a pivotal role when elections are near. Often the status quo depends on the acquiescence of the unorganized so organizing them not only provides resources to conservation but also undercuts passive support for opponents. If the lobbyist–public strategy is to be made more effective the unorganized will need to be organized and political nonparticipants turned into participants. The threat of action and a record of action, not poll data alone, moves decision makers (Dieter Rucht, 1999: 212; Bruce Miroff et al., 2002: 106, 108–9).

Mobilizing the unorganized can present special challenges and some advantages:

- Like southern California dry scrub at the end of a rainless hot summer, it's easier to start a fire under the unorganized than to control it. Yet major spikes in mass participation – from urban mobs storming palaces, prisons, and other hated monuments of the powerful to peaceful mass demonstrations demanding reforms or an end to outrages – have made history.

- The unorganized are likely to bring fewer political resources, being less educated and poorer. Those who have been excluded from politics, however, may have greater insight into how politics really works than those who enjoy more resources and see the political system as more benign. Not all of the unorganized lack resources, however; many middle-class individuals are not mobilized and have many skills.

- The unorganized are more difficult to reach because they lack formal leadership. Identifying social network hubs, opinion leaders, and media channels requires work. The experience of the US right-wing's mobilization of cultural conservatives over the last 30 years is a good example of what can be done, although they had enormous resources to identify leaders and channels, create leaders and channels, and otherwise invest in mobilization (Richard Viguerie and David Franke, 2004).

- Unorganized cohorts may have organizational experience from a work environment but are likely to need acculturation to political participation and continuing reinforcement.

- Those newly organized are less likely to have divided organizational loyalties such as the pull of membership in other NGOs.

- Unorganized cohorts who feel they have little vested in the status quo may be more willing to take political risks.

Prioritizing targets: the sympathetic and the powerful

Conservationists have generally given priority to groups who share their values, believing that such groups would require fewer resources to mobilize. Certainly this is an advantage. Increased membership has added financial resources and weight to the voices of some NGOs, but has not translated into a strong base of support willing to vote on conservation issues, let alone engage in more intensive action (Ronald Shaiko, 1999). Obtaining action beyond check writing from most of this base is not always necessary to make a difference, but much higher levels of mobilization are needed to fundamentally shift policy. To achieve higher levels of more intense participation, however, conservation NGOs will need to target more precisely and create new organizational structures.

Targeting more precisely means identifying and recruiting those among the sympathetic most inclined to political action. This is a limited pool in most societies.

Mobilization levels are low most of the time. The experience of other social movements suggests that to gain increased participation from most of those currently targeted would require making them part of a community centered on conservation (David Kertzer, 1988; Lewis Rambo, 1993; Doug McAdam et al., 2001; Amitai Etzioni, 2004). Other movements relied on preexisting venues such as Black churches in the US civil rights movement or universities and churches in the US antiwar movement of the 1960s. Conservation NGOs would have to create those venues by making a significant investment in them; there is for such a step (Martha Lee, 1995; James Jasper, 1998).

In the nearer term conservationists have more options in targeting when securing coalition partners. The history of the labor, public health, and other social movements shows that coalitions (and alliances) have usually been worth the investment and paid off (Mario Diani, 2003b: 106–7). The conservation movement has not been subject to similar systematic analysis, but anecdotal evidence suggests that campaign coalitions and longer term collaborations have added significant value (David Johns, 2003; Charles Chester, 2006). Groups can achieve together what they cannot alone through pooling resources, coordinating activities, and staking out a unified position. Coalitions have attracted new resources, including funding, media attention, and ultimately agency cooperation and broader political support leading to attainment of objectives.

Although good communication and cooperation among all conservation groups benefits the movement, entering into a coalition is much more of an investment and must be evaluated on its strategic merits – what help will partners bring, are there other partners that bring more without requiring more investment, can they deliver in time? It is always tempting to focus on groups that share similar values and this often makes sense, but only as a starting point. Similar groups usually have similar connections and networks and while the whole is more than simple addition indicates, access to more influential networks and political levers is unlikely to result from similarly situated conservation NGOs. More weight given to acquiring coalition partners (or quiet cooperation with) that bring greater influence, if successful, makes success more likely.

There are many challenges in creating successful coalitions. Among the most sensitive areas is funding. Potential participants do not want to see their funding diminished nor control over their budget transferred to a new, super-entity. Coalition agreements must be crystal clear about fund-raising activities, including how existing and new sources of funding will be solicited, use of one another's names in solicitation, and how resources will be allocated.

Coalitions among partners that are vastly different in size, experience, or other resources can lead to conflict over control. Larger organizations may expect deference. Smaller organizations – often more politically creative – can have more influence with partners if they are clearer about their own purposes and what they want from the coalition, and if they understand the motivations of partners' leaders. Dysfunctional leaders make an otherwise attractive organization a bad partner. In some cases it makes sense to work informally with units of an organization, bypassing top leadership, or to simply go directly to its membership and recruit. Many people belong to more than one NGO and NGOs even trade member lists. Recruiting without permission can generate bad feelings, however. Sometimes it's necessary to go up the hierarchy to gain cooperation, as when the provincial office of a land acquisition NGO opposed the purchase of biologically critical land when it became available. A visit to the NGO's national office, along with some persuasive words and documentation, secured funding. A focus on achieving greater influence in furtherance of achieving a goal, along with insight and good negotiating skills, are important assets for an NGO seeking to mobilize partners.

It is not possible for some groups – for example, NGOs engaged in direct action and those raising hundreds of millions for land purchases – to publicly be part of the same coalition. There's little benefit to the former, usually a cost to the latter, and the benefits of a coalition absent these drawbacks can be achieved with informal cooperation. Public political distance is needed for good cop/bad cop strategies to work. We will take a closer look at movement dynamics and NGO relationships in Chapter 21.

Nested hierarchies can be a useful means to manage large numbers of groups across a broad geographic region. In one case this involved larger regional conservation groups forming a coalition to work on landscape connectivity across 3000 miles of mountains. Each regional NGO in turn sponsored coalitions within their regions, inviting dozens of other conservation NGOs that were more locally oriented or more specialized.

Organized groups that possess very high levels of mobilization – by reason of wealth or political power, immediate access to other powerful people, the right messengers to influence those they lack immediate access to, by their understanding of the political process or knowing how to use the media or controlling media outlets – present significant opportunities for increasing conservation influence if they can be mobilized into alliances or recruited into NGOs (Tilly 1999: 262, 267). Such groups are regular participants in defining the political agenda, that is, what gets defined as a problem and what gets defined as an acceptable solution (Doug McAdam et al., 2001: 162; Thomas Dye, 2002). There are many advantages to focusing mobilization on more powerful groups and also some serious difficulties.

Not all organized groups are equally powerful, of course – some are of the elite, for example, in the United States those few thousand decision makers identified by G. Domhoff (1998), Thomas Dye (2002), and Charles Lindblom (1977). Other groups have greater or lesser influence with the elite based on their bargaining power and particular connections. Church leaders are not the equal of oil companies, and a local chamber of commerce or civic group is not the equivalent of the National Association of Manufacturers. Not all equally powerful groups are equally influential with all decision makers. In most countries elites have factions and the distribution of power among them shifts as a result of elections, economic evolution and cycles, and other factors.

Conservationists rightly consider decision makers – even sympathetic ones – as objects of pressure generated through mobilization. But in some instances decision makers as individuals and groups can be mobilized and made part of the pressure brought to bear on other decision makers. US mayors concerned with urban water supplies have lobbied for stronger laws that benefit conservation such as the US Clean Water Act and conservation provisions in farm bills. (Cities are mostly downstream from agricultural lands.) Key legislators are allies of conservation NGOs in many countries around the globe and spend their political capital to advance conservation goals. Some business leaders or sectors have values or interests that align with a conservation goal and this can form the basis for alliances; for example, a media mogul who likes wild places and whose immediate economic interests do not conflict with protection, or the insurance industry's interest in checking climate change. Such groups can bring other powerful players to bear on an issue.

Not only do alliances with powerful groups bring more resources to the table for conservation but also they deprive opponents of those resources, for example, depriving an elite of unity on a particular issue affecting biodiversity such as energy or agricultural policy. A divided elite contributed to apartheid's end (Elisabeth Wood, 2000). When polluting industries were divided over Clean Air Act amendments in the early 1990s they lost. They learned from this, unified in the next round, and won (Judith Layzer, 2006: 26–53). Ted Turner has used his western land holdings to help endangered species recovery in the face of opposition from powerful groups in the US west. Al Gore's role in countering industry propaganda on climate change cannot be disputed. These alliances create new personal connections between conservationists and the powerful that enable alliances on other issues to be explored. Conservationists could use connections, for example, to encourage the insurance industry, a major source of investment capital, to avoid sectors of the economy that are particularly destructive of biodiversity. All else being equal (i.e., rate of return), personal connections can make a difference.

Having powerful interests on the side of conservation comes with a price. Alliances with the powerful and wealthy have been justly criticized for leading the conservation NGOs involved to moderate their demands to avoid alienating their allies (John Dryzek et al., 2003). George Gonzalez (2001) documents several cases in which powerful and wealthy conservationists overwhelmed other conservation advocates in shaping goals and outcomes, such as supporting creation of a park but only if its size were limited and boundaries accorded with resource or development interests. Many conservationists oppose international trade regimes because they increase resource use and extraction thereby destroying habitat. Wealthy and powerful conservation allies have not hesitated to look after their trade interests and part ways with conservationists on trade issues. Conservationists similarly need to hold fast to their objectives. They also do well to remember that people who hold positions of power and wealth – especially great power or wealth – usually do so because they have fought for it and they want to keep it. Although a few wealthy and powerful people may be willing to make enormous sacrifices because of their connection to the natural world, for most the primary concern is getting and keeping power (Murray Edelman, 1988: 28). Because the greatest overarching threat to biodiversity is unending economic growth, conservation clashes with one of the most fundamental interests of power and wealth.

To be realistic conservationists must ask on what issues and on what basis they can form alliances with powerful organized interests. When the basis for the alliance is a quid pro quo rather than shared support for a goal based on interest or values, the trade-offs can be significant. The chips conservationists must use in payment or the compromises they are asked to make must be carefully weighed.

Prioritizing targets: groups that deserve more emphasis

Conservationists are doing better at targeting new allies, but need to do so more systematically and to devote more resources to it. Being more systematic means methodically evaluating the political landscape for potential allies relevant to pursuit of a goal. Some potential allies stand out.

Although the interests of public health and conservation cannot be conflated, many of the drivers that degrade public health also threaten wildlands and biodiversity. Public health professionals are organized and have influence on policy at all levels of government. They are in government, often have the ear of, and are sometimes decision makers themselves. Threats to public health often gain broad attention. In some cases actions that benefit public health will also benefit conservation and in those cases forging an alliance with the public health community can significantly add clout. Existing public health laws may give conservationists added leverage before courts and agencies.

Many economic interests are negatively affected by exotic species and are willing to take action to eradicate or limit them. (Many economic interests also benefit from exotics, such as cattle and sheep ranchers.) That farmers or fishermen, for instance, find exotics threatening to their livelihoods does not mean that the solutions they seek will benefit biodiversity, but they may. A common threat does not mean a common solution. Protecting agricultural crops from exotics usually does not necessarily make croplands available for biodiversity nor does it have positive ancillary effects. However, combining forces to pursue solutions that encompass the goals of both interests can result in enough enhancement of political influence to achieve results not otherwise possible.

The media are usually seen as a means to reaching others, not as a target themselves, but they should be considered a target because of the enormous multiplier effect of sympathetic media. (Chapter 11 contains a fuller discussion of this.) Consider how important groups might start to think differently if instead of hearing that "The wildfire has 'blackened' or 'destroyed' or 'devastated' 2000 hectares and threatens 20 homes," they instead heard "Fire today rejuvenated 20,000 hectares of forest and grasslands." Or, "Fires in this area are naturally recurring events important to the health of the region. The only homes threatened by fire were knowingly built in areas distant from fire-fighting teams or owners had not taken the advice to fireproof their homes by clearing brush around them." Such changes in the media have occurred as a direct or indirect result of campaigns, such as that around smoking and health in the United States. Media changes occurred, however, only when the ban on tobacco advertising in the electronic media removed the revenue stream from the tobacco industry (Bagdikian, 2004: 251–5). The changes in media reporting and the absence of ads contributed to a cultural shift in how tobacco use was viewed.

Although reporters and editors may resist being "educated," and although they are aware of media owners' interests and ideological preferences, and of the pressure to increase ad revenues, they still have leeway. Owners themselves are susceptible to change and even Murdoch's ideological agenda is tempered by profitability concerns. Conservationists have significant experience in dealing with news media, but have not made major efforts to gain entry into entertainment programming. Environmental issues have penetrated but conservation remains largely absent. What would be the impact of a conservation *Star Wars* on the conservation agenda?

The US right wing over the last three decades essentially bypassed established media by creating their own media (Richard Viguerie and David Franke, 2004: 322, 329, 330–2), which has become profitable as well as politically potent. The wealth of the economic and cultural right, including conservative churches with preexisting radio and television outlets and networks, made this possible. It also had a profound consequence: the mainstream media moved to the right to protect market share. Because conservation lacks the preexisting infrastructure and other resources of the right it will need to rely on the mainstream media in the United States and many other places for now.

Zoos represent a largely untapped ally. Most understand what is happening to the source of their exhibits. Many are playing a role in conservation by educating their visitors and supporting or undertaking in situ conservation projects, not just captive breeding. Zoos reach tens of millions of people every year. Visitors return repeatedly, bringing their children as they grow. Children's relationship with and empathy for animals can be profoundly shaped by the zoo experience (Kellert, 2002: 144–5). Increased cooperation between zoos and conservation NGOs can improve understanding of the plight of the natural world among audiences conservationists do not normally reach. Mobilization does not directly result, but zoo experiences that increase empathy make audiences more responsive to mobilization appeals.

Zoos are important cultural institutions. Zoo leaders are usually connected with other civic, business, and political leaders and can serve as brokers, connecting conservation NGOs with these groups. Zoos host events for a range of groups in the communities they serve; these can raise conservation awareness and legitimize issues and preferred solutions. Zoos are a source of funding for conservation science and planning and for species recovery projects.

The labor movement is strong in many developed countries and has played a major role in carving out democratic space. Only in democracies has the conservation movement thrived, although it is better tolerated in authoritarian regimes than social justice movements (Christopher Rootes, 2004: 622). Labor has not been a strong supporter of conservation, however, more often hitching its wagon to economic growth and focusing on environmental issues such as toxics (Brian Obach, 2004: 47–81). In the United States, where the labor movement is weak and economistically focused, it has often been an ally of big business, for example, major unions supported drilling in the National Arctic Wildlife Refuge. If nothing more, working with labor may head off unnecessary clashes and deprive conservation opponents of a populist ally. Some unions, such as those in energy and heavy manufacturing, may never be receptive or see a conservation-compatible future as desirable. In the longer term labor must play a role in the development of conservation-friendly livelihoods. In the near term those unions that are potentially more sympathetic to conservation – in the United States, for example, unions representing professionals such as educators – should be high on the list for consideration as allies. Members are highly mobilized, often span many social networks, and bring other resources. Unions tend to have a full political agenda, however. Some common ground exists in particular regions and at particular times – labor has supported some wilderness legislation and conservationists and labor have jointly battled the World Trade Organization (WTO) and corporations such as Maxxam that threaten communities and the natural world – but a quid pro quo arrangement is most likely.

In many parts of the world rural areas are the geographical landscape in contention. Remaining and recoverable wild areas and species generally lie outside cities and suburbs; hence rural communities' belief that conservationists mean to protect Nature at their expense – a view exploited by conservation opponents. In many parts of the world, however, there is a strong basis for alliances with important elements in rural communities. Economists (Ray Rasker and Ben Alexander, 1997; Ernie Niemi et al., 1996, 1997, 1998, 1999a, 1999b; Thomas Powers and Richard Barrett, 2001) have found that rural communities in the US west with intact wilderness are better-off than those that have sold or are selling their natural heritage. This has slowly been sinking in, as have findings that traditional rural industries such as farming, logging, and energy extraction typically add limited value to communities compared with alternative livelihoods. Farming, for example, may earn significant income for a rural area, but most of that money flows out of the area to buy tractors, petroleum, and chemicals, leaving a small net amount of wealth remaining in the community. In comparison, furniture making from small-diameter trees obtained from thinning keeps more of the income in the community. Rural civic leaders have been more responsive to conservationists on this account, and in some instances are taking steps without prodding. Amenity migrants come with their own costs – big houses, ranchettes, new roads, but tend to be more sympathetic to conservation goals.

Rural areas in the developing world are different than those in the developed world – more people are subsistence producers with fewer options; market forces often operate outside the law and involve superexploited labor. The capacity for alliances depends on whether rural groups engage in activities that degrade the natural world, such as logging, ranching, or farming in previously undomesticated habitat (e.g., parts of the Amazon) or whether groups seek to protect relatively wild area from industrial activities. Not all subsistence-based groups long present in a place and that oppose industrial incursions embrace wild creatures. Some have thoroughly domesticated and degraded the landscape.

Trial lawyers in the United States have significant influence and are not routinely engaged by conservationists. They share with conservationists an interest in keeping the courts open to people trying to hold accountable those responsible for medical malpractice, consumer product injuries, fraud, and pollution. These are often the same forces opposing conservation and checking them across a broad front aids conservation. Absent this check these entities simply factor in capped court awards as part of the cost of doing business and pass it along while continuing to inflict injury on people and Nature. Conservation litigators, such as those at the Center for Biodiversity or Earth Justice, are natural messengers to the larger bar, not just trial lawyers. This network offers enormous potential for conservationists looking to tap into new social networks that include critical political and economic players. Conservationists rely on litigation in countries other than the United States and before international bodies. International and domestic fora are entwined, as are fora dealing with environmental, trade, finance, and human rights issues (Linda Malone and Scott Pasternack, 2006); a broad range of advocacy groups are interested in maintaining access to and the authority of adjudicatory fora.

In countries colonized by settler populations that remained distinct from the populations already present (cf. North America and Latin America), the latter populations have maintained some control of a fraction of their former land base (e.g., the United States but not in Russian Siberia). Some of these groups are committed to "traditional" livelihoods seeking to protect wild areas and native species; and thus are potentially important allies. Many groups, such as the Nez Perce and Blackfeet, have taken the lead in protecting species and places.

Conservative religious groups are playing a much bigger role in conservation but their priorities are yet to be clearly defined. Several are interested in basing their work on conservation biology principles. In some respects their notions of "creation care" are closer to biocentrism than more mainstream religions' concern with the human environment. Still in the early stages, alliances with these groups hold the promise of bringing huge numbers of previously uninvolved people into conservation. Some larger NGOs are actively recruiting religious cohorts as well as forming alliances with them (Nigel Dudley et al., 2005), but many grass-roots groups are not. Globally there is a broad awakening of concern among religious groups about conservation, and these groups have enormous resources in many cases: connections with elites, the ability to persuade and admonish, and the ability to reach billions of people that conservation does not now reach. Religion has long been a means of enforcing societal ecological discipline, though not to the benefit of conservation in complex societies (Roy Rappaport, 1974, 1976, 1999). The possibility remains that religion can again exert a profound influence in restraining human ecological effects.

Both religious groups and some rural groups bring with them the resources of mature communities. They offer a template for community that conservation NGOs can emulate.

This shortlist of groups that deserve more attention is only suggestive and meant to reinforce the importance of mobilizing major new resources.

Other factors in targeting

Timing

Timing and ease of mobilization are closely related factors in mobilization. Unexpected opportunities or attacks may require rapid mobilization of resources and the rapidity with which forces can be mobilized becomes the deciding factor, other things being equal. Having resources (groups, connections, media materials) at the ready is one reason for mobilization in the service of organization building, for example, turning check writers into activists, and not limiting mobilization to campaigns. Existing NGO members, supporters, and volunteer staff are an obvious source of resources who can be rapidly brought to bear, including experts "on retainer," members at the ready to attend rallies or hearings, or extend a line of credit. NGOs are often surprised at the resources their members possess and that they are willing to put to use, such as their skills, time, and connections. Unfortunately most NGOs are not set up to effectively utilize member and supporter resources except in a few areas such as contributing money or letter writing and phone calling, and thus never call upon the additional resources. Direct action or activist NGOs are much better at involving people – and indeed depend on greater involvement by members and a sense of community – but NGOs with professional staff seem to find too much member involvement be intrusive. It can be if not approached strategically. For NGOs that do specialized work such as litigation it can be challenging, but other NGOs for not involving members more. Without deeper involvement, people's commitment atrophies resulting in high supporter turnover and a failure to respond to calls to action. This creates a situation where support shows up in polls, but not at the voting booth or in the streets or a boycott; this is the sort of shallow support decision makers know they can ignore. Successful movements are based on a strong sense of community and community is associated with deep commitment and endurance. As later chapters demonstrate, the group that plays together is better prepared to fight together.

It's also worth keeping in mind that some of the most effective NGOs never have had large memberships, but have been able to routinely call tens or even hundreds of thousands to street demonstrations and other action (Kirkpatrick Sale, 1973). Real connections count, as does message, messenger, and story.

Political geography

The ability to mobilize groups in the home district of an elected decision maker is often more important than mobilizing a more powerful group elsewhere. If elections are near, the importance of voters can outweigh more powerful but smaller groups inside or outside of a district. Even in authoritarian regimes strategically positioned groups – for example, a minority group in a border area or a media outlet in the capital – may have more clout than larger groups elsewhere.

What is needed

Targeting must heed the ability of groups to offer what is specifically needed. Some campaigns require large numbers of consumers, as when NGOs need to make good on a boycott promise. Hard-core activists may be needed to make a major media splash with a sit-in or demonstration. Those with advanced media skills may be needed to develop a professional quality video or public-service announcement. Scientists, litigators, retired business people, people with special connections or a high public profile are other examples of specialized talents that may need to be mobilized to influence decision makers.

In authoritarian regimes any political work can be physically dangerous. Some protection, in addition to political influence can be achieved by mobilizing support from a member or faction of the domestic elite, from international NGOs, or from influential foreign regimes. Such resources are scarce and relying entirely on outsiders can backfire (Clifford Bob, 2005: 5, 175, 185). Authoritarian regimes may point to outside help as an example of disloyalty or a threat to sovereignty as did former Russian President Vladimir Putin, justifying countermobilization and repression. A host of other issues attend reliance on INGOs in the best of circumstances. Such alliances are seldom between equals and they may have unwanted effects on the internal dynamics of the NGOs/movements in the authoritarian countries. In the final analysis, creating, maintaining, and expanding the political space in which to operate in pursuit of conservation goals must rest on internal capacity and strategies suited to the regime in question.

Other constituencies in authoritarian regimes often enjoy a special status that makes them good targets for mobilization. University students – as children of the elite or of those on whom the regime depends, such as professionals – may enjoy some immunity from repression (Misagh Parsa, 2003: 82). They also have access to communications technology and frequently enjoy prestige with other segments of the populace. Religious groups – as distinct from leaders who may be members of the elite – also enjoy immunity in many societies. But as recent events in Myanmar and the events in Tiananmen more than a decade ago made clear, there is no absolute immunity. Artists or public intellectuals with an international reputation may also enjoy some protection and even have access to domestic elites, although the execution of writer and NGO spokesperson Ken Saro-Wiwa by the Nigerian dictator Sani Abache and his oil company cohorts demonstrates such status has its limits.

Bases for alliances

The basis for a relationship with a potential ally is an important consideration in targeting. Coalition partners or allies may be willing to help conservationists in pursuit of a goal because they share common values (e.g., protecting a species), common interests (defeating a common opponent); or, on the basis of a quid pro quo (you scratch my back, I'll scratch yours). Partners that share values are the easiest to work with because of common language and purpose, although it's a mistake to underestimate the venom that can be generated among friends over differences in priorities. Alliances based on a common goal but differing motivations can work well when these differences are understood and expectations are clear. When they are not understood or clear the differences can disrupt cooperation because of mistrust or animosity. If conservation NGOs lose their clarity they can to be drawn into fights that don't advance their goals.

Much politics involves old-fashioned mutual back-scratching – the sort of amoral bargaining that leaves some conservationists cold because it runs contrary to passion-based politics. There are also conservationists who become so enthralled with being a player and fancying themselves masters of real politick that they forget their purpose. Quid pro quo relationships are an essential tool and require conservationists to be hardnosed in their bargaining. It also requires they have the resources other groups desire from allies. Conservation NGOs often don't have much to offer. Like Tennessee Williams's Blanche, they have depended on the kindness of strangers – not a good bargaining position.

The shotgun

It is not always possible to target precisely. There may not be enough time to identify and research prospective targets. An event or crisis may present itself and necessitate broad-brush outreach. If knowledge about a target is lacking or assumptions about them can't be tested, mobilization is more likely to result if predicated on the target audience being only mildly interested and unwilling to act boldly on their own initiative (Renee Bator, 2000: 528).

Available communications channels may not allow precise targeting. Smaller towns have one newspaper if any, and frequently no local radio. In larger cities, there may be no choice but to use media that reaches far more people than the targeted group. Urban radio allows for specific targeting whereas television and the mainstream press do not.

The failure or inability to target carefully can result in more than a waste of time, energy, and money. When messages reach the wrong people they can feed countermobilization. A message that inspires one group to act in support of conservation can cause others to mobilize against conservation. Opponents can be counted on to use whatever conservationists say against them, usually by distorting it; but there's no need to make it easy for them. Chapter 12 addresses how best to construct messages to maximize mobilization and avoid countermobilization.

The consequences of the inability to target can also be positive. Conservation messages falling on unintended ears have caused businesses and others to withdraw their support from anticonservation NGOs and their industry supporters, sometimes merely to avoid controversy. Media reporting, campaign materials, web sites, and other elements of outreach have recruited members or supporters from surprising sources, inspired others to create a new NGO in a different part of the world with a similar purpose, and generated alliances or other cooperative relationships. Messages unintentionally traveling far and wide can resonate with others and help transform a local issue into a national or an international one, increasing political leverage.

Targeting is important because resources are scarce, only some groups can be mobilized, and only some groups are worth mobilizing. Targeting is also important because people are mobilized as groups. People are reached through their groups, calls to action blunted or reinforced by intragroup relationships, and groups are moved by different sorts of appeals. This is why efforts at mobilizing 'the public' fail to produce desired results. Traits most widely shared are usually the least salient for mobilization. Salient motivators are attributes of groups much smaller than an entire society. Although most Americans say they want more wilderness (Campaign for America's Wilderness, 2003: 8, 23, 28), it doesn't appear that appeals on that basis work very well. Rather, appeals tailored to each group's salient feelings, attitudes, and values result in mobilization (Ronald Shaiko, 1999).

10 Understanding the targets of mobilization; and opponents

Some strategists argue that successful mobilization doesn't depend on a deep understanding of the target audience. Mobilization can be accomplished by relying on correlations. It doesn't matter why people act, it's enough to know that they do under definite circumstances, whether those circumstances are causal or not. Creating the circumstances results in action, just like stepping on the accelerator moves the car. The driver need not understand the underlying mechanics. But mobilization is seldom as simple as driving. If nothing else, the importance of interactivity among emotions, needs, views, material circumstances, and social relationships should dispel notions of Pavlovian simplicity; slight differences in any factor are multiplied by the interactivity, generating different outcomes. Mobilization is almost always creative rather than formulaic. Explanation and insightful intuition will prevail over a mere grasp of correlations.

Most conservationists readily admit that success in reaching audiences depends on knowing important characteristics of those audiences. But they often proceed without such knowledge because they can't afford to acquire it or they limit their focus to cognitive characteristics. At times they ignore information about target audiences and fall into the error of talking to others as they would talk to themselves.

The limits of the cognitive and the need for comprehensive understanding

Original research into prospective targets is expensive, but so is guesswork. Ignoring obtainable information on the attributes that distinguish and motivate various audiences wastes resources and makes conservation NGOs less effective (Mische, 2003: 276). There are many accessible sources of information about groups, especially in the more economically developed societies where commercial and political marketing is well developed. Data gathered directly by marketers are often closely held or expensive to acquire, but data gathered and analyzed by media organizations, polling companies, university researchers, and government are more readily available in many countries. When raw or massaged data are not available, findings usually contain enough detail to enable evaluation. Further research may be needed in some cases or reliance on judgment. As discussed below, research may be as simple as asking knowledgeable people about an audience or enlisting audience members to discuss the attributes of their group.

Mobilization of a group is more likely if NGOs possess a comprehensive understanding of the group: its social position and relationships, its interests, group dynamics, salient need-states, emotional predispositions, and the content of its sacred views, values, and myths.

Cognitive information is easier to come by because professional researchers, although they understand the importance of emotion in political behavior, almost always study the cognitive (Ted Brader, 2005: 21). In an otherwise cogent analysis of the factors that contribute to mobilization, William McGuire (2001: 22–48) focuses almost exclusively on the cognitive. It is important to understand the worldview of a group targeted for mobilization, but worldview is only part of the picture. Political behavior is shaped and catalyzed not just by beliefs and understandings. Most people do not make decisions and act as a result of critical reasoning but on the basis of nonconscious assessments that are influenced by experience, emotions, and needs (Antonio Damasio, 1994; Gerd Gigerenzer, 2007). Commercial and political marketers understand this, but social scientists and conservationists are stuck in people's heads. Repeated screenings of Monty Python's *Life of Brian (1979)* could help them see the error of this. Poor Brian Cohen starts on the path to political activism and ultimate martyrdom by following his hormones – he is attracted to Judith, a beautiful activist with the Judean People's Front (JPF). He is also motivated by strong anti-Roman feelings but not until he joins the JPF in pursuit of Judith does he embrace an ideology.

Mobilization is a common result when someone feels unjustly injured, that is, when the sense of violation is personal and strongly felt (Ronald Aminzade and Doug McAdam, 2001: 31, 36–7). Far fewer people engage in politics as a matter of material exchange and bargaining; for most it involves the projection of hopes, fears, or desires onto the larger political arena and its main characters (Murray Edelman, 1964: 4–6). Elites play to this, knowing that people generally lack the resources to accurately assess in real time the impact of policies and leaders on their interests; they also know that people are disinclined to make such assessments because of what such critical attention would reveal about themselves, that is, that they have let themselves be fooled (Carol Tavris and Elliot Aronson, 2007).

Conservationists share with academics a heavy focus on the cognitive and are heir to the Enlightenment tradition that exalts reason over emotion. Historians Eric Hobsbawn and Terence Ranger (1984: 8–9), in their study of invented traditions, found that those who were more highly committed to Enlightenment notions of rationality resisted creating or reshaping traditions in order to appeal to emotion in the service of political mobilization. A recent Wilderness Society (United States) handbook on mobilization advises activists to focus on values in their outreach, although the handbook itself appeals to activists' emotions (Bart Koehler, 2005: 43–6). In contrast, more opportunistic actors – usually on the far right – have made successful use of invented traditions. The implication is that resistance to emotional appeals comes from identification of the emotional with the irrational. Anecdotal evidence – from interaction with conservationists, who are often highly educated professionals or natural and social scientists – suggests that many consider delving into target audiences' emotions as slightly unprincipled or too personal. To mobilize on the basis of such knowledge feels like manipulation.

Conservationists are then left with trying to persuade people with cognitive-based argumentation that they should act. Persuasion usually fails because targets do not share the same values as conservationists and conservationists do not take the trouble to frame their information in the context of targets' values, especially their most basic values (the sacred). Persuasion also fails because conservationists ignore the relationship between views and underlying emotions. Because people are emotionally invested in their beliefs and their beliefs arise from emotional predispositions, views are difficult to change absent emotional openness or change. It's also likely that Stuart Ewen (1996: 409) is correct when he says that over the last century the opportunity for persuasion has declined as a result of the decline in genuine interaction.

Efforts to persuade that do not appeal to emotions and existing basic values, but instead try to alter them, are a bit like asking people to convert. Although conservationists are not usually arguing for major changes in worldviews, by not framing their arguments in terms of an audience's deepest understandings and values or by not invoking an emotional basis for action, they trigger the defense mechanisms associated with challenges to deeply held views. Overcoming such resistance is extraordinarily difficult (Lewis Rambo, 1993: 87, 172–3), and there is no need for it because by evoking emotion and proper framing people can be mobilized. In the long term fundamental change in people is important, but genuine conversion of groups is usually the result of personal or social crisis or catching up with a trend well under way. Absent crisis, fundamental change in worldviews occurs generationally or in groups that are socially marginal. Socially marginal groups usually do not bring significant political resources to the table unless a society is in profound crisis.

Cognitive appeals are also limited by the need to mobilize groups with a wide range of views (Helen Ingram and Leah Fraser, 2006: 107) and no single or closely related arguments are likely to work across the ideological range. Even if conservationists are focusing on a number of groups that have much in common, the common attributes are rarely the ones most deeply meaningful to each group and it is these that must be addressed if mobilization is to result. Recent electoral campaigns in the United States have demonstrated this point. Those campaigns that have disaggregated larger groups, for example, the industrial working class or white women, and addressed the attributes of the subgroups, have tended to prevail. Conservationists, for example, have long relied on data suggesting that women are more supportive of conservation, in significant part because they are socialized to be caring, feel empathy, and to act on it (Lynette Zelezny et al., 2000: 454). Looking more closely at women in the US northern Rockies, it is Caucasian females under 40 and over 65 who support conservation in large numbers; there is much less support among those in the 40–65 cohort (MacWilliam, 2002: 6). Conservationists need to find out why and tailor outreach accordingly. The greater the level of aggregation the more detailed and distinctive information is lost.

Salient attributes, of course, are not restricted to the cognitive realm. Far from it. Many observers have noted that successful mobilization hinges on the target audiences' identification with the NGO or movement seeking to mobilize them, and this is achieved by appeals that touch deeply felt emotions, offer belonging, as well as provide meaning in the context of the groups existing basic views (Doug McAdam, 1997: 155, 187, 286; Stryker et al., 2000: 10; Snow and McAdam 2000: 62–3; David A. Snow and Doug McAdam et al., 2001: 55, 118–9; Mario Passy, 2003: 26). Target audiences must feel that conservation offers them a purpose that bolsters their identity in the face of threats to it. People will not join if messages challenge their identity.

Because people mobilize as cohorts, an understanding of the mutual reinforcement mechanisms characteristic of a group is very helpful. Appeals that trigger mutual reinforcement among ordinary group members can be as effective as those that succeed in gaining the support of a group's leaders.

Material interests and social position

Some 60 years before Karl Marx and Frederick Engels (1998 [1848]) outlined their theory of class struggle and capitalism's apologists dismissed it, the US Constitution's primary drafter James Madison (1961 [1788]: 56–65), concerned about the stability of the Republic he hoped to bring into being, wrote that

> . . . the most common and durable source of factions has been the various and unequal distribution of property. Those who hold and those who are without property have ever formed distinct interests in society. Those who are creditors, and those who are debtors, fall under a like discrimination. A landed interest, a manufacturing interest, a mercantile interest, a moneyed interest, with many lesser interests, grow up of necessity in civilized nations, and divide them into different classes, actuated by different sentiments and views. The regulation of these various and interfering interests forms the principal task of modern legislation, and involves the spirit of party and faction in the necessary and ordinary operations of the government.

People who are similarly situated economically, geographically, or who find their material circumstances determined by their gender or ethnicity, or who have long been marked by a common history, frequently act as a group on those bases. They may respond in a highly effective way if well organized, or in ad hoc ways as individuals and cohorts.

Material circumstances do not translate directly into interest or organization, but are culturally mediated. Nonetheless material circumstances have a reality apart from social construction and they shape social construction, influencing and binding people even when they are not conscious of it. Material circumstances have shaped our needs and emotions through the evolutionary process. The objects of needs are overwhelmingly material and their presence, absence, and degree of scarcity affect experience and the emotions and values that arise from experience. The experience of need and emotion and the construction of belief are not reducible to material circumstances in any simple way and this is why groups must be understood in their particularity if they are to be mobilized. So although humans all (unsurprisingly) possess biophilic tendencies – Kahn's (1999: 112–3, 126, 163–4) research and review of attitudes across cultures found that children had empathy for Nature regardless of class or country. Empathy is weakened, suppressed, or enhanced differentially by later experience that is dependent on social position, among other things. Indeed, people can be socialized and acculturated to act against many of their interests as when workers vote for the candidates of big business, imagining themselves to be what they are not (Thomas Frank, 2004; David Brooks, 2005: WK14).

The role of the group in shaping the perception of interest cannot be overstressed. Robert Repetto (2006: 15) describes how peer pressure in fishing communities reinforces once-adaptive practices and keeps individuals from exploring new practices that would halt fish declines and maintain their livelihoods over the long haul. Absent supportive cohorts, most people are unwilling to act (Paul Stern, 2000: 416–7). Geographic dispersal does not diminish group effects. Ranchers across the western United States – a large geographic region consisting mostly of thinly populated rural areas – not only reinforce each other locally but throughout the region as well (Charles Davis, 2006: 251).

When an NGO understands the material circumstances and interests of a target group, how those interests are (mis) perceived by the group, and the group's internal dynamics, it possesses a good part of the basis for crafting a sound approach for mobilization. With this information conservationists can assess how their goals will affect the group in question and whether their likely response will be supportive, hostile, suspicious, or neutral. Conservationists can determine whether an audience's income or economic influence will be enhanced or diminished or whether their access to basic necessities be made more or less difficult. Other questions can also be answered. Will conservation proposals actually benefit a group's economic well-being in the long run, although the group does not perceive this? Will restoration of natural systems open new opportunities for those seeking to improve their circumstances? Will a new law increase one agency's power at the expense of another? What groups will be strengthened and weakened by realizing conservation goals? Conservationists can obtain significant leverage in their mobilization outreach by understanding the gap between interests and their perception. An appeal to the real interests of rural communities, for example, will open doors to more positive responses and undercut negative reaction if only in the longer run. This is an improvement over beating one's head against a wall by trying to appeal to misperceptions.

Understanding needs and compensatory needs in target audiences

Needs are primary behavioral motivators and closely linked to a group's material circumstances and definition of interest. The objects of needs are material. Basic matters of livelihood – food, water, shelter, warmth – depend upon obtaining, inter alia, calories and the plants and animals that can supply them. Safety and security are material – threats and safety lie in our relations with other people, other living things, and natural processes. Sexual needs, the need for belonging, purpose, to be recognized, to have a sense of efficacy depend on social position and networks and the larger social web of relationships. Within limits needs are also notoriously easy to deform, suppress, and redirect toward compensatory objects. People don't belong, don't feel loved or have difficulty loving, so they consume or seek status, wealth, and power. Societies through families and other institutions produce these deformations and encourage or require the focus on compensatory objects such as high consumption, status seeking, or onanism.

Although sickness results from the chronic deprivation of any need, not all needs are equally malleable. Acceptable substitutes for clean water and air are more problematic than substitutes for love or esteem. People may nonetheless adapt to the lack and resultant sickness if obtaining clean air or water is too risky or difficult. Their anger and frustration at the lack and its consequences may also be successfully directed at a scapegoat or turned inward. Hunger is easily directed away from nutritious food to junk and generates pleasure until the consequences catch up.

Some needs – and the manner of their deformation – are more important to conservation than others, including those related to livelihood, sexuality, belonging and purpose, and efficacy. Gaining information on these need attributes provides an enormous aid to mobilization. Knowing how a group experiences and defines its livelihood and security needs and how these are acted on, knowing their attitude toward sexuality and how sexuality is expressed, having a sense of whether individuals in a group feel they genuinely belong, knowing the strength of purpose of the group and feelings of efficacy, are as important as knowing whether a group likes animals, Nature, camps out, and so on.

Livelihood

Mobilizing around livelihood needs has been difficult for conservation because of the massive contradictions between conservation goals and the growth juggernaut; this has left the field to opponents. Some conservationists approach livelihood issues by arguing that no hard choices need be made, but that rings false to audiences. Others try to avoid the issue just as the US founders sidestepped the problem of slavery in drafting their Constitution, with tragic results. Still other conservationists call for material sacrifice to protect Nature, but this lacks broad appeal. Depending on how much an audience is invested in the view that endless economic growth and technological control are good, or their equation of consumerism with quality of life and markets with the best decisions, there are serviceable economic approaches. These are based on analyses showing that existing economic arrangements are not just an obstacle to effective conservation (Robert Durant et al., 2004: 510), but to people's ability to find connection with each other and maintain community – both essential to survival and well-being. Consumption beyond a certain level is not making people happy but the reverse (Bill McKibben, 2007: 34–8). Reestablishing connections offers much: it assuages mortality anxiety, lessens the compulsion to consume (Morris Berman, 2000), and allows people to see that ecologically damaging behavior is inimical to their best interests. Stephen Kaplan (2000: 505) and Raymond DeYoung (2000: 517) argue that people's need for a sense of competence can be increased by the challenge of living well by consuming less. Less consumption means less ecological degradation, more stable and reliable ecological services, and less reliance on complex and frail human systems. Less consumption also means less pollution and time spent working, and increased time for more interesting pursuits. Economists David Rosnick and Mark Weisbrot (2006) note that if western Europeans worked the longer hours of Americans they would emit 30% more carbon dioxide.

Because people are heavily invested in their manipulation of the world, appeals to abandon this approach to Nature in exchange for closer connection with Nature or a more reliable livelihood and less growth, won't work magic. Appeals must be compelling in other ways, for example, cast in vivid and enthralling stories. Elites, of course, stand to lose wealth and power if there is a shift in the status quo. Indeed, consuming less and placing limits on the human transformation of Nature into commodities may represent a more serious challenge to the status quo than events like the Russian revolution's challenge to capitalism (Ronnie Lipschutz and Judith Mayer, 1993: 266). Certainly some members of the elite do support conservation, but at what cost to their livelihood? Some economic sectors and individuals are less wedded to growth than others and not all kinds of growth are equally damaging to the natural world. Knowledge of these distinctions is important in crafting mobilization strategy.

Mobilization appeals to groups who lack necessities require an understanding of which necessities are lacking or scarce and how that lack is experienced. Strong economic safety nets provide a more positive context for conservation goals that necessitate limits on overall human appropriation of Nature. Such safety nets are absent in much of the world, including the United States (alone among developed countries), China (which abandoned the "iron rice bowl"), and much of the third world. Putting such safety nets in place will take more muscle than conservationists possess, but supporting such efforts can help conservation in some places if it does not increase total human consumption.

In authoritarian countries, where basic safety (as in freedom from repression) and income issues are prominent, conservationists have a direct basis for allying with other groups around a shared and pressing concern – creating a society in which political and economic relationships foster security rather than insecurity. Generalized repression of political activity, as distinct from enforcement of laws arrived at by a more or less open process, usually does not benefit conservation. ("Green dictatorships" are fantasies born of frustration, not real possibilities.)

Sexuality

Understanding a group's sexual attributes – the broad range of impulses that pull people toward each other in order to find pleasure and joy in touch, dance, song, sexual contact, and creative expression – is essential to mobilization. People who know they are animals, who have the capacity to live in their bodies and accept pain along with pleasure, who embrace their sexuality, have the capacity for empathy with other living things. For them "Energy is eternal delight." (William Blake, 1977 [1790]). Because they are in contact with their own needs they are better able to grasp that other creatures have needs and that ecological systems are living things not machines. Those who reject their bodies are frightened by their impulses and passions or who react to them with extreme discomfort and condemn them, or who deny their bodies, tend to attitudes the Puritans made a religion (Roderick Nash, 1982); they will fear and hate wolf and forest. Alienated from their bodies they believe the heavens will "torment Man in eternity for following his Energies."(William Blake, 1977 [1790]).

Because sexual energy and impulses are so powerful and thread through all of personality they are the object of social control. The so-called sexual revolution of the 1960s and the reaction to that revolution in the 1970s and 1980s are testimony to this power. Most groups are neither entirely comfortable with their impulses, nor sternly reject them; instead they live with the baggage of socialization and acculturation that burdens sexual expression with a range of internalized limitations that divide people against themselves and deaden them to some degree. Understanding how this energy is channelized and bound in a group is critical to assessing whether group is capable of being mobilized for conservation goals on the basis of sexuality and determining what messages will resonate or trigger recoil. It is not simply a question of whether ads such as PETA's mildly titillating nudes will work, but whether the group is capable of expressing the sort of profound passion that animates action for fundamental creative change.

Answering several questions will provide useful insight. To what degree does a group express sexual inequality, rigid rules governing sexuality, fear of sexuality (especially female sexuality), or disgust with sexuality? To what degree are power and wealth (domination) wrapped up with sexual expression? To what degree is sexuality well integrated with positive emotions and sensations such as intimacy, affection, joy, and the capacity to surrender to pleasure? The answers are usually not hard to ascertain because sexual attitudes are difficult to hide. They are expressed, if not always in a straightforward way, by behavior and in artifacts such as music, film, books, magazines, and by the rules elites and group leaders try to impose via laws, sermons, and texts, or by the absence of elite attempts to monitor and closely control sexual expression. Groups are often divided, hence the need to disaggregate. US Catholics, for example, use birth control in the same proportion as the general population notwithstanding the stern pronouncements of elderly male bishops and cardinals.

Belonging

The need to belong is strong, probably because it reinforces social cooperation – a highly adaptive trait for an animal with small teeth, no claws, but a clever brain. Knowing the degree to which the individuals comprising a group feel a part of it and share a strong sense of purpose is important in crafting the approach to mobilization. If identification with a group is weak, it makes sense to target individuals and cohorts that make up the group and to make appeals based on belonging and purpose. In the United States, which has a geographically mobile population, rapidly growing new towns provide no community, lacking a downtown or neighborhood sidewalks and houses with porches that might encourage interaction. In this milieu megachurches have blossomed by offering people a place to congregate, to find purpose, and to interact in small groups (Jonathan Mahler, 2005: 33, 36). Small groups organized by the churches around particular personal issues foster intense and transformational interactions that bind participants closely to the small group and the church. These churches also offer material benefits increasingly lacking in US society: day care, counseling, and athletic facilities. The lessons for mobilizing those who desire community are obvious.

If a sense of belonging and purpose is strong in a targeted audience – usually such audiences are organized – it makes sense to approach the group as a group and appeal to those attributes that are most salient to the group, thereby enlisting feelings of group belonging in addition to shared values, purposes, goals, or interests. It is also helpful to assess the purpose of the group; in many instances groups have quite pragmatic and narrowly focused purposes and there is a hunger for a larger purpose or a desire for engaging with recently awakened conservation concerns.

Efficacy

People need to feel worthwhile and this depends in part on their sense of efficacy. Self- or group-perception of efficacy is not always well tuned to reality; individuals and groups generally over or underestimate their actual and potential power to affect the world (Joseph Forgas, 2000; Joseph Forgas and Michelle Cromer, 2004; Carol Tavris and Elliot Aronson, 2007). Regardless, mobilization depends on an understanding of a group's estimate of its empowerment or deprivation. If a group feels empowered, then appeals to use their capacity in support of worthy goals will resonate. When a group's sense of empowerment is tied to control of Nature, however, it can be problematic; the control of Nature is antithetical to conservation's overarching goal of recovery of the natural world and letting most of it be. Milder interventionist feelings may be enlisted in support of active restoration efforts, however.

Disempowerment may be experienced as something general or specific. Many people who are sympathetic to conservation and who feel in general that they can affect things, also feel that conservation problems are too big and intractable (Deborah Guber, 2003: 52). Whether a group believes it is ineffective generally or only in some areas, identifying what it perceives as the main obstacles to its empowerment is critical because offering a way to overcome the obstacles is necessary to achieve mobilization. Scapegoating and tendencies to simplification are often present in such groups and dealing with these perceptions can be difficult. But often there is common ground between those doing the mobilizing and their target audience and this can form the basis for connecting and moving toward concerted political action. So although many rural people distrust conservationists and government as outsiders, they also mistrust corporations, including those that employ them (Ben Long, 2002).

Obtaining, interpreting, and using information

The availability of information about a group's need-states is uneven at best. Marketing and polling data can offer insights when available. NGO self-help is also available. Interaction with individuals from target audiences can provide important information and insight about a group if anecdotal information is not mistaken for more than it is. Brokers are also a major source of information. Caution must be taken not to simplistically type a group. The world is complex and humans categorize things to make life manageable, but categorization results in information loss. Groups and their members are not ideal types; both contain contradictions and tensions. Efficacy, for example, is not an all or nothing thing. Cultural conservatives are not the hyperrepressed crusaders their leaders call on them to be, nor are cultural progressives free of sexual hang-ups, basking in Reichian bliss. There is also tension among needs. Sexual expression may be sacrificed to more basic survival needs such as obtaining food and finding acceptance within the social group. The adaptation of societies to the larger world calls forth social structures that channelize needs, shape their expression, and seek to conform it to social structure. The channelization takes many forms including social suppression, psychological repression, and redirection of the associated energy to socially acceptable outlets. These details are important in determining how to best mobilize a group.

Among those conservationists who recognize that the overarching goal of ending the current extinction crisis requires fundamental social reform, there remains a disinclination to mobilize for such reform. This is unfortunate because such reform entails dismantling the deformation of needs and offers to reconnect people with themselves and the natural world. This can be quite enticing to many who feel alienated. As stated above, reconnecting with other living things and the larger universe can undermine the mortality terror that Sheldon Solomon et al. (2004: 127–146) and Kasser et al. (2004) argue is at the root of overconsumption and other social pathologies. Henry Thoreau (1964 [1854]: 335) understood this intuitively when he wrote over 150 years ago that "(a) man is rich in proportion to the number of things he can afford to let alone." Reconnection deprives compensatory pursuits of the energy they have commandeered. The unsatisfactory succor of fundamentalisms or schemes of control becomes apparent.

Most who embrace various compensatory needs, however, are not aware they have done so; they deny their dissatisfaction, and are heavily armored against the pain associated with the underlying deformation. It must be remembered that people have turned away from their genuine needs because they were too dangerous or hurtful at an earlier point in their lives and this turning away received approval and helped them survive psychologically. Such people fear where embracing their genuine needs might lead, or they fear the anger associated with their long frustration. As Jefferson wrote in the US Declaration of Independence

(1776), "experience hath shewn that mankind are more disposed to suffer . . . than to right themselves by abolishing the forms to which they are accustomed." Suffering, however, is not without its limits.

When conservationists understand how groups have adapted their needs to the social order of which they are a part, including the acceptance of compensatory distractions, they are in a position to tap into the longing and frustration associated with genuine need. Doing so has the potential to unlock tremendous energy, as other social movements have demonstrated.

Understanding emotion in target audiences

There are two aspects to a good understanding of the emotional structure of target audiences. The first is a group's mood or the emotions that permeate and dominate its "personality," defining members' reaction to need-states and the world generally, and underpinning its morality. Is a group fearful, angry, hopeful, anxious, ashamed? Do they possess a sense of place? A sense of well-being? Do they feel engaged in life and have a clear sense of direction? The second aspect of emotional structure consists of a group's reaction to particular events, messages, messengers, value statements, candidates, policies, and so on. Reactions are influenced by moods and predispositions but are also shaped by the moment, subject to the vicissitudes of trends, timing, and similar factors.

The complex interplay of factors that generate emotional responses is getting more attention from both practitioners and students. Cheetah biologist and advocate Laurie Marker (2006) noted that cheetah lovers are so emotionally attached to the animals that they don't understand the need to save the large ecosystems on which the animals depend. They can be difficult to mobilize to support habitat protection notwithstanding their empathy.

Joanne Weiss Reid and Karen Beazley's (2004) research on attitudes in Nova Scotia, and Craig Thomas's (2003: 3, 27, 258) analysis of bureaucrats' motivations, provide more systematic examples of the role of emotion and the need to take specifics into account. Reid and Beazley found that for many people sadness was primarily associated with the prospect of species loss; fascination and aesthetic pleasure were the primary reactions to the presence of other species. Fear that loss of species would hurt human chances for survival because of the interconnections was another widely shared emotion. People also feared government intervention to save species, and they desired policies relying on cooperation rather than regulation, that is, cooperative policy making that included them.

Emotions that were widely shared were not evenly shared. The mix of emotions present, their strength and centrality, varied among groups. The same emotion-based mobilization appeals are therefore unlikely to work on different groups. This point is driven home by Thomas's study of US bureaucrats' motivations. Motivations vary depending on an individual's role or position in an agency. Field staff juggle attachments to agency, profession, and local community. Professional staff have a strong emotional investment in adhering to professional standards and values, whereas line managers are moved by desires to maintain agency autonomy, stability, and predictability. Bureaucrats are not simply bureaucrats and approaches must be crafted that recognize differences. Thomas's study is also valuable for its analysis of how behavior is shaped by the interaction of emotional predispositions and the requirements of social roles.

Another study of bureaucrats from several countries found them discomforted by the recognition that the environmental problems they face are so serious that they require addressing profound moral questions – something outside routine administrative practice (Douglas Torgerson and Robert Paehlke, 2006: 314). Engaging agency staff in larger issues will require finding approaches that overcome this reticence. These findings also suggest that there are limits to what can be expected from agency personnel absent innovative approaches that address this discomfort. Unfortunately Paehlke and Torgerson don't provide information on how different kinds of bureaucrats experience this challenge or vary in levels of frustration, insight, and so on.

Which emotions?

The important emotions for politics are fear, anger, love and pleasure seeking, disgust, sadness, guilt, and shame. In the Chapter 12 on crafting messages many of these will be examined in the course of the discussion. In this chapter I rely on one emotion – fear – to explore how understanding the relationship between personality, emotional responses, and the objects, events, and experiences that evoke emotions is important in mobilization.

Fear has always played a central role in politics and may be the single most important emotion and mood in politics because it is relatively easy to evoke and to evoke with intensity. A mood of fear or general orientation of fearfulness – of life, of mortality, of vague enemies – makes a group less open to the world; they take a defensive posture. They are unlikely to have easy access to the empathy required to experience hurt and sadness at the injuries inflicted on the natural world because they are defended against the hurt that has caused their fear. Indeed, groups that are chronically fearful may feel threatened by Nature because fear generates a desire for control and Nature is the ultimate uncontrollable. Sherry Nicholsen (2002: 79) noted that genuine contact with another species changes personality and behavior, but fearfulness usually precludes the ability to make such contact unless the fear is focused on humans and Nature offers a refuge.

In contrast, fear as a reaction to something in the present – the approach of fire season or a fire – orients people to deal with the sources of the fear. Conservationists have made common cause with residents in some rural communities routinely threatened by summer fire by supporting programs to clear brush and take other precautions while educating residents about the importance of wildfire in the backcountry. The difference between fear in the present and a fearful mood lies in their accessibility to awareness and flexibility. In the former case a resident might skip clearing brush in a year when the spring and summer are wet; a posture of fearfulness is unlikely to be affected by a wet spring, though a dry spring may reinforce anxiety. Such differences (between mood or predisposition and an emotional response to the present) call for different approaches to mobilization and affect the likelihood of mobilization.

Predisposition and the present are not completely distinct. The presence or absence of a posture of fear will affect both the likelihood of a fear response in any situation and the intensity of the response. But it is more complex than this. Not all groups in which fear is a predominant emotion are focused on the same objects. Nor do they respond in exactly the same ways or with the same intensity, nor are emotions transferred to other objects according to the same pattern. The same is true of other emotions. The emotional response of materially insecure people to figures of authority tends to deference and the projection of strength and benevolence (Pippa Norris and Ronald Inglehart, 2004: 19), but not every figure of authority is the object of attachment. Early US baby boomers experienced virtually unlimited economic opportunities and they emotionally embraced dominant relationships and values, whereas later boomers faced a contracting economy, rejected mainstream norms as "inhumane" and "constricting," and embraced alternatives (Jack A. Goldstone and Doug McAdam, 2001: 217–9). The specifics of emotional responses are not only shaped by mood and the particularities of experience, but also by the symbols, stories, and metaphors used in communication (Karen Johnson-Cartee and Gary Copeland, 2004: 81).

Obtaining, interpreting, and using information

Obtaining information about the emotional attributes of a target audience presents the same issues as obtaining information about needs. Much of the best information is proprietary. More easily available polling data and public studies have holes in them (information most important to conservationists is missing), and people do not always report their motivations honestly or they are not fully aware of them. Developing close connections with politicos who run campaigns or marketing professionals can provide access to some of the most useful information available – these practitioners are focused on what works. Brokers who know a group or informants from the target group are good sources of information and if approached with well thought out queries conducted by NGO staff with insight and judgment, much can be gained. Often group leaders and those who influence a group's opinions are excellent sources, but their interests in the group often give them a limited perspective – they may be protective of the group or proud of it. All informants have blind spots about their group but these can be elicited by careful interviewing – a technique developed by anthropologists (James Spradley and David McCurdy, 2004). A combination of sources from inside and outside of a group is always preferable because it provides a combination of perspectives.

Analysis of the political and commercial marketing materials and tactics aimed at a group by those with good information about the group is another good source of information if information on what has and has not worked is also available. Sources include television and radio ads and programming, magazine ads and content, and other similar information from other channels a group relies on.

As with needs, emotions are often in conflict with each other, generating ambivalence or painful tension. The discomfort provides openings to groups because people seek relief from discomfort. Studies show that rural constituencies who often mistrust conservationists also feel a desire to protect Nature, even at some economic cost. What seems to override their conservation concern is not so much economics but a very strong negative feeling about "outsiders telling them what to do" (Karsten Heuer, 2004). If the approach to mobilization offers a way out of this tension and resulting discomfort it will be more appealing than otherwise.

Mobilization involves another area of emotional tension that when understood is a great aid in crafting the best outreach. Mobilization that requires little emotional and material investment from a person – a small check, signing a postcard – doesn't generate much commitment. Mobilization that has a higher emotional and material cost, although more difficult to achieve, generates much stronger commitment that is also more likely to last. Farm worker union organizer Cesar Chavez recounts a story of someone who had paid his first month dues but had then fallen behind two months (Richard Jensen and John Hammerback, 2002: 65–72). When Chavez knocked on his door on a rainy winter evening to collect the $7.00 he knew no one had any income. The man had only $5.00 and was heading to the store. Chavez said he would come with him and collect $3.50 for the union and the rest could go to groceries. It was difficult to ask for the dues but Chavez knew that if the union didn't require dues even from the poor, membership would not be valued. The person who paid his dues that day went on to become a major United Farm Workers organizer. For Chavez, who had little money himself, it reinforced two principles of successful organizing: there's never enough money to organize but you have to do it anyway, and people have a responsibility to help the cause.

Probably the single most important understanding conservationists need is of themselves: an awareness of their tendency to project their emotions onto targeted audiences. It is easier to project emotions than beliefs because humans tend to take emotions as universal. The lack of self-awareness extends to the act of projection itself.

Understanding worldviews, the sacred, and story

Conservation outreach typically stresses appeal to audiences' values and addressing values is certainly important. But values constitute only one element among many other elements in audiences' cognitive system or worldview. Although people are moved by emotions, needs, and circumstances, one of their most basic needs is to explain, understand, and give meaning to the world and themselves – a need required by the choices behavioral flexibility presents. The cognitive is partly driven by self-justification but it also provides another set of adaptive tools that serve the goals emotions and needs set, and it adds another level of behavioral regulators to the human repertoire. It enables consideration of the long term, evaluation of past actions and their efficacy, reflection on impulses before acting on them, consideration of a wider range of options in the pursuit of need objects and desires, innovation of tools and social organization, and manipulation of the world based on the capacity to imagine something different than what is.

Mobilization will be most successful when conservationists' appeals integrate an audience's needs, emotions, the many aspects of their worldview (not just values), including the sacred and other levels of understanding and prescription, and the stories and symbols that convey them. Enlisting need-states and emotion is not enough because people and groups must explain and justify their action to themselves and relevant others; they also imagine and plan. Evoking empathy or interest is incomplete without reasons for why it's good to follow the empathetic impulse, why feeling good about it is appropriate, why taking action is important and the cool thing to do. These additional, cognitive factors also trigger mutual reinforcement, which is heavily verbal. Thus, appeals that include a road map for action describing what should be done and why it's important and the right thing to do, how it will make a difference, and how it fits into the larger scheme, are critical.

To be able to craft integrated appeals conservationists need to understand the content of an audience's worldview (beliefs, understandings, prescriptions, values), (i) at all levels, (ii) the relationship between aspects of their worldview and the emotions and needs underlying each aspect, and (iii) the stories which convey the worldview.

With an understanding of the content and structure of a group's views and the stories that convey them conservationists can craft messages that fit. If messages and other aspects of outreach miss the mark they are likely to be ignored. If the values and views in an appeal conflict with the audience's, they are most likely to be rejected. Worse, the audience may feel challenged or threatened, triggering anxiety and hostility. Galileo's experience was extreme but it's a good reminder that people do not consider messages based on their merits most of the time.

The structure of the cognitive

A grasp of an audience's worldviews is important because effective appeals must fit into that worldview. Recall from Chapter 4 anthropologist Roy Rappaport's (1999: 263–7; 272–3) description of the four-tiered structure of worldviews and the regulatory aspect of each level. At the foundation of all worldviews are those views, unquestioned (i.e., sacred) and usually untestable, about ultimate meaning and why the universe is the way it is. The sacred serves to anchor and legitimate a worldview in the face of alternatives. The sacred provides authority to cosmological axioms, which are general propositions and principles describing how the material world is ordered and overarching values. These principles are not derived from the sacred, but are legitimated by it. The sacred and axioms are often interwoven in the foundational stories that transmit a worldview. It is from cosmological axioms that lower order understandings and rules are derived: the specifics of how the world works, the norms for individual and institutional behavior in daily life that establish social obligations and expectations, and so on. A fourth level of understandings and prescriptions addresses the identification and interpretation of events in the world and appropriate or required response, for example, to births, deaths, droughts, coming of age, and elections.

Sociologist Paul Stern (2000: 411–4) observed a similar structure to worldviews when he analyzed environmental views. He found basic and unquestioned value orientations toward the natural world (egoistic, altruistic, biospheric), each in turn giving rise to distinctive general notions of the proper relationship between humans and the natural world, and from these are derived specific norms that guide behavior in daily life.

The sacred and cosmological

Because fundamental views do not change easily it is essential to understand a group's sacred tenets and cosmological axioms so that outreach can be crafted to be compatible with them. The sacred and cosmological do change, but generationally (through socialization/acculturation) or in response to a major social or personal crises. If conservation messages call for changes in a group's lower order views – and they frequently do and must do so – they will have little effect unless nested in and "blessed" by existing higher-level views and associated stories. A few groups, as Roy Rappaport (1999: 444–5) noted, consider even lower-order rules to be sacred – unchanging and not to be questioned – and communication with such groups, if it requires any change in views, can be difficult. Such rigidity is maladaptive but people cling because of a high need for certainty; unfortunately such groups can be politically important.

Lloyd Burton (2002: 258–9), in his study of religion and wilderness, found that attention to the specifics of the sacred (and cosmological) is essential to effective communication. Does a group consider the wild an evil place, as Calvinists once did? What role does the Earth play in a group's creation story? Some Christian groups, for example, stress that part of Genesis that describes how Adam was made from the Earth (Adam comes from adamah, the Hebrew word for Earth). That provides an entrée for conservation messages. Many religions are place based, and unlike Abrahamic and other sky-god religions that emphasize symbolic structures as holy (e.g., temple and tabernacle), they regard the whole Earth as holy (a thing whose integrity must be protected at any cost) and some places as especially holy. Abrahamic and other western religions embrace two broad and very different notions of divinity (divinity plays a major role in the religious sacred) – one in which the divinity plays an overwhelming role, directing everything, and those in which humans exercise a much greater agency (David A. Leeming, 2002: 24–5). Such distinctions are not confined to religious sacred traditions. An Estonian informant explained a major cultural difference between his ethnic group and Russians thus: "We are all atheists; but I am a Lutheran atheist, and they are Orthodox atheists" (Pippa Norris and Ronald Inglehart, 2004: 17). Conservationists are familiar with the striking difference between humanists who hold a steadfast faith in reason and the capacity to manage Nature, and those who are eco- or biocentric and hold evolutionary and ecological processes to be wiser than humans.

In some respects a good grasp of a group's cosmological views is more important than the sacred because these axioms, although general, are more specific than the sacred, are directly about the material world, and are therefore easier to run afoul of in mobilization messaging. The lower order views that guide daily conduct are blessed by the sacred but they are derived from the cosmological. When conservationists ask not just for action from people, but action that breaks with past behavior and norms, the new behavior and norms must be convincingly derived from existing axioms. Matters are more complicated because axioms contain both understandings of how the world works – which may be based on knowledge or belief – and the meanings ascribed to these understandings. For a message to nest it must avoid conflict with both elements; this is a problem if a group's beliefs about how the world works is wrong on a point that is important to conservation, for example, that certain kinds of development are not harmful to a species.

A message that conforms to or evokes both elements will resonate better and such messages can usually be crafted because alternative lower order prescriptions can be derived from the same axiom as the extant ones. Thus, rather than challenging faith in progress – a deeply anchored axiom in many cultures – conservationists can redefine what constitutes progress; NGOs will succeed, however, only if the axiom and its existing derivative prescriptions are understood. Thus, if progress is closely linked to improvements in material security, it can be argued that policies bringing human behavior into accord with natural processes, protecting healthy populations of native wildlife, and generating less pollution rather than more stuff and more people, are better guarantees of that. Similarly, people (individuals and groups) have a sense of justice, which is a powerful motivator (Susan Clayton, 2000: 459) and is defined by some widely shared and many specific principles. Mobilization appeals that build on and extend existing principles of justice, such as applying them to animals and even ecosystems and demonstrating how they are complementary, can undercut resistance. Thus some religious conservationists argue that just as loving the divinity calls for loving one's neighbor, so it also calls for loving what has been created by the divinity: the world and all of its creatures and plants.

Justice offers a good example of the linkages between the cognitive and verbal, and emotion and need. Successful appeals to norms of justice evoke not just the energy associated with the emotional attachment to a norm, but also the emotional disposition and energy underlying the norm. "(J)ustice is rooted in a sense of fairness," and "such a sense is not the product of applying rational" thought, "but, rather results from emotional processing(.)" (Hsu et al., 2008: 1095; see also Jonathan Haidt, 2001). Further, people experience great pleasure in enforcing standards of justice (as fairness, not necessarily the law), even when it exacts a cost from them (Brian Knutson, 2004: 1246–7). Justice is also strongly social, binding the group.

When mobilization requires reinterpreting an axiom rather than deriving something new from it, or replacing an axiom, appeals to the sacred will be critical and so will an understanding of the axiom to be replaced and its relation to other axioms; even when axioms clash to a degree, they are part of a fabric. Understanding the structure of that fabric aids in design of a new piece. If the new axiom has a familiar form and story, including plot or characters, and mimics connections to other axioms, it will ease acceptance.

Although more open to change than the sacred, cultures conserve axioms even in the face of attractive alternative axioms justified in terms of the sacred. Only when axioms no longer explain what is happening in the world do they become susceptible to change. Because some members of a group are more open than others to new views it often makes sense to concentrate on them. They may not be the most influential people, however, and may even be marginal. But if a new overarching understanding and set of values offers advantages then others will be attracted when they see it works and smart leaders will try to stay in front. In the last several decades the historically dominant view within Christianity that humans should have dominion over Nature has come to be understood as stewardship of Nature, or Creation Care.

Bureaucratic personnel provide a good example of axioms that must be challenged. Bureaucracies, in addition to the constraints imposed by their political-economic context, resist setting goals that are not reducible to precise metrics (Douglas Torgerson and Robert Paeklke, 2006: 315–7). This makes looking at problems holistically almost impossible. Along with a heavy reliance on technical expertise that is presumed to be value-neutral, underlying values and the limits imposed by the deference to metrics are obscured (Douglas Torgerson, 2006: 16–9). Conservationists have not found a way around this axiomatic view and it may be that it is inherent in large-scale administrative organization. If so then bureaucracy as a primary instrument of policy will need to be replaced with other forms at some point. With what is not clear, but the capacity of volunteerism to solve large-scale problems is limited (Robert Durant et al., 2004: 496–7).

Thus far I have conflated analysis of religious and nonreligious sacreds and axioms. There are some important differences, however. Although secular worldviews routinely make use of religious symbols, stories, and ritual because of their popular familiarity, religious groups' views include other-worldly purposes and rewards that are important in motivating and guiding action (Ronald Aminzade et al., 2001: 158–9, 161, 174). Religious worldviews may provide for the intervention of spirits giving rise to a very different sense of the opportunities presented than that held by secular groups. Preexisting institutional structures also differ, along with associated commitment, loyalty, sense of community, and trust. Religious views may inspire greater risk-taking, but they may also prescribe a narrower repertoire of acceptable political activities. There are also important differences within each "community." David Wilson (2002: 163) found that in the United States, for instance, less theologically strict denominations have been more sympathetic to conservation but less likely to act. Those who belong to stricter denominations contribute more to their churches, and one might presume, if committed to protecting Creation, would be more likely to act because of the greater intensity of their beliefs. On the other hand both religious and secular worldviews often impute purpose to the universe, to forces of Nature, to events, and to the inanimate. Differences about what is imbued with purpose and what those purposes are vary, but not necessarily along religious/secular lines.

Day-to-day understandings and prescriptions

The map conservationists' compile a target audience's worldview should include all of those day-to-day level understandings and rules that are favorable for conservation mobilization and all those that present obstacles. The day-to-day includes views about what is good to eat, what is proper to eat, how much to consume, when to consume, who the status reference groups are (who are the Jones's one must keep up with), how many children to have, whether to vote, who to vote for, whether to engage in political action and what sort of action, what groups to contribute to, who to believe on issues, and what is of immediate concern such that it needs action, what sort of action is called for, whether the risks of acting are worth it, and whether the action demanded will make a difference. People generally do not act if they do not perceive that action will make a difference, if it is not of immediate and local concern, and unless it involves a deeply cathected principle of justice. Thus, views important to conservation may be directly so (e.g., those about consumption or the salience of conservation issues to political choices) or relevant because of their broad applicability (e.g., when is it appropriate to engage in collective political action). Some views are held more deeply (internalized) than others and have a bigger effect on behavior.

Day-to-day views are often contradictory. This is not simply a matter of norms and behavior parting ways, but of norms being at odds with others norms and the behaviors they rationalize; thus, a good life consists of lots of wilderness and lots of global travel. Conflicting norms are not difficult to derive from the same axiom.

It is possible in planning mobilization to know too much about a group and get bogged down in details or overanalysis, and lose sight of the overarching pattern to the views. The focus should be on the pattern and those details most relevant to conservation mobilization. At the pattern level, the overall degree of sympathy with conservation and inclination to act are most important. Important details included identifying views that can be directly appealed to for mobilization, their strength, and the stories they are associated with. Are there ambiguous views and stories that can be interpreted in a way favorable to conservation or do they need to be overcome? Not all views a group holds that are antithetical to conservation may be problematic if the action needed from the group does not hinge on that view; for example, efforts to block legislation weakening stream protection will not likely run afoul of a group's views that it's fine to own a second home in the woods.

When conservationists can build on or make use of existing views governing day-to-day behavior so much the better. More frequently, however, mobilization hinges on changing at least some lower-order views to achieve mobilization, such as those that incline a group to think action will make no difference, or that it's worth giving up a desired product to protect a forest or the ocean. Mobilization appeals can often be justified in terms of existing norms that people are not living up to. More often the norms need to change at least partially to generate mobilization. Accomplishing this requires demonstrating that existing views aren't working, providing compelling alternatives, and placing these alternative norms and understandings in the system of existing cosmological axioms and the sacred.

Alternative views are most persuasively conveyed through existing stories, including those that embody the sacred (myths). Knowledge of the relevant stories is as valuable as understanding the views at issue. So is understanding the roles available within the stories; it is often roles rather than the story itself that appeals.

The role of story, as we shall see in Chapter 13 is ubiquitous, but emphasis on story varies by group and culture. Stories play a bigger role in US political culture and pronouncements by expert authorities in German political culture (Francesca Polletta, 2007: 137). Germans are also more likely to turn to state leaders for authoritative stories.

Given the need to alter lower level views, it is reassuring to know that they are amenable to change and are indeed changing all the time. Change at this level is necessary for adaptation and even significant accumulations of change at this level usually do not call into question or require changes in higher order views because the latter are so broad (Roy Rappaport, 1999: 428). When accumulations of lower order change do alter axioms or even the sacred the change in these is usually imperceptible (generational), subject to denial, or is considered the recovery of lost truths. (The tension between adaptation and stability runs throughout culture.) The emotional underpinnings of axioms and the sacred are also slow to change. If the emotional underpinnings of day-to-day views are intensely held, change at this level is usually resisted even if the views aren't working well at explaining events or providing good guidance.

Mobilization thresholds and the salient activators of mobilization shift because of changes in a group's experience, such as the waxing or waning of its power, cultural innovation, drift, or diffusion. All affect in smaller and larger ways, ideology including values, a sense of what's possible, or who is considered friend and opponent. For groups newly mobilized politics can feel quite bruising; some groups are dispirited by a tough fight and others become invigorated. The length of time it takes to achieve goals can also be tough on the newly mobilized who lack political experience (Raymond DeYoung, 2000: 520). Mobilization appeals can take account of these factors and avoid unrealistic expectations. Even without causing a change in views per se, events may cause a shift in which threats or opportunities are emphasized by a group. NGOs must be sensitive to all of these factors because it means what worked yesterday may or may not work tomorrow.

The change mobilization requires in those mobilized (as distinct from the change those mobilized are intended to help bring about) is not equal. Altering product choice is one thing (buying a more fuel-efficient car), altering ongoing behavior is more difficult (devoting more time to activism) (Doug McKenzie-Mohr, 2000: 546). Changing deeply seated views on efficacy is tough – people who are suspicious of their ability to affect political change need to have the possibility of change demonstrated; being told it is possible is not convincing. Many Americans firmly believe they can know a candidate's personality via television. It does little good to tell them this is nonsense and that they should be skeptical and look at a candidate's record on the issues. That takes effort. Conservationists are better off to pitch their candidate on personality to groups focused on personality. Encouraging voters to be become critically thoughtful and skeptical is a much longer-term project.

Obtaining, interpreting, and using information

Information on a group's views is relatively easier to come by than information on their emotional attributes and needs. For sacred and cosmological propositions mythological and equivalent texts are available, though these are subject to interpretive nuances over which wars have been justified, if not always fought. A group's popular literature, music, film, jokes, and so on can usually be mined. Members of groups can be sought as informants as can a group's leaders with the caveat that leaders may have a less-than-perfect grasp of those they lead, or they may have an interest in putting a particular spin on their group's views. Marketing publications such as the *Journal of Consumer Research, Simmons-Study of Media and Markets* and *MRI (Media Research Inc.)* are available in most university libraries. Conservation-friendly public relations and communication firms such as Pyramid Communications, the Public Media Center, and many others are good resources; such firms often do pro bono or reduced fee work for smaller NGOs; these firms are committed to the cause, see it as an investment, and want NGOs to be more effective.

Although groups that are the target of mobilization frequently disseminate and consume voluminous amounts of material on their views (e.g., books, newsletters, broadcast programming) it cannot be taken at face value. Debates between factions, differences between leaders and followers, and deliberate dissembling mean that all representations of views on behalf of a group or about it require interpretation before outreach on them.

In assessing an audience's worldview it is helpful to keep in mind the difference between knowledge about the world and beliefs about the world. A group's understanding is always a mix of the two. Knowledge – tested, verifiable explanations and theories – is always incomplete. Those with great knowledge in one area may be quite ignorant in others. Because of the need to explain phenomena people fill in the gaps in their knowledge with surmises, looking to what is plausible and what fits with existing knowledge, values, or principles. Asked about the decline in newspaper readership during a time of war when people would be expected to show an increased hunger for information (information is the basis for knowledge), researcher Kathryn Bowman at the American Enterprise Institute noted that Americans don't look to factual information to inform their understanding, but consult their values (John Young, 2003). This is not a novel observation and has caused intellectual authorities to comment that people are entitled to their own opinions, but not to their own facts. Unfortunately people aren't listening to such authorities. If one was inclined to support the US invasion of Iraq then Hussein had weapons of mass destruction (Steven Kull, 2003, 2004, 2006). If sexual activity outside of marriage is bad then only abstinence prevents disease, condoms do not. All groups fill in their knowledge gaps with conjecture; some more than others, however, seek knowledge rather than rely on surmise. The point for those engaged in mobilization is that groups do not distinguish between knowledge and belief. Although some groups are much more open to learning than others, efforts to enlighten a group can put them on the defensive and derail mobilization.

Channels of communication

It is also important to identify and understand the means of communication by which a targeted group receives information. Public relations firms make it their business to know this and many manuals (Sandra Beckwith, 2006; Jacobson, 1999, 2006; Kristen Grimm, 2006) have been written on this subject, although their information is often necessarily generic. Different channels of communication in different societies possess different dynamics. Advertiser-driven media are both shapers of audiences' frames of reference and play to existing frames (Ronald Bettig and Jeanne Hall, 2003; Ben Bagdikian, 2004). Reporters, anchors, columnists, editors, and owners all possess an ideology and agenda – a mix of commercial, political, professional, and career interests and goals. Many established channels of mass communication have culturally distinct industry norms and expectations. These include political biases, conventions of self-censorship, a greater or lesser reliance on novelty and drama to gain market share, differential preferences for stories about the personal rather than structural, and the degree to which they pay attention to context (W Bennett, 2002).

Opponents

Serious opponents and potential opponents deserve the same effort at understanding as potential partners and allies. Too often conservationists are surprised by groups that emerge into opposition, or by the vehemence and lack of principle in their opposition. They should be outraged, but not surprised. Conservationists have also missed opportunities by ignoring some groups based on assumptions about them, when with a better understanding they could have kept a group neutral or even won some sympathy from them.

Understanding opponents must go beyond their stated positions and obvious interests. As with potential supporters, knowing opponents' circumstances, emotional predispositions, need-states, organizational structures, and leaders will improve the chances that conservationists can anticipate their behavior and effectively deal with it. Particulars count with opponents not just with allies. A business that extracts or utilizes resources may behave differently depending on whether it has diverse sources of supply. A rancher fallen on hard times who regards livestock-raising as primarily a business will likely welcome being bought out. A rancher in similar circumstances who considers ranching his holy patrimony will resent such an offer and go into debt to buy ammunition to shoot wolves. Some groups may oppose a particular conservation proposal, goal, or NGO, while others don't like conservationists, outsiders, or Nature. Conservation is to some a scapegoat for unwanted change (e.g., from automation, exhaustion of a resource, consequences of cut-and-run business policies). Some conservation opponents lack good political and economic understanding; they don't know who their real friends and opponents are and are easily manipulated by those who play on their ignorance and fear.

Good intelligence is essential. Knowing opponents' strategies, resources, what sort of fight they can mount and how long can they sustain it, their leverage with officials, communities and the media, their access to experts, their claim to icons like the cowboy or threads of dominant mythology, their degree of coziness with the keepers of sacred stories, provides real strategic benefits. In particular, a good knowledge and understanding of the stories and symbols they plan to use and how they plan to use them makes countering them less onerous. Much of the battle over "hearts and minds" consists of efforts of those on opposing sides of an issue to frame the issue favorably by advancing the most compelling story. A realistic assessment of opponents' political sophistication and their weaknesses such as internal divisions or vulnerability to scandal, their record with communities they claim to be protecting, is as important as knowing their strengths.

In the case of opponent organizations, especially large hierarchical ones, knowledge of leaders and their record provides important insight into likely policies, strategies, and tactics. Past behavior insofar as it can be used to reveal motivations and patterns of thinking is often a good guide to the future, but conservationists must avoid the mistake of professional militaries the world over who are always preparing to fight the last war. Although some opposition leaders may continue relying on what has worked in the past, even as it becomes less and less effective, most are more innovative although patterns and continuities exist.

As the conservation movement challenges the status quo in increasingly significant ways its more serious and well-heeled opponents will commit more resources to countering it and will utilize more sophisticated or draconian tools and approaches. This is not a likelihood but a certainty. Those clinging to wealth and power have filled prisons and graveyards without remorse throughout human history (see section on Repression in Chapter 19).

11 Messengers and channels for mobilization

Once upon a time in the United States there were three television networks that dominated the broadcast media and a handful of personalities trusted by its citizens to give them information about events with which they had no direct experience. Press deadlines were twice daily, tied to morning and evening newspaper editions and the nightly news broadcasts. Then everything changed. There were more networks, dozens of channels feeding niche markets, and gaggles of talking heads competing with each other for advertising dollars 24 hours a day. With more channels came far fewer owners. Talk radio programming blossomed like thistles in a country lawn and broadcasters replaced journalists with entertainers. Newspaper circulation shrank and literacy declined (Lois Romano, 2005: A12).

Media changes have occurred in every country, although in many parts of the world big cities still have dozens of newspapers and literacy is growing. But all forms of media and the other means of communication on which mobilization depends have been transformed, including news and infotainment magazines, books, niche opinion journals, the pulpit with its admired civic and religious leaders, music, concerts, theater, cinema, and social networks. Corporations have started think tanks to compete with older, semi-independent ones, in order to lend credibility to greed and concentration of wealth, and public intellectuals find it difficult to compete with blowhards. Interest groups continue to provide members with political information but it is increasingly difficult to compete with the noise. The internet has made sharing information easier but for uncritical surfers it is difficult to distinguish reliable or peer-reviewed information from disinformation or foolishness.

Despite these changes the centrality of messenger and means of communication to mobilization remains. I treat them in the same chapter because they are closely entwined. In one-on-one communication they are the same. In other forms of communication the authority of the messenger is tied to the channel – the pulpit, the anchor's chair, the social network hub. It also works in reverse, with charismatic personalities energizing new or old channels. US talk radio came to prominence in part because of its cadre of entertainer-messengers (Richard Viguerie and David Frank, 2004: 321, 331–2). Messenger and channel must be suited to each other, just as they must be suited to the audience. Groups receive information via several channels, but one or a few are usually favored. Mobilization depends in large part on using effective messengers and favored channels.

Messengers

Most of us remember a fable from our childhood wherein a powerful person killed a messenger bearing bad news. The obvious lesson was that killing the messenger did not alter reality – except for the messenger, of course; and the recipient, who, in rejecting the message lost a final opportunity to avoid disaster. The story raises some questions for conservationists. Would a different messenger have been able to persuade the recipient to listen? Is there even a message without the right messenger?

The messenger is an essential part of the message and the baggage they bring disposes recipients to listen carefully, to distort the message, or to reject it outright. The wrong messenger can start recipients' knees jerking, short-circuiting the receipt of information (James Combs and Dan Nimmo, 1993: 145). Conservationists have long admonished each other about the importance of the messenger when dealing with ranchers, other rural audiences, and indigenous groups. This is good advice, but it applies to every audience. Michael Gunter (2004: 143) found that even when conservationists talked to each other the messenger mattered: higher significance was given to messages delivered by messengers of higher status.

Messenger attributes

Effective messengers possess several types of attributes, the particulars of which vary among audiences. Thus, messengers must be trusted or otherwise credible but what makes them so depends on the audience. Trust depends in part on an audience believing they know the messenger's character. In one-on-one communication – as when the leaders of nongovernmental organizations (NGOs) are bargaining or when personal networks are involved – the messenger is often personally known or their record is known. In mass communication effective messengers must also convey a transparency that allows an audience to believe they know them; on this trust and credibility rest. The relationship between audience and broadcast journalist, editor, political or economic leader, or cultural luminary, usually takes time to develop, but some messengers have an instant rapport with an audience. Early insights by Sigmund Freud (1959 [1922]: 12–3, 25–32, 47–57) into relationships between political leaders and the led have largely been born out by Murray Edelman (1964: 4, 7, 16–8) and others (Wilhelm Reich, 1946 [1970]; Theodore Adorno et al., 1950; Bruce Mazlish, 1990; John Jost et al., 2003) and apply to messenger–mass audience relationships. A mass audience "knows" a messenger when they project their emotions and needs onto the messenger and the audience experiences the messenger as a suitable object because the messenger embraces the projection as adoration and mirrors back reassurance; there is a fit.

Mass messenger–audience relationships are unequal even if dissembling is not involved because the messenger (or his/her handlers) always knows the audience better than the audience knows the messenger. The most successful messengers are those who understand or intuit the audience's expectations and conform to them without being seen as doing so.

Those for whom politics is the pursuit of interest rather than symbolic reassurance are less susceptible to the thralls of the quality of the messenger, but even the most hardnosed are not immune. Because the sway of a messenger's attributes depends on the issue area, those who are pragmatic bargainers on behalf of their interests, may be less skeptical and more accepting of the messenger's influence in other areas of life.

It is easier for audiences to do a reality check on local issues and events than on more distant ones, but local messengers are just as capable as national messengers of dissembling and manipulation and frequently get away with it.

Because growing a new messenger into a trusted one can take time, enlisting an already trusted messenger is an easier path. Conservationists working in the rural northern US Rockies only got their message heard when, after years of trying to reach locals with their message, they found messengers among the local business community. When new messengers must be relied upon careful attention must be given to the attributes that will make them most likely to be accepted.

The social position of a messenger is another factor in being "known." For example, conservationists who have lived in rural southern Oregon for more than two decades find that they are still deemed untrustworthy outsiders by large segments of the community. They are known as "outsiders" because of certain social attributes: education, cultural liberalism, dress, etc. Information and policy suggestions they provide is therefore suspect. Evidence-based arguments erode this view very slowly if at all. In these same communities, recent arrivals may be quickly accepted as insiders if they possess the right attributes, in which case their views are not discounted.

To be known does not require, as noted above, that messengers be of similar social standing, and they are not, in the case of mass communication messengers. Indeed, in matters of public affairs most people find elite opinion leaders more persuasive than others (Karen Johnson-Cartee and Gary Copeland, 2004: 113). A study of commercial campaign spots found that the facial expression and body language of those who hold power evoked strong emotional responses (Ted Brader, 2005: 11). This is one reason technique has become so important and leaders' countenances are matters of practice, practice, practice. Some leaders, of course, are naturals.

Not just any member of the elite resonates with every group. To be considered reliable a political, economic, religious, media, and other opinion leader must be perceived to share or have at heart the interests of those who look to them. In the creation of the Steens Mountain wilderness many in the rural southeast Oregon high desert country looked to the big ranchers for guidance. When the big ranchers supported a protected-area proposal the community followed. Trusted leaders are by definition looked to, they are believed to have interests similar to those looking to them, and they are assumed to have the interests of the larger community at heart. Generally groups recognize that those they look to for leadership are not of their background, but they believe their leaders share important things with them that make the leader an insider no matter how distant they are in geographic space or social position. It harkens to the fit – the audience can credibly project onto a messenger what it needs to project, and the messenger is attuned enough to mirror it back.

Obviously "knowability" is not the only attribute that lends credibility. The messenger must be the right age, gender, socioeconomic status, ethnicity, and have the correct demeanor – all elements of what Renee Bator (2000: 529) and William McGuire (2001: 24–5) summarize as physical attractiveness and general likableness. For US audiences this includes a pretty face and makeup for women, a good smile for both genders, and more formal dress. Attractiveness is important because it is associated with positive character traits, including the notion that the messenger genuinely has the audience's interests at heart (William McGuire, 2001: 24–5).

Effective messengers need not be real. Smokey Bear and countless other fictional or animated bearers of information have been singularly successful in mobilizing audiences. Characters from films, books, television shows, and commercials peddle products, candidates, and causes successfully. Some individuals are remarkably versatile in that they have more than one persona available to act as a messenger. Political and economic leaders have their carefully cultivated public persona, which sometimes they calculatingly shed in favor of their equally carefully cultivated "private" persona. On rare occasions they may even portray a glimpse of the private person behind the private person. Celebrities have similar options. They may utilize popular characters they have portrayed, they may utilize their public persona, or the "real" person behind both roles and public image. Audiences may identify with any or all of these characters. In still other cases characters come to personify an institution or organization, evoking the values and images that constitute a brand.

Sometimes the messenger is a phantom, as when groups act to correct errors in the claims made by advertisers on a billboard, unfurl a banner from a high building, or monkey wrench. It is just as important to attend to persona in these cases as it is when creating a commercial, speaking to a rally, or engaging in civil disobedience. An attractive or sympathetic face is needed even if the face must remain literally anonymous. Think of Zorro (the Fox) or the rabble dressed as

Native Americans who seized and then destroyed private property by throwing it into Boston harbor in 1773. If care is not taken to craft a compelling persona then it will be created by opponents, by a media that seldom regards with sympathy those who challenge the elite, or by audiences who feel threatened and project unattractive attributes onto the messenger.

The effectiveness of real, fictional, animated, or phantom characters depends on a developed (and attractive) persona. To achieve the development needed a messenger must usually be embedded in a story at two levels: the messenger must have an implied or explicit story about themselves, which is related to but not subsumed by the story the messenger uses to deliver the message about the issue (Evan Cornog, 2004). This aspect of messenger effectiveness is too important to leave to chance and professionals know it. This is explored thoroughly in Chapter 13.

Presentation and setting

A messenger's manner of presentation is often much more important than other elements of communication, such as consistency and reasonableness of message. People often don't remember enough to catch inconsistency, and are not very critical of it in any event. A respected community leader, apparently speaking frankly and not talking down, using stories and metaphors that resonate, will be effective even if he said something different last week. It is likely to be accepted even when contradicted by obvious aspects of reality, for example, "We've got to log the big trees in the backcountry to prevent forest fires near town." Tendencies toward critical evaluation are easily blunted by self-interest, groupthink, and efforts to minimize dissonance. When several influential messengers address an issue and call for similar action they reinforce each other's credibility (William Brock, 2006: 62) and shut out contrary messages. Fast talkers, notwithstanding the popular view, are considered more knowledgeable by most audiences and produce more attitude change (William McGuire, 1989: 46).

Mass communication messengers usually associate themselves with important symbols that possess their own legitimacy, such as flags, monuments, scenic vistas, accouterments of high office, and nationalism. These symbols are closely associated with positive message reception. Creating a setting for message delivery with such symbols – a newsroom anchor desk, a group of admiring children, or military hardware are concrete examples – allows the messenger to claim the symbols. It's not always possible to control the setting; a demonstration or rally may not go as planned due to police interference or the weather. The other aspect of setting – the audience's – is usually not susceptible to the same degree of control. But it is possible to take their setting into account. For example, much is known about where and how people watch television and what sorts of camera angles, music, and visuals create drama and vivid images that in turn evoke emotion and garner attention. In the case of some public events such as invitation-only rallies, setting is part of the choreography that builds excitement, structures, rituals or collective rhythmic actions that bond people, and generates interactivity that reinforces the authority of or the pleasure taken in the messenger.

Those who discount the importance of the messenger should consider the US presidency of Ronald Reagan. Millions embraced the affable if increasingly demented president even as they rejected his wars that killed tens of thousands of Central American peasants, his billions in tax breaks for the wealthy that tripled the national debt, and his dismantling of health, safety, and conservation programs and rules. An increasingly avaricious upper class, chafing under already weak democratic constraints, could not have found a better messenger to help further weaken those constraints. There are also very effective messengers who have served a nobler cause such as Nelson Mandela.

Channels

Channels of communication are the means by which groups share information (broadly defined) within themselves and with each other, often with the intent to influence members or one another. All groups of people have preferred channels for receiving information and information via these channels is more likely to have influence than the same information conveyed by other means. Because mobilization is a process and because groups typically rely on a cluster of channels, several channels are usually important in mobilization. Some channels are better for gaining the attention of a group, while other channels are better for moving a group to action or reinforcing mobilization. Social networks are important in mobilization because people act as part of a cohort, but it is first necessary to reach the cohort or someone in it and that may be achieved by a specific mass media channel, for example. Different channels are required when seeking to reach members of a group and when seeking to reach their leaders.

Channels have inherent strengths, weaknesses, and costs apart from whether they are relied upon by a group targeted for mobilization. Television, for example, reaches many people, doesn't communicate complexity very well, and is expensive. Channels include informal social networks and formally organized networks, one-on-one and small group meetings, intraorganization newsletters and meetings, canvassing, street theater, demonstrations, cultural events such as talks, concerts and exhibits, the specialized print media, mass print media, mass electronic media, film, and others. A good knowledge of what channels are most effective with a target group and how the channels work alone and together, contributes to mobilization success.

Channels: mobilization and scale shift

A primary task of mobilization is to bring conservation NGOs to a political position from which they can influence decision makers at the scale at which problems are solvable. Mobilization is not simply about gaining more resources in a simple, additive way. Achieving influence at the right scale requires gaining access to channels of communication that reach those groups needed to influence the right decision makers – channels are not usually available to organizations operating at smaller scales. Access may be gained by reframing local and regional efforts in a larger geographical and explanatory context, by innovative action that grabs major media attention, by inspiring efforts elsewhere and linking up with them to create a large-scale presence, or by directly creating the capacity for action at larger scales by creating new formal networks, coalitions, alliances, or organizations.

In 1991 a small group of prominent conservation biologists and advocates, recognizing that existing protected areas were becoming islands in a sea of development and were failing to stem the loss of species and wild places, proposed that protected areas be made larger, new ones created on the basis of science rather than beauty or recreation potential, and that protected areas be reconnected at a regional and continental scale as they once had been (Reed Noss, 1993; Michael Soulé and John Terborgh, 1999). This group – the Wildlands Project – lacked access to the channels of communication necessary to influence decision makers at the continental and national levels. The most prominent founders of this NGO engaged other scientists and advocates to argue for this scale shift and to create a grand conservation vision based on it (David Johns, 1999). Initially considered radical by more timid national groups and as too big by many grass-roots groups, this analysis of the problem and proposed solution took root as its necessity and practicality was demonstrated by scientific corroboration, escalating species loss, and the development of regional conservation plans. In the course of a decade access to important channels of communication was gained and large-scale conservation moved beyond a handful of courageous NGOs and foundations. It was embraced by staid conservation groups, one or two of whom claimed to have invented it. Several governments have been significantly influenced as well.

The Yellowstone to Yukon Conservation Initiative (Y2Y) is another example of scale shift enabling access to new channels of communication. Founded by the Wildlands Project and Canadian Parks and Wilderness Society, Y2Y brought together dozens upon dozens of NGOs working along North America's mountain spine. Most of these groups worked on a relatively small geographic scale and had difficulty gaining the attention needed to influence decision makers. Larger NGOs in the region – usually associated with iconic areas like Yellowstone or Banff parks – recognized that even these protected areas were not large enough to maintain their biological integrity in isolation. By creating a network embracing NGOs throughout the region a framework for cooperation was developed. More important was the creation of a vision of the region as a biological whole and that saving a part required saving the whole. This vision not only invigorated participant NGOs but also captured the imagination of funders, governments, and the media; access to new and influential channels of communication was opened (Charles Chester, 2006).

The Siskiyou Regional Project took a different path gaining access to new channels. This NGO, which works for permanent protection of the Klamath–Siskiyou region of southern Oregon and northern California, created a national constituency for the region by means of an ambitious strategy and a master storyteller. Although the Klamath–Siskiyou is a spectacular region of mountains, wild rivers, intact fish runs and extraordinary botanical endemism, it is not an icon nor is it likely to be transformed into one. Through a compelling road show featuring storyteller Lou Gold support in almost every US state was created, including some very influential people and important media. Combined with deliberate efforts to mine social networks and use them to reach the White House, access to these new channels nearly secured a national monument and laid the groundwork for legislation.

Opening new channels

Mobilizing powerful new constituencies is a top priority for conservation if it is to become a more potent political and social force. Many conservation NGOs, especially those most committed to ending anthropogenic extinction, do not have access to reliable channels connecting them with powerful groups. One-on-one communication is extraordinarily important in reaching such groups and grass-roots conservationists are rarely part of the right social networks. Many NGOs lack a good understanding of channels and their use. In the remainder of this chapter the strengths and weaknesses of various channels are examined along with the means for accessing them with available resources. The discussion is necessarily generic, but concrete examples are provided. There is no cookbook for fashioning irresistible mobilization communication; the world is too messy and contingent for that.

Social networks

The targets of mobilization are almost invariably groups, organized and not. Groups, such as the Business Roundtable, Sierra Club, or White females aged 30–39 living in the US Rockies, are not the same as social networks. Although some people's social networks may be confined by a group, for example, a church congregation, networks generally span groups and include people's school and work cohorts and those with whom they share a common recreational or political interest. Networks are bounded and typically constrained by class, gender, ethnicity, and other factors that structure interaction. In Britain many working-class social networks are associated with sports teams and middle-class networks are associated with university days (Hobsbawm and Ranger, 1984: 287–8, 297).

Some networks evolve into organizations as when business leaders or activists decide some joint tasks require more formality. Networks of conservation leaders lend themselves to the creation of umbrella groups or coalitions (Oliver and Meyers, 2003: 183–4). The Y2Y started as a network and became an organization many years later. Although some networks become organizations, all organizations have their roots in networks and then go beyond them.

The power of networks lies not just in social relationships – that is, ongoing interaction – but in the shared frames of reference and stories (Florence Passy, 2003: 23). When the Zuni appealed to Native and White churches to help them protect sacred land from a coal mine it was effective because religions hold some things to be off limits to commerce (Robert Cox, 2006: 269). The resulting support swayed elected officials who pressured regulators. Shared frames of reference are an attribute of social networks and organizations; with networks the frames are informal whereas organizations have ideologies that are the product of leaders and intellectuals.

Social networks communicate news, values, information, knowledge, appropriate behaviors, cues on how to feel and react, as well as specific recruitment and action appeals. Social networks allow for both accidental and deliberate concatenation of resources. The designated driver campaign received expert creative support because New York executives associated with the Ad Council had Hollywood connections that came to their aid (Jay Winsten and William DeJong, 2001: 292). Ad production and programming expertise maximized the impact of the ads.

Social networks are a primary means of reaching into a group. Mobilization follows social network connections, with network cohorts rather than individuals deciding to become involved or drop out (Lewis Rambo, 1993: 36, 108; Pamela Oliver and Daniel Myers, 2003: 178). Civil rights activism (and tactical innovations like sit-ins) spread through the network of Black religious leaders during the mid-late 1950s (Sidney Tarrow and Douy McAdam, 2005: 132–4) and 1960s recruitment of White supporters spread through the north through networks embedded in colleges. These networks formed part of the basis for the reemergent women's, free speech, and antiwar movements.

Using networks for mobilization requires an understanding of their general properties and the specifics of a particular network, for example, knowing that network hubs are important, and knowing who the hubs are in the network being targeted. Networks are not uniform fields. Relationships among social network members are either strong or weak. Strong ties are characterized by emotional closeness and more frequent interaction. A person typically has strong ties with 15–20 other people; they form a nonrandom cluster of high interactivity within a network (Buchanan, 2002: 39). From a mobilization perspective strong bonds are important because they mean more influence and define the cohorts that act. Recruitment – or resistance to recruitment – mostly proceeds via strong ties.

Strong ties can also limit mobilization communication because in many cases an individual's closest ties are with others who share mostly the same ties; thus some strong-tie clusters are nearly closed (Buchanan, 2002: 44). Members mostly interact among themselves, with only partial overlap with other clusters. The degree of overlap in clusters shapes how mobilization proceeds although there is data suggesting that acquaintances (weak ties) are more influential than close friends in some areas (Rob Walker, 2004: 75). Indeed, if mobilization proceeded only via strong ties it would require direct contact between many more clusters and mobilizers.

Weak ties involve less emotional closeness and less frequent interaction (although coworkers interact frequently without being close). Weak ties are important because most interaction between clusters is via weak ties. Weak ties spread messages rapidly (Buchanan, 2002: 42–3). A message dependent solely on strong ties might pass between 50 people in order to get from person A to person B; by means of weak ties it takes only five or six intermediaries to literally span the globe. An individual's weak ties include a wider range of people who are part of many other clusters and networks. Weak ties are thus important within and between networks. (Networks can be visualized as partially overlapping circles, as in a Venn diagram.)

Although weak ties transmit messages rapidly, people respond less strongly to most types of messages via weak ties, with some exceptions as when there is a preexisting interest on the part of the recipient.

Some people interact with far more people in the networks of which they are a part, they are involved in more clusters within a network, and they are involved in more networks. These people are network hubs or cybernetic centers and for the reasons just noted are most effective at getting the word out. They are also effective because others look to them for the latest news, for "inside information," and for cues concerning desirable action. Careful attention to identifying and focusing on hubs – and not mistaking irritating busybodies for hubs – makes mobilization more economical and more likely to succeed. When there are more links between clusters within a network (more people acting as hubs, people acting more effectively as hubs, or more overlap among clusters), mobilization is more likely to take off; absent those links mobilization often fails to reach critical mass (Buchanan, 2002: 205).

Hubs or network leaders are particularly important in more authoritarian cultures because network members look to them for leadership (Jeffrey Broadbent, 2003: 204–29). In these cultures appeals to group loyalty carry greater weight in moving network members to action than in more egalitarian cultures where substantive appeals (and peers) have greater weight (Doug McAdam, 2003: 287).

As just noted, hubs are disproportionately located in the overlapping areas of our metaphorical Venn diagrams. A professional conservationist who is a network hub may still mostly associate with other conservationists but is more likely to have maintained links with people from university days who went into communications, became wealthy, or who know the prime minister or president. All hubs are not equal, however. The university they attended will determine the networks they have access to, for example, an Ivy League school in a major urban center provides access to elite networks that a rural school in a small country or province cannot.

All NGOs rely on social networks for mobilization, but they vary in terms of the networks they have immediate access to. How to reach networks an NGO does not have access to? Organizations with high visibility gain access from that visibility (Florence Passy, 2003: 28, 42). Their presence in the media (including the internet), at events, through diffusion of their literature, attracts notice from people in a variety of networks. NGOs with close ties to the state or other elite organizations can utilize the connections they afford – power and wealth attract attention – and elite organizations have many channels through which they communicate to other elite organizations and subaltern groups. In addition to channels of communication elites control important social and political leverage.

Brokers – people who specialize in bridging networks and organizations – are another means by which NGOs can connect with key people in previously inaccessible networks and groups. These key people – who may be hubs or formal leaders – can then influence their networks and group.

Increasingly social networks consist of people who have never met. Although telephone and mail make possible interactions among people who have not met, the ability to communicate rapidly with large numbers of people via E-mail, multipurpose cellphones, and similar devices is new. The ease and velocity of electronic communication have altered communications locally and globally. Most of this communication is horizontal; participants link with each other as desired, bypassing a central gatekeeper or other authority such as an editor. Charles Tilly (2004: 121) and David Meyer and Sidney Tarrow (1998: 12) note the central role of electronic networking in organizing mass protests on very short notice. Some of these protests have had profound consequences such as toppling governments. These networks (academics term them "smart-mobs") depend heavily on weak connections and observers question whether these have the ability to sustain political action as do NGOs. Social movement leaders have long warned about the weaknesses of spontaneity – the lack of follow-through, the inability to adhere to a strategy, the vulnerability to manipulation by opponents and demagogues, and the inability to respond effectively to mass repression. It is one thing for organizations to rely on electronic communication as a tool, quite another to base interaction solely on such connections.

Brokers and organizers

One of the most effective means of reaching a group in the absence of network ties is a broker. Brokers are people whose business is to connect groups that are not connected. They make introductions. Brokers help ensure that contacts are best suited to the purposes for which connection is made and they may facilitate inter- action. Brokers usually act on behalf of a group seeking to connect with another (Doug McAdam et al., 2001: 157), though in some cases individuals take on this task independently to further goals they support. Several Americans who received news of and supported the Zapatista rebellion used their media and other connections in North America to generate high-profile coverage and enlist the backing of social justice NGOs (Sidney Tarrow, 2005: 114–7). Brokers may be NGO employees, a contractor closely associated with a movement or type of interest group (e.g., civil rights, conservation, labor), or a professional without strong political sympathies.

Brokers are particularly important in cross-cultural settings (Burton, 2002: 132–4). Effective ones have an understanding of the worldviews of all the parties being brought together, such as Native American traditionalists, conservationists, and recreationists working on a land-use plan. Without such understanding initial meetings may not materialize let alone produce mobilization.

Many conservation advocates have good brokerage skills but lack the broad contacts or the necessary understanding of some groups. In some instances brokers need to credibly present themselves as neutral parties to the target audience in order to get a hearing.

Brokers, as social scientists say, "minimize transaction costs" because they have skills and existing connections – it's cheaper to use a bridge already built. As with so much else in politics, not all brokers are equal. Each broker's connections vary even within the area. The choice of one rather than another will open some opportunities and foreclose others.

Brokers are worth cultivating even if a particular purpose doesn't exist in the present. By the same token professional brokers are only a good investment when the goals and target audiences are clear and the prospective broker's record can be evaluated against the task required. Brokers are particularly important when conservationists need to reach a powerful group's leadership. One-on-one contact is critical and this is often the only way to arrange it. Fortuitous contacts are too infrequent to be counted upon.

Reaching an audience through the right channel – in this case brokers – does not guarantee mobilization. Mobilization requires more than contact as we noted in Chapter 8's discussion of strategy. The targets of mobilization must be moved to take desired action – as members or supporters of the NGO seeking their action, or as allies or partners. Moving people to action usually falls to the NGO, not to the broker. Some brokers, however, do more that make introductions – they excel at the other aspects of mobilization and it is often worth engaging them if a NGO is working with a new type of group.

Door-to-door canvassing and the public square

Canvassing is a very labor-intensive means of reaching out to new groups, but offers one-on-one contact that can be persuasive even between strangers. Recent US studies show canvassing is highly effective in getting out the vote compared with leaving door hangers, leafleting, direct mail, or other means (Donald Green and Alan Gerber, 2004: 11–22). Greenpeace USA has periodically used door-to-door canvassers in neighborhoods likely to be sympathetic. Whether this is effective beyond acquiring new members who are mostly check writers is unknown. There is evidence from Canada that unless visits are repeated annually memberships lapse. I know of no comprehensive study of how canvassing compares with other means of outreach in generating support other than voting in an upcoming election.

Canvassing is well suited to elections because of district-based voting, but less well suited to other forms of mobilization because neighborhoods do not often define groups that are important to conservation. Canvassing faces obstacles in many urban areas because it is difficult to gain access to buildings or to those areas where the most influential people live.

Taking advantage of the public square where they still remain – shopping malls, for example, have replaced the public square in many towns and generally ban politicking – by leafleting or setting up a table permits one-on-one interaction. It is less labor intensive than canvassing. The effectiveness of leafleting and tabling depends on the willingness of people to approach, talk, or take material and read it later. Much hinges on the attractiveness of material handed out, displays, and messengers. Materials must be vivid to evoke interest and emotion and not be threatening to the intended audiences. Humor can help undercut threat reactions, but not all messages are compatible with humor. Finding those with the right personality for this work can be difficult.

Working the public square, unless in association with an event that draws specific audiences, does not allow precise targeting, making it difficult to bring together messengers, messages, and materials in a manner most likely to be effective. Larger cities have numerous public places that attract more homogenous crowds enabling more precision with materials.

Canvassing, leafleting, and tabling are better suited to increasing awareness than mobilization per se. They lay the groundwork for mobilization, making audiences more likely to heed further messages by acting.

Events

Events include concerts, film exhibitions, conferences, speeches or presentations, street theater, and demonstrations to name a few. Each has its own characteristics as well as shared attributes. Events may serve many purposes and some of them simultaneously such as increasing awareness, deepening understanding, affirming and increasing support, enhancing solidarity and identity with conservation, and initial recruitment. Events serve as a channel to those who observe or attend them. Many events, such as concerts or film screenings are entertaining and involve little risk but may also attract those with more interest in music or film than conservation. Street theater and similar actions provide entertainment as well, but targeting specific audiences can be difficult (as with leafleting the square), and some audiences find them threatening. In its early days Earth First!ers' dressed up in animal costumes to leaflet and do skits in front of federal offices adding some wry humor to their serious points and defusing observers' fears. The movement that helped bring down Serbian dictator Slobodan Milosevic relied heavily on street theater during times when demonstrations were too risky. Using clowns and farcical antics, critical statements were made quickly and with biting humor before the police could arrive. These actions kept spirits up and people ready for action but did not afford much opportunity for recruitment. Normally obtaining contact information from observers and following-up by other means is essential to mobilization. Relying on observers to follow up is less effective, but those who do take the initiative are likely to be more serious.

As with leafleting, events like street theater can take advantage of other's events such as concerts, sporting matches, public hearings, and speeches (open or closed) by political leaders. Matching the NGO action with event's level of gravitas or with a counterpoint is usually most effective. Thus, an entertainment event is best matched with entertaining skits or music to draw attention. But a political leader making a ceremonial or celebratory appearance might most effectively be met by confrontation or satire.

Some NGOs have arranged to install semipermanent exhibits and displays that permit contributions to be made in international airport terminals; departing passengers are asked to donate their unused currency rather than exchange it. Airports, train stations, stadiums, and similar venues are prospects, but closing such deals takes good connections and a willingness to forgo controversial messages.

Speeches by conservation luminaries, with some notable exceptions like Jane Goodall or the late Jacques Cousteau, usually do not draw those groups that conservationists most need to reach, but can increase levels of mobilization among the faithful, making activists out of the merely sympathetic.

Parties hosted by sympathetic nonconservationists are a good means of bringing together conservationists and those not usually exposed to their message first hand. Parties rely on social networks but do not leave it to the network to carry the word. They provide a congenial method for NGOs to share information about issues and identify new supporters. Although labor-intensive parties are a good means of enlisting supporters' enthusiasm. Parties are also a good tool for increasing support from the already involved; they can be used to honor past contributions and ask for increased levels of action and commitment.

The great strength of events is the opportunity provided for one-on-one interaction in settings that range from the relaxed to the intense. Audiences can put faces to a cause, increasing the likelihood of a stronger and more positive response. They frequently provide opportunities for ritual-like involvement that create bonds with a cause and coparticipants. Live music and song associated with dance is particularly forceful but must be linked to a purpose and not just the pleasure of movement. Concerts, films, speeches, and storytelling events all evoke strong emotion if properly staged. Storytelling in the right room with the right light and a small circle of people can be profoundly moving. A compelling film followed by audience interaction can build on the response to the film and engage the energy made available.

Event-based mobilization is more likely when people are asked to proclaim their commitment by publicly signing a petition or other statement, verbally proclaiming support, chanting slogans, or sharing a satirical joke with others. Like the call to be saved at a revival, the right mixture of opportunity, encouragement, and choice is important.

Events have important downsides that must be weighed given limited resources. They can be expensive even when much is donated. Breaking even or taking a loss may be fine if other important goals are achieved. Events may offer little immediate return, being only the first step in a longer process of mobilization. Many types of events require significant initial outreach via other channels to make people aware of them and entice them to attend. Demonstrations and street theater can lend themselves to countermobilization and repression. Many events, of course, are not organized primarily to reach direct observers, but to gain media coverage and reach a much broader audience. The dynamics of events organized primarily with the media in mind will be discussed below.

A major benefit of events, leafleting, tabling, and other outreach that brings conservationists into contact with new groups is the grounding it provides to conservationists. These activities offer conservationists a better understanding of how others think and feel about wild places and creatures, where it fits in their lives, and provides insight into what steps are needed to make life on Earth more important to others.

Direct mail

As with events, direct mail is a multipurpose channel: it mobilizes new support, provides existing supporters with information, and asks them to do more. Smaller NGOs barter lists to prospect for new supporters and some new members result. Recruiting members who belong to other conservation NGOs has its purposes, but it does not grow the movement by reaching new groups. NGOs also aim direct mail at audiences identified as receptive to conservation by marketing research or polling data. Despite the targeting a 1% return is considered success and mobilization is generally restricted to smaller contributions. A few people mobilized this way will over time grow into major donors. Some may become activists if the organization has a role for activism; too few do.

The demographics of direct mail are narrowing; groups born after the post-World War II baby boom are not responding. Anecdotal evidence suggests that web sites and other electronic channels have replaced direct mail as means for reaching these age-groups but it is unclear to what extent people search for the web sites of groups from scratch or from what channels (such as a brochure) people find out about web sites that in turn lead to mobilization.

Direct mail remains important in mobilizing older demographics and many value-based groups and in keeping them mobilized. Increasingly, however, mail is combined with electronic communication. NGO newsletters and action alerts are now available by E-mail in addition to post.

Direct mail continues to be successfully used along with E-mail in electioneering. Conservationists still have much to learn from those who engineered the massive direct mail campaigns that helped bring the US right to power in 1980 (Richard Viguerie and David Frank, 2004: 322, 343, 346). Viguerie's direct mail organization financed mailings for clients so it would not be beholden to party or corporate donors. Viguerie and the client had ownership of the list of those who responded. With each campaign Viguerie added to his lists of responders. He thus had more to offer each new client, and with each new client he gained another list. By maintaining ownership, he was able to offer potential clients lists of voters by geographical area and by issue area such as abortion, gay rights, property rights, and defense. Mail could then be precisely targeted by issue and election district down to the precinct level. The payoff was significant and US conservation has suffered as a result.

Viguerie noted other factors that contributed to his success. He recognized he was in the business of movement building and that it is a long-term process. He recognized he could have the biggest influence on the Republican Party if his organization remained under his control and did not merge with the party. This allowed him to help only those candidates who furthered his goals; the party was simply a means to an end and he was not obliged to help all party candidates. His control of valuable voter lists meant he could effectively help the party on his terms, pushing the party further to the far right. This in turn invigorated voters on his lists and helped to insure their disciplined response to appeals. Direct mail also allowed him to bypass established media and Republican Party gatekeepers.

Though skepticism of Viguerie's claims of complete independence from corporate largess and Republican power holders is appropriate, his success in mobilizing millions of Americans cannot be doubted. He helped to amalgamate antichoice, anti-1st amendment, antiaffirmative action, anticonservation, and antilabor factions and to utilize this amalgamation of voters and contributors to push candidates and the political debate toward greater subservience to powerful business interests and the ideologically medieval. (Nowhere else in the developed world do grownups debate the veracity of evolution or argue that the Grand Canyon is a few thousand years old.) Although the economically powerful were the overwhelming beneficiaries of this effort – the medievalists always played a distant second when it came to delivery on their policy goals – the usefulness of Viguerie's strategy for electoral influence should not be dismissed: keep control of the lists, don't become subservient to party leaders, and bargain hard.

Print materials

Printed materials have always played a central role in conservation mobilization. They are included in direct mail, distributed as leaflets and brochures, provided to the media, and presented to decision makers. Conservationists are readers and are predisposed to the printed word and photos. Despite their often stunning aesthetics these materials increasingly do not evoke enough interest to be read and absorbed in societies where they must compete with electronic media and high levels of commercial noise. Many groups lack the financial and expert resources to make their materials stunning or to craft different versions for different audiences. Color materials are usually too expensive to print and distribute in saturation quantities. Getting materials in the right hands can be difficult and the most valuable audiences are overwhelmed with mail and other materials to read, including those given to them by friends.

Despite limitations printed material is an effective mobilization tool with a variety of audiences. Much depends on matching the quality of the materials with the audience (Doug Tompkins, 1998). Printed media must tell the audience that their importance is understood by the mobilizing NGO and that the NGO knows what it's doing. High-quality material lends even a small NGO much needed solidity with audiences. If attractive enough the recipient will keep the material and return to it. Printed material is tangible, enabling NGOs to give people something physical that they can take with them, share with others, and reread. Printed materials are well suited to storytelling with words and graphics. Tangibility and quality can translate into welcome happenstances. A successful actress came across a high-quality conservation magazine in the Rome airport. As a result she made a significant contribution to the group that published it, arranged for a high-profile article in Vanity Fair, and hosted a benefit party in Los Angeles.

Although many printed materials must include information that quickly dates them, it is often possible to create lavish documents that do not include quickly dated information and to marry them with cheaper materials that include the time sensitive.

By themselves printed materials allow for minimal involvement through return cards and envelopes for donations, petitions to elected officials, and sample letters to the media or others. Printed material can, however, direct people to other channels such as web sites, physical addresses, or phone numbers. To result in mobilization printed material usually needs to be combined with other NGO action including earned and paid mass media. Smaller groups, less able to obtain media exposure, can rely on a high-quality interactive and easily navigable web site. Those with more resources can maintain sites designed for a variety of audiences, giving some groups special access to particular portions, such as the press, partners, and so on. The press and many demographics are good at using the web to gather and vet information.

The net has also been used to spread disinformation. Anticonservation groups – unable to make points on the merits – have created web sites that masquerade as the official web site of conservation NGOs wherein they post materials intended to stir hostility.

On a per capita basis the creation of a good web site and e-mail alert system is a good investment, but probably better for maintaining mobilization and regularizing income flows than for initial mobilization.

Mass media

An effective presence in the mass media is necessary for the conservation movement (not each NGO) to frame the political debate and agenda (via message propagation and repetition) and to mobilize new constituencies (Karen Johnson-Cartee and Gary Copeland, 2004: 143). Although some larger NGOs have magazines aimed at nonmembers and available at newsstands they do not achieve saturation levels of communication nor do they enjoy adequate circulation among influential groups. A few conservation NGOs produce television, video, and audio programming, but these reach limited audiences as well, though some have made a difference such as the original Cousteau series. In the long term it is essential that conservationists own and control commercially successful media outlets, much as US conservatives have come to do. In the meantime they must develop strategies to obtain better access to the full range of the mass media and obtain coverage and programming that supports rather than ignores, marginalizes, or denigrates conservation values and goals.

The mass media consist of many channels, including print, radio, television, books, film, theater, billboards, and others. Each channel type is distinctive and includes many specialized subtypes. The print media, for example, consist of newspapers of general and niche circulation, general news magazines and magazines aimed at narrow markets, opinion journals, and academic and professional journals. Media ownership in many countries has become more concentrated although outlets have become more specialized and aimed at narrow audiences. The objects of mobilization have preferred channel types (e.g., television and web) and preferred outlets (e.g., specific programs and web sites).

Conservationists confront several common problems in using the mass media for mobilization including concentrated ownership and outlet fragmentation. Concentrated ownership by conglomerates that have many interests averse to conservation means that both earned and paid media are more difficult to obtain. Media owners seldom meddle in minor issues but they do not hesitate to veto news and other programming that might injure their business interests, those of friends and advertisers, or their favored causes (Ben Bagdikian, 2004: 154–5, 198, 216, 238–42). Where media ownership is less concentrated (in 2004 London had 12 daily papers, Paris 33, Tokyo 31) and ownership is more politically diverse, conservationists and other critics of the status quo have an easier time.

Where media are dependent upon advertising revenue those who pay the bills have significant influence. Many advertisers are quite aggressive in trying to influence the programming their ads are directly and indirectly associated with. This may amount to no more than seeking programming that puts people in a "buying mood," for example, comedy, adventure, and escapism; serious and controversial programming does not sell products. Some advertisers insist that programming avoids criticism of their favored interests and values.

Smaller NGOs rely heavily on earned media and even the largest NGOs depend on it. During brief periods and on certain issues earned media have played a major role in advancing conservation goals. Along with other factors they have contributed to a sea change in attitudes in developed countries and many other countries (James Jasper, 1999: 70); but the change is not deeply rooted and has not altered fundamental behavior. In particular instances – such as when a local newspaper story about an animal rights activist resulted in numerous people contacting her and led to a network of activist NGOs being founded across the state (James Jasper, 1999: 76) – it may provide a real boost, but not consistently. The cost of paid media varies but it is almost always too much for smaller NGOs and larger groups cannot achieve saturation levels that opponents can – some of whom own media. Radio is certainly cheaper than television or the national press, allowing some sustained paid access in some markets. Outlet specialization makes targeting easier, but often more expensive because desired outlets are pricey and more outlets are needed to reach all targeted groups.

Conservationists will never have the financial resources of their opponents as a whole, but they can compensate for this by being strategically smarter in their use of media and emphasizing alternatives such as brokers. Conservationists are more creative and must use this to develop more emotionally evocative stories than opponents and they must seize the political opportunities generated by the blowback from the human destruction of natural systems. Good personal relationships with journalists and editors count for much. Advice in this area is provided by several sources and need not be discussed in detail here (Sandra Beckwith, 2006; Kristen Grimm, 2006; Susan Jacobson, 1999, 2006). Responding to media requests quickly, providing story angles and visuals in the right format, understanding the media outlet's audience, and encapsulating complex issues in concise statements or otherwise making issues newsworthy, is excellent advice. What does bear stressing is the premium on providing good information, a compelling context to explain its meaning, and being trustworthy. Being likeable also counts for a lot.

Good relationships only go so far – until they bump up against media owners who oppose conservation goals. Even absent direct ideological control over programming – and it exists – reporters and editors regularly avoid stories that would hurt parent company interests (Robert Cox, 2006: 172–3). When General Electric – NBC's owner – was fighting efforts to make it clean up chemicals it had dumped into the Hudson River, NBC news offered little coverage and the president of NBC lobbied against the plan before government officials. Reporters and editors are quite sensitive to unspoken cues from on high and self-censorship becomes second nature.

The "he-said-she-said" approach to coverage gives credibility to positions that deserve criticism if they deserve any mention at all. By providing equal time to industry hacks or, for example, the Bush administration's doctored "science," audiences receive a distorted version of reality. The US media's coverage of climate change followed this pattern for two decades, up until 2007. Those opposed to limiting carbon emissions reinforced and exploited this institutional predisposition to delay policy change (Environmental Working Group, 2003). In an earlier era apologists for tobacco companies were given a very long free ride as well. Although US government scientists became adept at leaking findings there are only so many leaks the media can report and only so much room for stories on the environment or stories about the suppression of science.

Silence – the failure to cover a story – is the most common method the media use to shape the perception of reality. Silence may result from an outlet's political or commercial agenda, from political pressure, because a story is too complex (many are complex) or because it's not judged as newsworthy (timely, emotionally appealing, dramatic; Robert Cox, 2006: 164, 175). Biases can be general. The US media, for example, are more willing to entertain calls for individual change than social change because the former are less threatening to the status quo (Laurence Wallack and Lori Dorfman, 2001: 393). Despite this the proponents of "Buy Nothing Day" have not found many takers for their ads encouraging individual shoppers to refrain from heading to the malls on the day after Thanksgiving.

Media workers are networked in such complex ways that the notion of "herd" mentality does not adequately capture the interpersonal and professional dynamics. It is true that newspeople ignore mobilization campaigns or the problems they are trying to address until one of their numbers – perhaps bolder or hungrier – decides it is news, and then they follow (Meadow, 1989: 257; Pamela Oliver and Daniel Meyers, 2003: 188). The media in one country may influence the media in another to take up a story. So obtaining initial coverage is a priority, but conservationists should also be prepared to take advantage when a story does appear and encourage its diffusion by having a chorus of groups calling for further coverage and commenting on the initial coverage. This chorus is important in advancing the preferred story line or framework for making sense of the issue, which in turn determines how groups will respond. Offering the preferred storyline by packaging it as a compelling story with a good hook can overcome many obstacles.

It is worth identifying the "bold and hungry" among reporters and opinion leaders or network hubs at other levels within the media – those who by reason of personality, interest in conservation, or place of employment (an independent or innovative outlet) are more likely to break stories on issues that are usually ignored by others and that others look to for their lead, be they editors, publishers, station managers, or owners. This understanding of media social networks is invaluable. In the United States, for instance, broadcast media take their cue on what's important from New York and Washington, DC newspapers. Once the progressive Mexico City paper *La Jornada* took up the Zapatista cause in a serious way and provided major coverage, other news organizations followed its lead and looked to it as a major source for what was going on in Chiapas (Sidney Tarrow, 2005: 114).

The success of the Zapatistas in gaining coverage and global support and thereby limiting the repression that would have otherwise come holds important lessons. Preparation was important. Launching their rebellion on the day NAFTA came into effect provided an important news hook. Building ties with sympathizers in the capital who would stage large demonstrations guaranteed to independently attract the media was another. The Zapatistas also understood that relying on earned media as a central campaign element required constant tactical innovation in order to keep them interested. Contacts, strategy, material and organizational resources, leadership, and messages were in place when the rebellion was launched (Clifford Bob, 2005: 44, 185). A campaign, let alone a rebellion, is always a gamble, but knowing the channels that reach the right groups lessens the gamble.

Other problems with earned media are more difficult to overcome. Movements themselves ebb and flow for a variety of reasons from winning a major victory that causes support to dissipate to being drowned by bigger events. Media audiences have limited attention spans and with media focus on capturing audience share – especially well-heeled audience share – the media look to what's new. (The electronic media have also contributed to the ever shrinking attention span.) It is possible to address cyclic and fatigue factors (Verta Taylor, 1989; Robert Cox, 2006: 166–8). The media generally have become more accessible to activists as activism has grown, making greater reliance on earned media realistic (Doug McAdam et al., 2001: 333–4, 337). But a simultaneous inclination on the media's part to downplay grass-roots challengers to the powerful and characterize direct action as illegitimate remains a hurdle (Pamela Oliver and Daniel Meyers, 2003: 177; Lance Bennett, 2005: 220–1). Overcoming negative branding generally requires "approval" of the group or cause by members of the elite (Ben Bagdikian, 2004: 154–5), for example, climate change. Such approval may not extend to all groups and the media are quick to drive a wedge between acceptable NGOs and those that represent a more serious challenge (Jules Boykoff, 2007: 216–47).

Because in many countries news coverage is just another cost center expected to show a profit, drama has become more important because it draws audience which increases ad revenue (James Combs and Dan Nimmo, 1993: 78). The success of conservative talk radio hinges on the fact that Limbaugh, O'Reilly, and others are entertainers, and only secondarily politicos (Richard Viguerie and David Frank, 2004: 339–40). What passes for drama, of course, varies by audience. But it is drama that garners audience share and revenue that offers a path around corporate ideological censors.

Old-fashioned political censorship – the kind enforced by violence and threats of violence – presents different problems for conservationists seeking to use the mass media for mobilization. Legal challenges before domestic and multi- or international tribunals offer one approach and various human rights groups can assist (Linda Malone and Scott Pasternack, 2006), but this can create blowback. Pirate media can be effective but makes conservationists a higher profile target for security forces. Reliance on foreign media also provides opportunities but as discussed at greater length in the section on Repression (Chapter 19), it subjects conservationists to treatment as "foreign agents."

A final media hurdle worth noting is the lack of formal science training among reporters who do not specialize in science (Robert Cox, 2006: 174, 181–2). This lends itself to inaccuracy in portraying expert opinions and is compounded when reporters want "both sides" or the drama of conflict. The best response is to try and provide drama in the story provided to reporters so they don't need to go looking for it or invent it. The facts do not speak for themselves. It is also useful (and justifiable) to compare many experts opposing conservation with the medical doctors (MDs) who shilled for the tobacco companies; and to gently remind reporters that their predecessors uncritically accepted such experts and ought not to repeat the same mistake. Reporters can also be encouraged to find the drama in going beyond the he-said-she-said format and actually assessing claims by comparing them to the evidence. These matters are discussed in detail in Chapters 12 and 13.

Important mass media particulars

The various forms of mass media – print, television, radio, film, and music – have unique aspects including how they interact with other institutions in different societies. These present conservationists with differing obstacles and opportunities. My examples are from the US media is not because it is typical but because that is where most of my experience lies and the subject of much of the literature. Indeed the United States is atypical among developed countries in its low press readership, a press largely restricted to elite views, and media provincialism.

The US press is relied upon primarily by the upper middle and upper classes, although it is still vigorous in many other countries. There is a press hierarchy; those with a major presence or outlet in Washington, DC or New York are at the top, followed by those in other major cities, and then by those outlets serving smaller markets. The press is more accessible to conservationists whether earned or paid, in part because newspapers cover more stories than electronic media. Coverage in the top tier, national newspapers, and newsweeklies is more difficult to obtain because space is limited. Some regional newspapers still do investigative and analytical stories of high quality, often putting the New York *Times*, Washington *Post*, and *Wall Street Journal* to shame. These stories may have significant policy influence through their effect on important nonelite groups but are ignored by the elite.

Because press coverage is so influential in setting the tone of elite political debate in the United States and because it shapes what stories the electronic media cover, it remains an important target for influence in and of itself, and as a means of reaching important audiences. On the positive side journalists are doing more in-depth stories on the natural world, seeking to understand what is driving events and trends (Robert Cox, 2006: 170–1). On the downside because of the rise of major media outlets (mostly electronic) that have little regard for basic reality and indulge jingoism, superstition, and scapegoating, neither press coverage nor full-page ads influence the agenda as they once did; they are only part of the game and decision makers know it.

The regional and local media remain important to conservation because of the role they play in defining issues at those levels and their influence on the careers of officials, including national officials in their home districts. Press trends at this level are generally not good with some exceptions. More papers have been bought up by the big chains and in-depth coverage and news generally have declined because it costs more to provide and can create controversy inimical to ad revenue (Ben Bagdikian, 2004: 198). When conservation is covered the story biases vary with the general disposition of the readership according to some studies (Robert Cox, 2006: 173). Corporate political views are sometimes imposed, but conservation is considered unimportant and left to local editors and reporters so long as profit goals are met.

Chain ownership in the United States contrasts with much of Europe, where major political constituencies have their "own" widely read press. A much greater diversity of stories are covered and wider range of views are part of the debate. This lack of total corporate dominance is one reason climate change has been taken seriously longer in Europe. The United States, for example, has no mass circulation equivalent of the UK's *Guardian*.

The press has the capacity to generate strong emotional responses, but generally lacks the impact of electronic media or film. Conservationists can address this by emphasizing those elements of journalism that touch people deeply. This does not require abandoning principled efforts to convey information, but by being more creative about how it is done. Visuals, for example, are more evocative than words (Ted Brader, 2005: 29). The 19th century political leader "Boss Tweed" did not care about what journalists and editors wrote – his constituents couldn't read – but he raged over the cartoons and caricatures, ordering his minions to stop them in any way they could (Carnog, 2004: 189). Indeed, some of the best press coverage of conservation issues comes from cartoonists such as Tom Toles, who not only have a better grip on reality than many reporters and editors, but also seem to enjoy greater freedom to directly and undiplomatically speak the truth. Unfortunately cartoons are not enough to mobilize people. But supplying photos, symbols, and other visuals rather than leaving it to reporters to do that work will get better results.

Conveying drama in the press is not difficult and is the primary method of evoking interest and emotion. Visuals do this but so must the storyline. Conservationists will generally find reporters receptive when they can provide a storyline that offers drama. Investing the time to develop a storyline that frames the issues, a compellingly phrased opening, and a sprinkling of tag lines doesn't mean reporters will incorporate them, but they usually have a positive influence on the article.

Although full-page ads in top newspapers may no longer decisively inject elite debates with views from the grass roots, such ads are effective in other ways. When people sign onto a newspaper ad they are more likely to follow through on their commitment with other action (Renee Bator, 2000: 357). Such ads are more effective when they offer readers a means of responding, such as contacting an NGO to obtain a bumper sticker, decal, or some other means of publicly identifying themselves with the cause. What is essential to achieving mobilization through impersonal means such as print ads is involving people by making them feel the issue is personal and of direct concern. It is more important that the ad triggers an involving response than that the response be politically significant. Once involved politically meaningful acts can follow.

Conservation has relied heavily on books and shorter printed materials such as articles and pamphlets. Most all of this writing is either directly political and cerebral, or evocative "Nature writing." Both have been effective in reaching limited audiences, and in some cases these audiences have influenced much larger audiences over time. Rachel Carson's *Silent Spring* had such an effect. Works by Terry Tempest Williams, Gary Snyder, Dave Foreman, David Quammen, and Paul and Anne Ehrlich, have played a major role in energizing and inspiring people and educating the educators. Such books have an important multiplier effect, but the effects remain circumscribed.

One method of overcoming this limitation is to better use fiction. Nonfiction offers story and drama, but novels are more widely read and moving (and more often made into films that reach even more people). Tony Hillerman, Peter Mattheissen, Barbara Kingsolver, and James Welch have written powerful books in which Nature is a major character. Daniel Quinn's *Ishmael* influenced a generation of college-educated Americans and many others. How to better use this channel? Best-selling conservation novels can't be cranked out on demand and writers are generally not looking to politicos for direction on their next book. But novelists, screenwriters, and others take their cues from real-life stories. The more conservationists can get their stories out, tell them well, the more likely they will find fertile ground among writers. Social networks can be used to make writers aware of the issues, and more importantly, aware of characters who are living a good story. There are other methods of bringing attention to conservation. *Ishmael* was in part the result of Ted Turner offering a huge prize for an environmental novel, suggesting that establishing regular and high-status media conservation awards would motivate. As Nature increasingly bites back and it emerges onto center stage more and more material will be available. By itself the material may not speak clearly. A main task of conservationists is to help give Nature a voice other humans can understand.

So-called think tanks annually emit tons of propaganda that is not worth the dead trees it is printed on. Although some of the older ones produce legitimate analyses, many do little more than provide intellectual cover (however thin) for the interests of wealthy and powerful benefactors and are best seen as interest groups rather than research institutions (Andrew Rich, 2005). They have enabled official US denial of climate change, cloaked resource wars as battles for democracy, called the rollback of environmental laws and regulations a principled stand for freedom, and made the Enron, Worldcom, and the subprime mortgage fiascoes possible. Of the two lessons for conservationists, neither is that conservationists should mimic this behavior. Rather, conservationists must publicly raise the profile of the funding sources for these think tanks and connect the dots with think tank findings. Secondly conservationists must take to heart the need for people to explain and justify themselves and to have a purpose. Conservation can provide purpose and justification for protecting Nature without dissembling – something those who destroy Nature cannot do.

Although the US elite read (and watch television) they are aware that groups they must contend with in their societies rely on electronic media for information and entertainment. Not all of these media take their cue from the elite press – especially the far right media of Murdoch and some religious broadcasters. A major goal of US conservatives was repeal of Fairness Doctrine that required equal access for opposing views on broadcast media (Richard Viguerie and David Frank, 2004: 322, 331–2, 344–5). Repeal occurred in 1987. One thousand talk radio stations were on the air within six years and outlets such as Fox News grew to reach wide audiences by playing to those who were angry with the status quo but clueless about who was responsible for the status quo.

The loss of the Fairness Doctrine is far from the only reason conservation has suffered in the electronic media (Robert Cox, 2006: 166–8, 189–90). Coverage has declined over the last 15 years (with the possible exception of climate change) and at least one report suggests that heavy exposure to television retards proenvironmental attitudes across the political spectrum.

The electronic media (broadcast, cable, and to some extent internet) are so important to conservation because of their power to mobilize audiences by evoking passion. Not only can they convey information in real time, but their ability to present human voices, music, faces, bodies, and other images vividly and dramatically is unsurpassed (Brader, 2005: 30, 159, 175, 181–2). Television has habituated people in many societies to visual images thereby making them more important to story. If in the United States an image machine has replaced the old party organization in electoral campaigns (Evan Cornog, 2004: 93, 150), the visual has become a major part of image. (In reality party organization still plays a major role in getting out the vote and preventing some voters from participating.) Influential media effects do not occur automatically – those crafting the messages must get it right. Identical messages have markedly differing effects depending, for example, on the music and images used. The effect of music, images, and other elements is particularly important in ads which are typically too brief to develop a substantive argument. (This partly explains the repetitive half-hour ads on television that hope to nab channel surfers for a few moments.) Most news stories are no longer than an ad spot and the same rules apply.

Because opportunities are limited it is important that conservationists "get it right" and obtain the best talent they can whether it is crafting an ad or pitching a story. Conservationists cannot outspend opponents; they must out-create them. In the 2006 US Congressional elections business outspent labor by more than 10 to 1; conservation cannot even approach labor's spending. Former US President and reformer Theodore Roosevelt (1911) likened the influence of business on politics to the influence of slaveholders in an earlier era and called for the removal of business from politics. That call was made 100 years ago and its realization remains distant. In addition to being more creative than opponents, conservationists can try to enhance their status through elite connections and spokespeople. It's an old game. If one has status or influence it's easier to get coverage because the media benefits by providing it. This in turn adds to influence, and so on. Conservation has not reached that critical, synergistic mass.

The internet makes much programming available on demand, giving them a longer life and potentially greater repetition effect. Some programs or segments are picked up and shared by social networks. Internet availability enables NGOs to better research media representation of issues and rhetorical effects; it also makes hanging people with their own words and images much easier.

As with writing and fiction, conservationists are not doing well with nonnews programming. The treatment conservation receives is often favorable but it is not pervasive enough to raise concern. Entertainment programming is so important because of the role it plays in establishing dominant stories, frames of reference, and role models. For the two generations before the coming of TV the cinema played a similar role: Americans got their idea of the 19th century based on Westerns rather than from history books – the west was settled by people in white hats facing fierce resistance from unreasonable savages. Scottish attitudes toward nuclear power have been heavily influenced by *The Simpsons* (Murdo MacLeod, 2002). People approach entertainment with expectations of pleasure, not of being informed so their receptivity is higher and their critical faculties less engaged. When people experience pleasure, are touched emotionally, or simply titillated, they absorb a story's meaning and message nonconsciously. When stories are repeated enough there are lasting effects on views and experience.

One of the earliest examples of the political power of entertainment pro-gramming on television (Edward R Murrow's documentary reporting came much earlier) was a Peruvian soap opera *Simplemente Maria* (Arvind Singhal and Everett

Rogers, 2001: 343–56). The broadcast of the series beginning in 1969 had a major profound and unexpected effect on large numbers of people. The program featured Maria, a migrant to Lima from the countryside, who takes literacy classes, learns to sew, and succeeds in bettering her economic situation. At the heart of each episode are three role models and the same pattern of interactions and consequences: a positive role model, a negative one, and a role model that changes as a result of interacting with the other two. Each time the positive or transformational role model does the right thing they are immediately rewarded. The audience ratings were not only extremely high, especially among the urban poor, but the program caused huge increases in literacy class enrollment and in sewing machine sales. Results were similar when the program was shown in other Latin-American countries. The lesson was not lost. A Mexican TV producer, using the Peruvian program's formula, refined it and created similar programs aimed at affecting family planning, gender equality, and other behavior. They were commercial hits and had a significant measurable effect on behavior.

A TV program in the Philippines focusing on safe sex had behavioral effects similar to those experienced in Peru and Mexico. The effects were amplified when other channels were used such as print and broadcast ads, buttons, and songs by high-profile rock stars. In 1984 a South African doctor created a TV series called *Soul City*. Aimed initially at reducing behavior associated with HIV transmission, it came to address a number of other issues. The 13-part prime-time TV series ran simultaneously with a 60-part radio series. The TV program enjoyed very high ratings – about half of the entire TV audience. As in the Philippines, supplementary campaign materials were used, encouraging viewers to discuss the issues and take action, which increased the likelihood of behavior change.

All these programs focused on individual and interpersonal change, not mobilizing people for collective political action. Would they work as effectively on collective political action? Conservation calls for both personal and collective action. The decision to act, whether individual or collective, involves some of the same elements – a call, the response of cohorts based on needs, emotions, and appeals to purpose, and an assessment of risk. Political action, however, because of its public character, less immediate and certain reward, and higher risks, demands greater energy and commitment.

Convention holds that the mass media are good for creating awareness, providing information, and shaping the political agenda, but that mobilization requires face-to-face channels, that is, organizers (Robert Hornick, 1989: 309). Some recent campaigns demonstrate that mobilization can occur through the media without field agents. This result was due to a very thorough understanding of the target audience, a very clear message, saturation level message repetition, and good sense of what could be expected from the target audience given the risks and their resources. Anecdotal information indicates that people may seek out an NGO after media coverage to join or become active (James Jasper, 1999: 76). Using multiple channels contributes to mobilization via mass media (Robert Hornick, 1989: 329). The US right wing successfully relied on a combination of direct mail, talk radio, cable TV, and the internet (Richard Viguerie and David Franke, 2004: 330–1). Church-based networks and pastoral exhortation (though illegal) also played a major role in these partisan mobilizations.

How can conservationists better access the nonprint mass media? Marine conservation biologist Randy Olson got frustrated with how scientists were communicating and went back to school to learn how to make films. His first film, *Flock of Dodos*, about the intelligent design attack on evolution, has been widely seen by general audiences. If other conservationists pursued work in the media it would be a boon to the movement though it would take time to show results. More immediate access and influence are needed. Existing connections need to be cultivated and new connections deliberately sought. Many in the industry are sympathetic, but to be persuasive conservationists will need to show that conservation issues can be interwoven into commercially successful programs and film. Conservation practice offers much raw material for stories from the mythic to the personal: characters, conflict, drama, and the occasional comedy. Once a few successful programs and films are made others will follow. Commensurate with the difficulty are enormous potential rewards, such as deepening people's sense of urgency and efficacy, creating new role models, and transforming the dominant mythology into one friendlier to Nature.

Entertainment is not confined to fictional programming, as Nature documentaries demonstrate, but they reach a limited audience. Gore's *An Inconvenient Truth* has been viewed by millions and the factors that contributed to its mainstream status are important: Gore's stature as a very public member of the elite, the long struggle to raise awareness of the issues, and the significant threat of climate change to the human enterprise. Not all of these factors are easily replicable around each conservation issue, but even less successful films can make a difference.

Checking the extinction crises will require permeating the world's cultures with conservation messages in the way that the first *Stars Wars'* trilogy permeated many of these cultures. Film is especially suited to the mythic, but also to lesser stories. The draw of *Star Wars* was the combination of its mythic qualities – a cosmic battle between good and evil – and the personal roles it offered to audiences, allowing them to find themselves in the myth. Both cinema and some television programming build on other programming and film, filling out the elements of a full worldview; for conservationists this holds the promise of a new template for understanding Nature and over time contributing to cultural transformation. Film in its theatrical venue is particularly influential. The darkened collective setting and simultaneous viewing across large geographical areas by tens of millions of people provides a ritual-like experience; the experience continues to reverberate in social interaction about film content and its experience. The film *Day After Tomorrow* greatly increased awareness of the climate change, albeit by scaring people with a scientifically implausible story line.

DVD format gives film and some television programming a longer life, but the small screen and lack of simultaneous viewing lessens impact. Much popular film is aimed at achieving blockbuster status by appealing to adolescent audiences, but films aimed at adults have enjoyed improved commercial success so it is possible to reach influential audiences this way.

There is much room for innovation in mass communication as the Zuni demonstrated in their battle to stop a coal mine (Robert Cox, 2006: 268–72). Like most conservation NGOs the Zuni faced limited resources, one of which was the refusal of billboard owners to sell them space. The campaigners instead painted their billboard ad on the sides of a large truck trailer and drove it around major cities, the state capital, and other important sites. The truck not only worked as a billboard but it gained media coverage because it was innovative. The Zuni also made use of a traditional channel of communication – sending runners with their campaign message to other tribes, to the state government, and to coal company officials. This also leveraged media coverage. Successful use was made of radio as well by running ads in the languages of their audiences – Apache, Hopi, Spanish, and Navajo.

A chorus of channels

Channels of communication are not so discrete as they seem from the foregoing. They are interactive and are not just complementary. NGO outreach is most effective when it mimics the interactivity of well-entrenched patterns of social influence, that is, through "complementary and reinforcing" pathways (James Dearing, 2001: 305). This includes personal interaction; the greater its extent between mobilizer and audience the more likely mobilization because the process is experienced as collaboration rather than outside-inside or top-down. (Neil Bracht, 2001: 323). Put more simply, media outreach combined with grass roots organizing is an effective combination (Laurence Wallack and Lori Dorfman, 2001: 395)

In selecting channels there are other factors to consider in addition to determining the cluster of channels a target group relies on. Not all types of messages can be carried by all channels with equal effectiveness. Television ads and news do not generally convey complex issues well. Documentary or dramatic programming can do a better job but will usually not be seen by as many people. For simple messages television can be quite persuasive (William McGuire, 1989: 47). Print media can convey complexity better but is often less effective because it lacks the capacity to evoke emotion through body language, voice, or music. Mobilization based on argumentation has built-in limits in any event because the critical faculties of many groups have been dulled by too much television watching, which reduces the patience necessary for in-depth analysis and processing of information (Jane Brody, 2004: F7).

Different channels are more or less capable of obtaining audience attention and messages that require high attention will fare poorly on channels that attract low attention. Attention is a matter of the time and focus. If people have time to focus on a message, the more intellectual and emotional processing it receives and the more enduring its influence (Renee Bator, 2000: 530, 531, 533). Mass media ads, as opposed to some programming, are usually only capable of attracting low attention. To capture high attention a message must be exceptional; vividness, captivating images, memorable characters (the messenger), or a poignant situation may attract high attention even when the medium normally doesn't. Personal testimony often garners high attention and makes a lasting impression when other types of presentation do not (David Meyer and Sidney Tarrow, 1998: 224–5). But these factors cannot always overcome channel infirmities; not all messengers, for example, are recognized in every medium. Anyone who has heard Karsten Heuer speak of following the Porcupine Caribou herd across the tundra for weeks, or Amy Vedder speak about her work with mountain gorillas, knows the power of intensely personal voices. But a voice that moves people in the course of a half-hour presentation in a low-lit room may not work as well on a minute radio spot, or the op-ed page.

Low-attention channels, such as paid ads in electronic media, simply cannot accommodate messages that require a great deal of processing – they need to hit quickly and deeply or be repeated over and over and over. It's pointless for conservationists to complain about a low-attention medium's inability to carry their full-dress arguments and information. In low-attention media or circumstances messages must be straightforward, brief, and as emotionally evocative as possible. This can be done with misleading. Those mobilized do not need to immediately understand the issues in great detail; they can gain an understanding over time. By insisting on needless complexity conservationists are limiting the channels available to them and are writing off important constituencies.

As noted, low-attention messages work, but require sustained repetition. Some channels accommodate this better than others because of their cost or ability to consistently reach people. Much of this discussion is moot when conservationists lack the resources to use the best and most effective channels. Resources include not just money to buy the access but the expertise needed to select channels carefully and use them in a way that maximizes mobilization through proper staging. Because there is much science in mobilizing groups does not mean the need for art is absent; quite the contrary. Nor is perfection required; getting the channels right need only be good enough, that is, better than opponents.

12 Mobilization and messages

For all of the wizardry and fetish of technique the substance of message remains central to successful mobilization. This is true when talking one-on-one, in materials distributed to a mass audience, or speaking with and through the media. Whether messages result in mobilization depends on several factors such as messenger and channel. But these factors do not diminish the need for messages to:

- resonate with all aspects of personality – needs, emotions, reasoned cover, and the sacred (James Combs and Dan Nimmo, 1993: 145; James Jasper, 1999: 71; Paul Stern, 2000: 419) – and at the level of social interactivity. To resonate they must address the specific personalities and subculture of the audience, strike them as compelling, and be part of an enticing story (see Chapter 13), only then do messages stick, penetrate, and ultimately become integrated into people's personalities and shape behavior in a lasting way;
- be temporally sequenced to raise awareness, frame the issues, and call for action;
- do a better job of resonating with audiences than opponents' messages;
- be tuned to audience responses to earlier messages, opponents' messages, and events that change audience receptivity.

This chapter will focus on the most salient principles that govern message resonance with a target group's needs, emotions, and worldview. Other aspects of crafting mobilization messages and their temporal sequencing are also examined.

A reminder about the obvious

Messages must be crafted that *resonate with the intended audience, not with the message propagator.* Conservationists still forget that what brings tears to their eyes does not bring tears to the eyes of others. What makes others fighting mad is not what makes conservationists fighting mad. Conservationists share many personality attributes with other groups, but other groups do not act on those shared attributes but on those aspects of their needs, emotions, and beliefs that are most salient to them (Riley Dunlap et al., 2000: 427). Conservationists, for example, rely on charismatic megafauna and stunning landscapes in their messages. These creatures and landscapes do have wide appeal, but they have generated mostly shallow support. News coverage of wildlife issues has never been significant when compared with coverage of water and air pollution (William Paisley, 2001: 19). Even when the audience is other conservationists assumptions of similarity cannot be relied upon. Those nongovernmental organizations (NGOs) in one country that were successful in gaining support from NGOs in another (usually richer) country did a better job of matching their requests with the target's interests, views, and goals (Clifford Bob, 2005: 22, 193).

Conservationists often emphasize inspiring people with their messages. Inspiration is important because it stimulates the energy people need to overcome their inertia and the external obstacles to mobilization. If conservation is to go beyond saving a few places along the path toward global ecocide it must feed muscle as well as the spirit. Indeed, conservation is about the most practical thing in the world – the human relationship with the world on which its livelihood depends. Messages must address the practical in a variety of ways or the resulting mobilization will be neither deep nor lasting (Lewis Rambo, 1993: 76–86).

The practical aspects of messages provide guidance on what must be done and how it can be done to reach conservation goals. The practical side also includes guidance on the path toward livelihoods more compatible with conservation. Guidance sometimes will provide explicit direction: take this specific action to sway an agency or other decision maker. When possible guidance should enlist target audiences in the process of crafting the best means to reach well-defined goals, and avoid the sort of false solutions that are typical of civilized societies, for example, add more fertilizer to depleted soils (David Ehrenfeld, 1979: 107).

Resonating with needs

Nature, including human nature, is material; even the spiritual is anchored in the body. Survival depends upon meeting needs and for this reason informed appeals to need-states generate the strongest responses (Wilhelm Reich, 1970 [1946], 1961 [1949]; Karen Johnson-Cartee and Gary Copeland, 2004: 133). Conservation's opponents understand this, which is why they describe conservation as a threat to people's need to feed and house themselves. It is a potent appeal in societies like the United States and much of the third world where economic insecurity is a major means by which elites manage large segments of the population.

Mobilization efforts on behalf of biodiversity conservation – as compared with appeals to protect clean air and water for people – have generally failed to successfully tap into needs. Even campaigns to protect clean water and air have not elicited major responses absent immediate threats or a major incident of degradation (Decision Research, 2002: 208). Tepid responses or nonresponses are the norm to threats distant in place or time and examples such as Easter Island carry little weight (James Combs and Dan Nimmo, 1993: 61). Slowly developing threats and those that are causally complex are equally difficult to mobilize around. Even when conservation-friendly alternatives to the painful boom and bust cycles of resource extraction are the focus of messages it has been a tough sell. People, of course, can survive in the face of much biodiversity loss. Claims to the contrary have provided conservation opponents with ammunition to undermine conservationists' credibility.

Positive messages about the ease of transition to conservation-friendly liveli- hoods have not fared much better because they ring hollow. Not every community can transition to a new economic base by attracting engineers, architects, and bond traders who work by computer distant from urban centers. Loggers and miners generally do not become bond traders and they justifiably do not want to flip burgers for minimum wage. Appeals to the precautionary principle – protecting all biodiversity to maximize future options, or preserving wildlands and creatures for the great grandchildren – are not successful in the face of urgently felt needs.

Based on the specific characteristics of each group to be mobilized, messages will be most successful at mobilization when they appeal to pain avoidance, and pleasure seeking, physical bonds with the natural world, needs for belonging, love, esteem, and justice. Because esteem and justice needs have strong emotional and cognitive components, respectively, they will be discussed in later parts of this chapter.

Pain and pleasure

Avoiding pain is closely linked with maintaining biological integrity; hence it is a powerful motivator. For those groups who are in touch with their own needs and the relationship between such states as hunger and thirst and pain and pleasure, a basis exists for understanding that other creatures feel hunger and thirst in ways similar to them. Appeals to empathy based on needs may resonate deeply. Messages that portray the consequences of habitat loss on feeling creatures and offer opportunities to protect them and prevent pain also offer the opportunity for those acting to experience pleasure in doing good. When fund raisers such as Andy Robinson and Kim Klein tell conservationists they are doing people a favor by asking them for support this is what they are talking about. It is no accident that the great religions that have arisen in the last 3000–5000, have a major focus on giving meaning to and relieving suffering (Connie Barlow, 1997: 70). Thus their broad appeal and the ubiquity of the golden rule.

Some people experience places as being alive. Attachment to these places usually has roots deep in childhood or in an epiphany that transcends the geographical mobility that weakens human connection with place. These places are part of individual or group identity. Injury to them is painful and threatening. Mobilization appeals based on either basis will resonate, but the economic relationship of the audience to the place must also figure into the message.

For audiences that do not feel compelled by economic pressure to consume the places they care about appeals can be straightforward. Such messages should focus on raising the salience of connection to place over competing aspects of identity. These messages can undercut groups wanting to block protection of place who often rely on appeals that privilege the alleged injury to humans that protection would cause. Messages that reinforce and strengthen empathy with Nature – situations in which entire species and systems are at risk – should strive to demonstrate that humans have alternatives to destroying habitat and depend on a healthy natural world.

Many rural audiences confront potential conflicts in their relationship with place and messages must account for this. They often feel a very strong attachment to wild places but their material security is frequently tied up (or perceived to be) with activities that consume the places they love. Because of the destruction of these places is often "only a bit at a time" it is easy for rural audiences to embrace denial and accept arguments that they are not destroying what they love. Messages must expose the contradictions. To do so they must initially bypass the rationalizations by focusing on the pleasure that comes from the audience's connection with place and other creatures and how it nurtures them. Then messages can be effectively advanced with cognitive explanations for a more conservation-oriented relationship with Nature. Mobilizing rural audiences is less about basic identity change – something very difficult to achieve – than a change in emphasis on aspects of identity.

A major challenge confronting conservationists is that ties to the natural world are increasingly abstract. Fewer and fewer people have the direct experience of wild Nature that makes it a big part of their identity and creates strong bonds of pleasure (Richard Louv, 2005), so other bases for messages must be found. Conservationists need not rely entirely on pleasure taking in Nature for mobilization. One reason cohorts are so important in mobilization is the pleasure they take in each others company while doing things together; this can be a successful impetus for mobilization (James Jasper, 1999: 73).

Messages, sexuality, and physical bonds

Nature in humans is sexual and there is enormous potential for energizing target audiences by appeal to sexual impulses. These include the longing to touch, to hold and be held, to dance, and to interact in ways that foster the experience of physicalness. The connection between Nature within each individual and Nature without is why the undulating river's sensuality resonates with human sensuality; why rippling water soothes and white water stirs blood and adrenalin. For groups that are at home in their bodies and have good access to the joy and sorrow they bring, a strong basis for connection between sexual impulses and the natural world exists. Empathy for others, including other species, begins with the experience and acceptance of the individual's own needs (Harold Searles, 1960; Dorothy Dinnerstein, 1976). Grounding in finite, needy, and mortal bodies provides both access to the energy associated with need-based impulses and a potential willingness to protect the expression and satiation of needs in other creatures. Grounded humans can recognize that only in a caring relationship with other creatures and Nature as a whole is it possible to fully experience and maintain their own well-being. The desire to embrace and feel the embrace of a living world is the "secret" attraction of hiking, Nature safaris, camping, float trips, some aspects of zoos, and television Nature programming. To be connected with one's own and Nature's sensuality transforms assaults on Nature into a personal assault. When abolitionists in the 1820s and 1830s came to experience slavery as a personal affront they acted, bringing their daily lives into accord with reform (Michael Young, 2001: 99–114).

To realize the mobilization potential of tapping into sexual needs and energy requires that conservationists be well tuned to the audience. There is much sensitivity about sexuality and if the messages are amiss strong negative responses are likely.

Sexuality is contested terrain. Elites and other groups in every society seek to harness it by shaping how it is experienced and expressed. In some cultures sexual energy flows like a river, finding expression without conscious effort and giving delight and sometimes pain. But damming, diverting, or confining this river to a narrow channel is a means to effective social control (Wilhelm Reich, 1970 [1946]). For as long as elites have existed they have tried to direct this energy to serve their world-building (and Nature-destroying) compulsions (Dorothy Dinnerstein, 1976). The pervasive suffering that has characterized hierarchical societies through the millennia has rendered people amenable to this control because suffering is at root physical. In such circumstances the false promise that suffering can be avoided if one escapes the body and finds refuge in an ethereal, split-off spiritual life has great attraction.

Those who embrace their own sexuality tend to rejoice in the sexual in Nature, taking delight in bees' performance as plants' sexual organs, in the struggle of salmon to spawn, in copulating bears and the aerial dance of birds. For them Nature is not to be dominated or controlled, but to be embraced as a source of joy and the satisfaction of needs. Most people, however, are only partially and ambivalently at home in their bodies. They lack to some degree the capacity to express, embrace, and empathize with their own needs and thus with the needs of other creatures.

17th century North American Puritan Cotton Mather documented with approval his alienation from his body and disconnection from Nature.

> "I was once emptying the Cistern of Nature, and makin Water at the Wall. At the same Time, there came a Dog, who did so too, before me. Thought I: 'What mean and vile Things are the Children of Men, in this mortal State! How do our natural Necessities abase us, and place us in some regard, on the same Level with the very Dogs!' My Thought preceded. 'Yett I will be a more noble Creature; and at the very Time, when my natural Necessities debase me into the Condition of the Beast, my Spirit shall (I say, at the very Time!) rise and soar, and fly up, towards the Employment of the Angel.'"
> (Alan Heimert and Andrew Delbanco, 1985: 327).

The Puritan disconnection from self and thence from Nature was expressed in their treatment of the wild as evil, the home and work of the devil, and as something to be controlled and subdued (Roderick Nash, 1982: 34–40). Similar views regarding female sexuality and Nature predominated among other groups in the early modern West (Carolyn Merchant, 1980).

Much of the world's population lives in conditions that foster similar but less extreme views (Pippa Norris and Ronald Inglehart, 2004: 22–4). Material deprivation in childhood is highly correlated with hostility toward sexuality and especially female sexuality. Women and Nature are experienced as inherent threats to proper order and are to be controlled.

Discomfort with sexuality need not give rise to puritanical responses. Victorian-like responses are more frequent. If the puritanical welcomes the rigid suppression of sexual energy as good, the Victorian experience is a more consciously divided one, in which sexuality is condemned outside of a narrow range of acceptable situations but indulged in with some pleasure in those situations nonetheless. Those who sanctimoniously condemn a breast exposed on prime-time television secretly rent porn flicks and enjoy them, albeit with guilt later on. In the Puritan sexual energy is turned on itself; in the Victorian sexual energy is expressed in self-contradictory forms. Among other groups in modern societies a psychocultural narcissism is characteristic (Christopher Lasch, 1978). Sexuality is divorced from emotion, and is driven by efforts to distract from an inner emptiness through sensation seeking and the desire for approval from others. For yet other groups sexuality and Nature are highly romanticized. Sexuality is draped in expectations of true love, perfect union, and perfumed genitals. Nature is similarly experienced as something desirable, but only when free of discomforting predation, biting insects, and dirt.

These categories oversimplify; individuals and groups combine elements of healthy sexual expression with the puritanical, Victorian, narcissistic, and romantic.

The life of one prominent American conservationist makes this plain. Prize-winning historian Edmund Morris (1979) described Teddy Roosevelt (TR) as morally rigid and obsessed with his purity of sexuality. Faced with arousal while courting his wife he would engage in strenuous physical pursuits such as riding at a gallop for hours. Like many a Puritan he took pleasure in killing wild things and seemed utterly without empathy for the bleeding creatures he gunned down for "manly" sport. Morris puzzled over this given the pleasure TR developed for spending time in the wild and his actions as president to protect wild places. Yet this contradiction is not so difficult to explain. The natural world was the one place TR could partially escape his moral rigidity, at least in manageable doses. While he could not accept his own sexual impulses as they were, he could find some joy in Nature "out there" and in overtly nonsexual physical exertion. He alternately embraced and killed the split-off part of himself, just as some seek out prostitutes while publicly condemning and punishing them. Murderous hostility can result from being brought too close to one's rejected impulses and is a means of maintaining the bifurcation between need and its natural expression.

Those who lack physical passion or who have made an enemy of it by turning passion on itself can be mobilized for conservation, but are not good candidates for messages that tap into sexual energy. For those who embrace sexuality and have a significant degree of physical empathy there is great potential because of this need's capacity to bind people with unrivaled strength to objects of satiation and empathy, including the natural world.

Warnings of pitfalls should not make conservationists overly nervous. They are not embarking down an untrodden path. Conservationists already make appeals to sexuality when their messages appeal to the physical joys and drama of the wild and its creatures. By paying better attention to what works they can be more effective. And although mobilization for political action is different than peddling beer or a car based on sex (it involves greater commitment), commercial marketing offers some lessons from a wide range of audiences (for a topical survey see Tom Reichart and Jacqueline Lambiase, 2002, 2005).

The physical connection between humans and other creatures and wild places has also been a staple of conservation literature and broadcast programming. Integral to the writing of Terry Tempest Williams and Gary Snyder are the joys of the sensual and the embrace of Nature as lover. This is one of the great appeals of their work and why it is read beyond the conservation choir. Nature, after all, is the real thing – both the source of broadly sexual need and the source of the natural objects that satiate it. Conservationists are not offering compensatory indulgences or artifice, but connection to the living universe.

Given the tension most groups experience concerning sexual needs it is more likely that messages appealing to it will be successful if they are implicit rather than explicit as is the case with ideologically based appeals (Murray Edelman, 1988: 24,

49–50). Polemics create resistance and sexual polemics add discomfort, increasing resistance. Appeals to the sexual connection between humans and Nature will be better received when they are embedded messages that show situations in which closeness to Nature offers a fuller life through fuller experience.

Messages and belonging

Humans depend on group membership for survival and group survival depends in part on the quality of the strength of social bonds and cooperation. Evolution has not left the matter of group membership primarily to rational calculation of the benefits. The impulse to belong is hardwired; its expression is flexible but nonetheless strong. Humans belong to families, bands, neighborhoods, professional societies, NGOs, interest groups, political parties, nations, global empires, and countless other grouping to which they feel affiliation. The evolution of language is entwined with our social impulses. Language allows more information to be shared, cooperation to be more complex and to extend over great geographical distances. Humans define themselves in part by the groups to which they belong, although some groups are more important than others.

Mobilization messages can appeal to the tug of the group on two bases. One basis is that of community per se. When people do not feel well connected or satisfied with their membership in the groups to which they belong they are open to appeals from others. Another primary basis of appeal is to a desire for association with a larger purpose. Those who are satisfied with their group memberships may still lack association with a group that offers the sort of purpose conservation NGOs do.

Being part of a community and the community's purpose should both figure in appeals to belonging. Messages are more effective when they take account of people's preference to join groups where they will feel at home – either groups with backgrounds similar to theirs or groups possessing attributes they aspire to have. Audiences need to believe they will be welcomed and valued. Belonging messages should also aim to stimulate mutual reinforcement for mobilization among cohorts. Stimulating discussion among cohorts about the message or the NGO that issued it achieves this. So do messages that remind people that mobilization includes continuity of cohort association in addition to new opportunities for association. When mobilization of a large group proceeds through its leaders continuity needs less attention. In cases where people have outgrown or feel uncomfortable with existing associations they may seek escape rather than continuity; messages must be sensitive to this (Lewis Rambo, 1993: 83).

When purpose is linked to belonging in messages it can focus on specific goals or higher purposes (or both). When conservation speaks to the latter it offers people the opportunity to be part of something that is meaningful beyond anything else – to be part of saving life on Earth by halting the sixth great extinction. Such an achievement will echo through all time. A higher purpose does not require that the scale action requested be grand – it can be local rather than global so long as it is linked with great purpose. Great purpose offers redemption for oneself and the human species. Charismatic leaders are a good means of personifying high purpose to some audiences in addition to attracting those who desire to live vicariously through them.

For most groups messages need to communicate that conservation is socially approved because people want to belong to something that will gain them approval. Most people do not want to be on the leading edge. If conservation is portrayed as a growing and effective movement, but not arrogant, the message of approval is communicated with a sense of efficacy thrown in.

People do not live only on high purpose. Belonging is foremost a practical matter (survival) and mobilization appeals based on belonging in conjunction with daily activities should be part of the message. Survival per se is usually not on people's mind, but the benefits of being part of group of capable, connected people often is. So is the desire for friendship, beer-swilling, hiking, talking, finding a mate, hanging out, and everything else that vigorous social networks provide.

For some target groups mobilization in support of conservation carries risks ranging from the disapproval of friends and family to repression. Even when risk taking is attractive – as it often is to the young or those fed up with abuses of power – messages should recognize the risks and offer both reassurance and an understanding of why the risks are worth it. Reassurance comes from messages that link conservation mobilization with a group's existing standards of justice or people's desire to be part of something heroic – a movement of Davids standing against Goliath. So does explaining the link between belonging and high purpose. Mobilization costs in other ways – it asks for time, expertise, and money. The costs of belonging should also not be ignored in messages because it tells people that mobilization is worth it. (Stressing "sacrifice" is different and is discussed below.) When people pay a price for joining a group it increases their sense of commitment (Lewis Rambo, 1993: 116; Richard Jensen and John Hammerback, 2002: 68; Carol Tavris and Elliot Aronson, 2007: 15).

Appeals to belonging are also a means of mobilizing institutions in support of conservation. Many US religious congregations took action on behalf of Central American refugees created by US President Reagan's war on peasants in part because of appeals to their sense of belonging to a larger purpose and a larger church of believers that included the wars' victims (Sharon E. Nepstad and Christian Smith, 2001: 166–9). The appeal to belonging – to being part of a community defined by high moral purpose – leads congregations to experience what was happening in Central America as something immediate and personal. They also saw themselves as being uniquely responsible and well placed to act given that it was their government backing the terrorists who were creating the refugees. The risks of support were real but small.

The advantages of mobilizing institutions are significant. They bring resources to bear such as organization, labor, money, and skills that are much greater than when an equal number of unorganized individuals are mobilized. All the strengths and weaknesses of coalitions and alliances discussed in Chapter 8 apply; resources must be invested, leaders can be fickle, interests diverge in the heat of battle, and tactical compromises must be made. In cases where strongly held values are shared and there is regular interaction among groups, a new and common identity may emerge.

If conservationists have been mediocre in their messages addressing need-states such as belonging, they have been worse with follow-through. In Chapters 17–19 the importance of community to sustaining mobilization will be discussed. Without a *community* of tens of millions of people conservation cannot succeed.

Messages and emotions

Fear, love, anger, joy, defiance, disgust, shame, and other emotions are the proximate cause of most human behavior. Recall that the words emotion and motivate come from the same root – to move. Needs frequently work via emotions as when threats to safety make people fearful. The imprint of culture on how needs are experienced and expressed uses emotional responses. Emotions underlie values and thoughts and humans invest emotionally in their worldviews. The emotional structures humans develop from their experience means they are rarely totally in the present; the past shapes present encounters. The emotional structures of individual and group do not constitute a unified whole, but often pull in different directions, creating distinctive internal individual and group dynamics (Jeff Goodwin et al., 2001: 15). Individuals and groups are each characterized by particular emotional predispositions – emotions or moods prevalent in the psychic background – and by particular emotional reactions to certain situations, for example, a new job causes anxiety in some and excitement in others.

The complexity of human interactions intuitively suggests that groups are more difficult to understand than individuals, but experience indicates otherwise. People in social networks and organizations accommodate each other and by seeking reinforcement for views and behaviors, attributes are stabilized and made more predictable. This is good news because mobilization is primarily aimed at groups.

Fear and messages

Fear is probably the most powerful and ubiquitous of all political emotions, tied as it is to basic needs such as physical safety and integrity of identity. The adaptive value of fear is obvious – much in the world appropriately generates fear, such as having to share the Earth with people who are obsessed with wealth or power. Fear's capacity to motivate is well understood by those seeking to manipulate others. One of Hitler's top deputies, Hermann Goering, said in an interview during his war crimes trial that it was easy to gain popular support for war – people just needed to be told they had been attacked (Gustave Gilbert, 1947). Fear works the same in a dictatorship or a democracy, he said. Political campaigns of all sorts rely on fear to mobilize people. Oil company apologists warn of economic calamity and job losses if fossil fuel use is curtailed. More realistically the specter of catastrophic climate change fosters fearful images. Fear reactions, then, may or may not be well grounded.

Conservationists often successfully appeal to fear but have sometimes over-played their hand. Although the extinction crisis is real and dramatic, its conse-quences are more akin to the fall of Rome, with the consequences unfolding over decades and centuries rather than in the minutes it takes for apocalypse to occur in a film. It is the film standard many audiences require before they will listen to messages and only rarely is that standard met by reality, for example, events such as the Exxon Valdez spill. Opponents of conservation have exploited these occasional overstatements and used them to attack conservationists' credibility. Opponents also make their own overstatements, usually regarding negative eco-nomic consequences of proconservation policy choices.

By improving their own fear-based messages and crafting better responses to opponents' fear-based messages, conservationists can greatly improve mobilization results. One way fear messages work is by gaining priority attention over other messages and causing people to seek out and consider new information rather than rely on routine and habit (Ted Brader, 2005: 13, 61–2, 131). Fear is a dis-comforting emotion and people seek to diminish discomfort. Those with a sense of efficacy and greater flexibility seek the source of discomfort and will reconsider support for candidates and policies. Those who believe that reconsidering choices won't make any difference are more likely to deny their fear rather than act on it (Ted Brader, 2006: 58–9). Denial is also the likely response from those who have a high need for certainty and who are surrounded by those similarly invested in a candidate, party, or set of policies (John Jost et al., 2003; Carol Tavris and Elliot Aronson, 2007: 15, 19, 63).

Messages aimed at those willing to act on their fear are most effective when they clearly identify what is feared and what steps can be taken (such as becoming mobilized) to alleviate the emotion. Messages aimed at those inclined to denial must show how they can make a difference and how the action called for fits with their central myths; both undercut resistance to the message. Then messages can tackle industry fear-mongering about conservation-caused job loss with the raw material that economists have compiled for many regions showing Nature protection is economically positive (Powers, 1996; Ray Rasker and Ben Alexander, 1997; Ernie Neimi et al., 1996, 1997, 1998, 1999).

For those who feel closely connected to the natural world the fear of its loss is palpable; messages to this group based on fear have long resulted in mobilization. As biological losses mount, however, depression and despair can result instead of mobilization. The despair pitfall can be avoided by balancing threats of loss with reports of victories and reminding people that their actions do make a difference even though the difference may take time to show itself.

Most audiences conservationists need to reach do not feel very connected to Nature. Fear of a diminished wild is not experienced personally enough to generate mobilization. To make such losses personal conservationists must *clearly and credibly* link them to human well-being. Material well-being carries greater force than psychological well-being, but both are important. Former Secretary of the Interior Bruce Babbitt (2005: 44–6) attributes success in gaining support for Everglades restoration to the link between water and public health, rather than to conservation. Protecting public health does not necessarily move policy toward conservation goals – sometimes they may be in conflict – but mobilization and changing policy is a process, and where the two can be linked it often makes good sense to do so. Over time NGOs can build emotional ties to the movement and biodiversity.

Conservationists can improve their own messages by an examination of industry's fear campaigns. Conservation cannot match industry's capacity for saturation level messaging nor their reliance on propaganda (disinformation, scapegoating, the propagation of self-serving "solutions," and exploiting the human propensity for wishful thinking). But by understanding that people also, for example, fear their dependence on oil, on large, profit-seeking corporations, and on problematic political regimes in countries supplying oil, they can craft messages that undercut opponents. As the oil-supply crisis worsens saturation-level industry messages will ring distinctly hollow and conservationists need to be ready to push alternatives, included less energy consumption. Political opportunities are often the most important political resource in effective mobilization. Often a crisis is needed to make clear to groups that there is much to be afraid of if they wait too long to mobilize in support of alternatives to the status quo.

The pervasive fear of mortality and its causal links to conspicuous consumption and profligate reproduction represents a daunting challenge to conservationists (Ernest Becker, 1973; Sheldon Solomon et al., 2004). It is also one that cannot be overlooked because of the direct consequences for species and habitat destruction. There is precedent for breaking the connection between fear of mortality, consumption, and reproduction – or rather, preventing it from being made. Historically many cultures have moderated death anxiety behaviors through identification with Nature and the endless process of death and birth (Paul Shepard, 1990;

Morris Berman, 2000). Religion and philosophy traditionally address mortality and conservationists will be most successful if they work with those who have been crafting messages in the area for a long time.

In wealthier societies consumption is the dominant means of ameliorating mortality anxiety; in poorer societies religious notions of immortality predominate along with values that denigrate earthly existence and keep birthrates high (Sheldon Soloman et al., 2004; Pippa Norris and Ronald Inglehart, 2004). Such distinctions are obviously important in developing messages. For religious audiences it is useful to recall the important role "the wilderness" plays in many religious traditions – as places of refuge and where prophets find their voice. The increasing centrality of "creation care" in many religions provides a new basis for cooperation on conservation issues. In this view the obligation to love the Creator includes the obligation to love one's self, one's neighbor, and the rest of Creation (Peter Illyn, 2001). The task of conservationists is not to offer connection with Nature in lieu of belief in immortality, but to work with religious groups to build an understanding of what needs to be done to secure creation. Nonconservation differences need to be put aside. Groups do good things for many reasons.

For the nonreligious the main path away from death-inspired consumption and edifice building lies with starting to build emotional bridges to nonhuman life using what is at hand – parks, zoos, gardening, a focus on urban wildlife. Although mobilization for collective political action will not result immediately from messages that move people toward connections with Nature, to the degree that individual choices (less consumption) relax pressure on habitat conversion, the task of conservation is realized. Creating art, like an urban garden that attracts wildlife, also offers a path away from anxiety and toward the distant past when there was no word for salvation because there was no need for one.

Love and messages

Love of other creatures and wild places can be a powerful motivation. Love offers pleasure and purpose but can also be frightening because loving entails emotional openness to the pain associated with injury to or loss of the loved. Encouraging loving action in a time of great loss must overcome some inherent resistance, but there is also great appeal in acting to protect what is loved.

E.O. Wilson (1984) among others has argued that humans are predisposed to love at least some elements of Nature. This notion – the biophilia hypothesis – is based on observations of human preferences and on the argument that it makes sense people should care about the world that made them and sustains them. A more precise understanding is needed, however, to craft messages on this basis. The real challenge of love-based mobilization lies in making it intensely personal and intimate – the kind of love that moves people to take the risks associated with action. Fear is almost always personal and immediate, whereas people sometimes confuse ideas about love with the emotion.

Love as emotion is closely associated with the need for love and the desire for love in return. Nature certainly gives much to those who intimately connect with it, but for most mobilization targets – those who spend little time in the wild or even a local park or zoo – the return is thin. This becomes an additional source of resistance among some groups to love-based mobilization. Evoking love in the young release enormous energy, but they have little toleration for frustration. Older audiences are more likely to persevere in the face of adverse companion emotions, being more discerning about what they love, and perhaps less passionate.

Love of other people is usually a stronger motive than the love of Nature. But people badly injured by war, domestic violence, or an authoritarian upbringing have found their way back to themselves by means of loving the wild rather than people. For most audiences, however, messages should stress the promise of mutual affection among movement cohorts in addition to appealing to an audience's love for Nature. Many are attracted by the prospect of close human interaction and bonding – that goes beyond belonging – to love (James Jasper, 1998: 417; James Jasper, 1999: 73).

Love is not only an emotion and a need but a way of behaving that entails nurturing the love object in its growth and development (M. Peck, 1978). Commitment means behaving toward what is loved regardless of the inevitable ups and downs of felt love. This distinction is useful in approaching mobilization because it is loving behavior not the emotion that conservation seeks. It is love as emotion, however, that brings lover and love-object together; when passions cool, Peck argues, relationships either end or something longer lasting is built based on commitment to loving behavior.

Joy, happiness, harmony, and pride

Strong appeals evoking joy, happiness, and a sense of well-being or harmony have been successfully used by conservationists. But when linked only to the natural world they do not stir enough passion in many groups. Those who can be recruited based on the joy of working to ensure there will always be whales breaching the ocean's surface, or the sense of well-being that comes from immersion in smells, light and sounds of the deep forest, are probably already in the conservation movement.

The approach typically taken by marketers is to link these feelings and the product they are selling with something already strongly cathected. A new beer, no different than countless others, becomes a best-seller when linked to, for example, a sense of adventure, manliness, a desire for freedom. Thus many Americans embrace conservation when the message is linked with notions of national heritage; together these evoke well-being and joy. The Grand Canyon and Yosemite were protected in part because advocates successfully portrayed them as national treasures linked with American greatness and national pride (Dave Foreman, 1998–1999: i–4). But many groups don't find joy in Nature, and many more are ambivalent. Nature in the US emotional repertoire is both exalted and feared, the object of reverence and rape. When this applies to an audience messages should focus on tipping the balance toward reverence by associating conservation with other revered objects that evoke well-being. In an effort to stop the lucrative Caribbean parrot trade conservationists succeeded in identifying parrot protection with anti-imperialist nationalism; this was enough to overcome economic temptation because the feeling of national pride was personal (Butler, Paul R. 1992).

Elisabeth Wood (2001: 267–81) found that supporters of the El Salvadoran insurgency were mobilized, despite the risks, by their pride and desire to "defy and repudiate" the repressive authority of a brutal government. There are certainly many people who view human behavior toward the natural world as unjust and defiance of or countering that injustice evokes pride and satisfaction. The challenge for conservationists is to evoke the same sort of immediate and personal response when Nature rather than people are unjustly injured. Principles of justice are usually considered ideological because they are the subject of much rational discussion and argumentation (and pontificating), but at root the justice is emotional (Jonathan Haidt, 2001). Although some aspects of what people consider to be just appear to be hardwired – consider the very general but almost universal golden rule – justice is also shaped by individual and group emotional predispositions so that for some the pettiness of an eye for an eye is justice and for others justice involves going to the root of problems and trying to heal people. Regardless of the content, people find satisfaction in acting against transgressors, even when risk or costs are involved (Brian Knutson, 2004: 1246–7; Dominique de Quervain et al., 2004: 1254–58). That the intensity of the motivation increases with the satisfaction obtained makes appeals to justice attractive for conservation.

Disgust, anger, hurt, guilt, and shame

Many emotions that cause discomfort in addition to fear can move people to mobilization. Indeed, political movements *do not exist* without emotions such as anger and outrage. Disgust, guilt, shame, and sadness are also important motivations. If too intensely aroused for too long negative emotions can lead to burn out, numbness, or pointless violence. Psychiatrist Harold Searles (1979: 240) likens our psyches to Rene Dubos's description of ecosystems: they can adapt to heavy pollutant loads but only to a point and then they become less functional and predictable as the load increases. In politics this translates into bad decisions. To be effective in mobilization negative emotions must be disciplined (Ted Gurr, 1970; Johnson, 1982). Target audiences can become irritated and resentful of images and messages that evoke intense emotion but are not followed by an outlet for the energy evoked. Anger and outrage, for example, serve mobilization if they have concrete objects such as specific adversaries who behave badly (Jeff Goodwin et al., 2001: 16–7). It is not enough to arouse disgust at clear-cuts or revulsion at wildlife slaughter. Messages must also suggest immediate opportunities for simple actions that are convincingly efficacious enough to transform the negatively experienced feelings into something more positive. In this way the ebb and flow of intensity can be maintained within productive boundaries.

Appeals to anger have been a mainstay of conservation mobilization; there is no shortage of outrages perpetrated on the natural world by decision makers everywhere. Anger, however, is a derivative emotion, arising from hurt when something cared for is injured. Often people are unaware of the underlying hurt – the first emotion they feel is anger. For those groups that experience hurt it can serve as a basis for mobilization. Whereas anger generates protective and defensive energy – the basis for striking out at the injuring entity – hurt lends itself to a more considered response, a deeper evaluation of causes, and action to remove those causes. Hurt can also lead to withdrawal from action, however, in order to focus on personal healing rather than on healing the injury to Nature.

Sadness is another emotion deriving from hurt. It is associated with loss, as when a forest is razed or a population of creatures annihilated. Loss changes the world for those who suffer one, just as the death of a close one does. Sadness is often initially incapacitating. It also possesses the potential to transform as it did former British Columbia Premiere Mike Harcourt when he first saw the massive clear-cuts in the heart of Vancouver Island. Messages evoking loss must offer a path through loss and grief to action that prevents further losses. The grieving process itself ultimately restores the capacity for action.

Those who bring excessive idealism to politics may experience sadness (and anger) from the loss of faith in human goodness. This is part of political maturation and is liberating for many rather than debilitating. Reality is almost always a better basis for politics than delusion because it leads to better decisions. For those suffering from this crisis of faith there are good therapeutic options for keeping alive a willingness to engage conservation politics such as immersion in the complete work of Gary "The Far Side" Larson (1998) and Monthy Python (1979).

Guilt as a basis for mobilization is overrated even when audiences are supposed to be terribly guilt-ridden. Guilt – serious regret about something done or an opportunity foregone – is easily tolerated and people do not act to relieve it absent other emotions, such as attachment to a place or species (Biodiversity Project Staff, 2002: 63).

Shame is feeling bad about oneself rather than about an action. If guilt is about the sin, shame is about the sinner. For that reason shame is a much more powerful motivator. For the same reason it easily overwhelms and debilitates. All cultures and subcultures rely on both emotions in regulating behavior, but to widely varying degrees. Effective evocation of these emotions requires a good understanding of their place in a particular society or group.

In (sub)cultures that rely heavily on shame for controlling behavior its use must be carefully crafted. When shame is deeply ingrained its evocation is extraordinarily painful. At the first hint of shame-based pain one is likely to resort to denial, projection, or other compensatory reactions; or become depressed. Denial, projection, and reaction are bases for political mobilization, but in the service of authoritarian causes. Shame – feeling unworthy, inadequate, and powerless – makes people susceptible to surrender to an all-powerful, other-worldly parent who offers forgiveness, solace, and faux power, in exchange for adherence to their absolute moral authority, that is, the earthly leadership of the all-powerful parent (Arlene Stein, 2001: 115–31). Shame-based appeals are most effective when conservation messages hint at the emotion and simultaneously offer a means of avoiding it. Thus, to destroy Nature or to stand by while it is destroyed is shameful, but to act to protect it is worthy. This offers a way to feel other than powerless as well. Many who feel deeply ashamed are unable to escape the authoritarian trap.

Anecdotal evidence suggests that groups for whom shame is not deep-seated can be effectively moved to action by messages that evoke it. This is so because the sense of shame is not overwhelming and therefore does not generate powerful resistance and denial. It can be tolerated and acted on in a way which promises relief from the experience. By living up to one's responsibility to protect one's home (Nature) the sense of being unworthy of one's home can be overcome.

The overarching lesson for emotion-based messaging is the one so well illustrated by *Simplemente Maria* (Chapter 11): messages must tie pleasure and positive emotions to desired behavior, negative emotions and social disapproval to undesirable behavior, and demonstrate the possibility of transformation. This makes the problem and the desired action personal and immediate.

Humor is not an emotion but it always evokes some emotion – pleasure, embarrassment, amusement, and so on. From a mobilization perspective it is a tactic used by mobilizers to evoke emotions in an audience such as trust in the speaker by means of self-deprecating remarks that demonstrate a lack of arrogance. Such remarks convey that the cause is serious but the speaker does not take his/herself too seriously. Making light of the issues themselves may be judged inappropriate by audiences and undercut the likelihood the organization will be taken seriously. In some circumstances it is useful to poke fun at opposition leaders, their hypocrisy, preposterous claims, and stumbles; indeed, showing them as inept and foolish may be more effective than showing them as self-serving sleazeballs. It is better to attack leaders of the opposition than the rank and file to avoid appearing unfairly critical, though sometimes opposition followers may be chastised as misguided if it is done gently as Randy Olson does in *Flock of Dodos* (2006). Humorous deprecation of opponents evokes pride that one is not among them and is instead on the right side of things. Satire is for insiders.

Messages and the cognitive

Humans are far from the calculating animals posited by Enlightenment true believers and modern economists. The vast bulk of all emotional and cognitive processing is nonconscious. Despite this humans have a profoundly felt need (in addition to an adaptive necessity) to explain and give meaning to the world. To varying degrees people need to believe that their explanations and meanings are the correct ones and not created by them but by a divinity or inhere in the order of the universe. Because human cognitive and behavioral flexibility allows so many alternative views of the world and because for individuals to function through time and human groups to cooperate based on shared views, a particular alternative must be selected and adhered to. Self-interest and socialization cannot carry the burden of ensuring adherence, thus the evolved need to believe the alternative adhered to is right, proper, and in the very nature of things rather than a mere human creation (see Chapter 4 for a fuller discussion; Roy Rappaport, 1999). The most basic tenets of all worldviews are unquestioned, that is, they are sacred.

Messages, the sacred and cosmological principles

Messages are more likely to result in mobilization if:

- they are framed explicitly or implicitly within the context of an audience's sacred propositions. If at all possible messages should also be framed explicitly within the context of the basic principles (cosmological axioms) which concern the operation of the material world;

- they build on or otherwise incorporate existing views and rules of a group governing their day-to-day lives;

- in order to achieve mobilization day-to-day rules must be overcome (e.g., the desired behavior runs counter to norms) then the desired behavior must be shown to be compatible with the sacred and with axioms. This may require reinterpretation of an axiom, emphasis on a different or neglected aspect of an axiom, or application of an axiom not normally applied to the behavior at issue;

- when an axiom must be overcome appeals to the sacred are convincingly made that demonstrate that the existing principle is no longer working to achieve sacred purposes.

Conservationists rely heavily on appeals to the cognitive, especially values, but infrequently take account of the sacred views of the target audience. Because political mobilization requires a change in how people allocate their time, money, and other resources it must overcome more resistance, and the energy associated with the sacred helps achieve this. Failure to consider the sacred in creating messages also increases the possibility messages will miss their mark or run contrary to an audiences, sacred sensibilities. Even inadvertent challenges to the sacred are commonly experienced as threats. They cause messages to be rejected, the source of the message to be categorized as suspect, and may result in countermobilization.

By framing messages in terms of a group's most fundamental and unquestioned assumptions about purpose and meaning, receptivity is increased. Messages should be framed in terms of the sacred whether or not the action sought by NGOs runs contrary to existing day-to-day views or behavior of target groups. If messages do ask a group to act differently or to alter some of its views – to become politically active when they have not been, to make conservation a priority in selecting a candidate, to consider wilderness or a troublesome species worthy of protection – then it is essential that they be framed in the context of the sacred. When messages are successfully linked to the sacred the tremendous energy invested in the sacred becomes available for mobilization. This kind of energy is needed to overcome social inertia and to achieve significant policy change. In the 18th and early 19th centuries US abolitionists held rallies on July 4 in order to make appeals to the principles of equality enshrined in the Declaration of Independence (Francesca Polletta, 2004: 175). Such appeals were only as effective as the legitimacy of the Declaration. During the same period religious groups that regarded sin as a form of slavery to the devil were responsive to appeals that labeled slavery a sin and therefore something that should be abolished (Michael Young, 2001: 111–2). Oil industry leaders have long understood the power of tapping those regions of the mind that religion speaks to (Stuart Ewen, 1996: 385). Eco- and biocentric conservationists have also framed their arguments on the basis of a strong belief in the intrinsic value of all life but because this view of the sacred is not widely held it does resonate with most groups.

Messages solidly framed in terms of a group's sacred views also tend to diminish the anxiety associated with mobilization. Continuity with the sacred reassures in the face of the unfamiliar aspects of mobilization.

Framing messages in the context of the sacred is not so difficult because sacred views are usually quite general and in any event the sacred *legitimizes* views and rules about daily conduct; the latter are not *derived* from the former. But messages must be compatible with the purposes and meanings of life established in sacred propositions. Thus, messages seeking to mobilize groups in support of setting aside a substantial area of land or ocean for a species or natural processes might be framed, depending on the group, in terms of being required by a divinity; being part of a group's historical mission or by reason of the group's special social position or consciousness; being required to realize a group's deepest humanity through the exercise of compassion; or being required by a sense that justice or broad self-interest that does not permit degrading biological processes on which all life depends.

The greater challenge is framing mobilization messages in the context of the target group's cosmological axioms – those general principles about how the world works including overarching principles of justice and value (the descriptive and prescriptive are entwined). The views and rules that govern daily life are derived from these general principles so the logical relationship makes a fit potentially more difficult. Americans, for example, have long believed they are a chosen people with a special mission and not limited by the historical forces limiting others (Richard Hughes, 2003: 155–9). This view sacralizes the axiom holding that Nature is to be conquered and civilized. As the experience of Nature changed through the 19th and 20th centuries notions about how to treat Nature became more ambivalent; some groups maintained the view Nature ought be subdued, others abandoned it, and others held that it was okay to subdue it in human centers but that Nature is potentially fragile, the source of many goods when intact, and that some areas should not be under the collective human thumb. Axioms evolve, but not evenly among groups; messages must account for this variation.

Some axioms are amenable to a new interpretation or gloss. A US Senator seeking to designate a new wilderness near a fast growing city justified his proposal with a new twist on the axiom that growth is good and wilderness locks up land. He argued that wilderness protection was needed to make growth possible and that without more protected areas as an amenity growth would slow (Bruce Babbitt, 2005: 83–5). Such arguments can have the effect of perpetuating notions that growth is good even as they help to win a particular battle, and those consequences must be weighed.

In other situations messages may need to appeal to an axiom that is part of a group's worldview but is generally not applied to conservation issues. Thus, when mobilization for wilderness protection confronts the phalanx of well-anchored beliefs in growth or unrestricted private property use, appeals to principles of justice that normally govern human relationships might be used: all creatures' homes deserve protection. At least two studies have found that appeals to justice for other creatures, rather than to the value of biodiversity, generated strong support (Susan Clayton, 2000: 459–62; Michael Gunter, 2004: 16). Appeals to axiomatic principles of justice also allow people to read into them specific attributes of justice that resonate most strongly with them (McAdam, 2003: 276). Not all groups or individuals within them will accept this extension of justice principles to the natural world.

Sometimes a worldview contains axioms that are directly in conflict with each other. Thus, some (US or Canadian) groups simultaneously hold people ought to be able to live wherever they want including wild places, and also hold that wild Nature is essential to human freedom and the unique American or Canadian character. Such inconsistency rarely troubles most people, but messages that cause people to make choices between the two axioms will bring the conflict into consciousness and this can be a source of resentment against the source of message. But it also causes some to consider their priorities and leads to mobilization or abandoning opposition to conservation.

There is often no choice but to challenge an axiom held by a group because the group is important to mobilize and the action sought from the group runs contrary to an important principle. To displace an axiom or diminish its influence is not easy. They are usually as well entrenched and resistant to change as the sacred, but they do evolve and most adherents recognize that axioms change and do not consider such change a challenge to the sacred. When an axiom is challenged by calls to mobilize, those messages must do several things to be effective: they must explicitly appeal to the group's sacred, they must *make plain the ways in which the existing axiom is failing to explain the world and give good guidance* and in particular demonstrate that the axiom no longer serves the purposes of the sacred, and a new axiom must be offered (Rogers M. Smith, 2003: 155). Thus, the view that Nature must be subdued and conquered has lead to the destruction of life-support systems diminishing the quality of human life and needs to be replaced with one that directs people to find solutions that sustain most natural systems free from the effects of human degradation. Or, more satisfaction and fulfillment is to be found in Nature than in dominating Nature.

If messages can build on the existing understandings, values, and rules that govern a group's daily behavior so much the better. But a wide range of views and rules governing daily behavior either rationalize or encourage ecologically destructive action, for example, having the latest toy is good, or destroying "inconvenient species" will have no negative ecological consequences. To obtain mobilization in support of conservation goals will require changing such views or diminishing their salience.

When a message calls for action contrary to day-to-day views or rules it must appeal to the sacred and axiomatic *and* it must demonstrate how the day-to-day is failing and provide an attractive alternative. People will cling to old views even in the face of reality and some pain to themselves. A path forward must be offered along with practical advice for navigating it. Justifications must also be provided for the new view that makes it reasonable, proper, and advantageous. For example, messages encouraging protection of predators comes up against long-standing views of them as problems – views heavily invested with habit and fear. As with axioms, sometimes a view or rule not applied to conservation can be invoked. Imagine someone decided you were inconvenient or in the way, or wanted to take your home because they could make a bundle from it? It wouldn't be fair. It's not fair when we do it to other creatures. And we don't need to. Here's why; here's how? One of the deciding factors in the success of messages that must overcome existing views is the appeal of the stories used to convey the messages (see Chapter 13).

The immediate and local

To result in mobilization messages must be of immediate and local interest. This doesn't mean people cannot be mobilized in support of action addressing matter in places distant or well into the future, it simply means the matters at issue must be brought home both emotionally and cognitively by the message. An analysis of international conservation effort found that successful appeals did just this by stressing justice for Nature and wildlife rather than the concept of biodiversity protection (Michael Gunter's, 2004: 16). Concepts of justice, as noted above, are strongly cathected. But appeals to justice are obviously inadequate by themselves as people mostly ignore injustice distant in place and time. The challenge for conservationists is to first catalyze emotional connections and then provide explanations that make sense of them. Sometimes the explanations help evoke feelings as well; they always reinforce them.

Although a mugging down the block will usually have greater impact than the sinking of an ocean liner with all passengers on the other side of the world, people do act in response to nominally distant events when they are affected by them. There are many social processes at work that push groups to see (and experience) the larger scale as local. Conservation messages can build on the consequences of these processes by finding ways to link other changes in scale-thinking to conservation (Ronald Aminzade and Doug McAdam, 2001: 31).

The economic forces that make conservation necessary are driving home the view of many that economic decisions in New York or central Asia affect people in Panama and Tanzania (Alfred Crosby, 1986; Sing Chew, 2001; Michael Williams, 2002) and that such decisions are not in their interest. They might agree with Thomas Jefferson (1992 [1817]) who wrote that "Merchants have no country. The mere spot they stand on does not constitute so strong an attachment as that from which they draw their gain." Although the same economic system that injures people also injures wild creatures and places the connections have not been used extensively in conservation mobilization. A few groups, focusing on the economic factors driving climate change, have raised the issue with regard to arctic species. Many groups speak of the threats to species posed by climate change, but few fully address the sources of the problem at the scale of the problem. Certainly some NGOs do not want to offend supporters of globalization.

Some messages don't do a good job of making the problem (e.g., climate change) appear that action can make a difference, which detracts from the problem's immediacy. When messages ask for money and provide little guidance for action or a means of becoming personally involved there is little inclination to see an issue as local and urgent. There is no mention of steps people can take to help create habitat linkages (attend a lobby day, pull fences), raise energy standards (write your legislator or volunteer for his/her campaign), and slow trade with countries that are building coal-fired power plants (boycott Wal-Mart).

The large-scale forces generating many conservation problems do not always manifest locally but messages can hit audiences with local immediacy when the linkages are made, provided the emotional response to the problem is there. In crafting the message elements calling people to action, it is helpful to keep some things in mind. The conservation movement cannot imitate globalizing agents. Global thinking and institutions require "a simplification too extreme and oppressive" to have ecological merit (Wendell Berry, 1993: 19). The simplification and reductionism imposed by governments and business works for them because their task is simple: turn Nature into commodities for a profit and exercise control over people and territory. Their approach also creates unintended consequences, hence the constant state of crisis. Protecting ecological integrity is not amenable to such reductionism because it must take account of the complexity and uniqueness of place. A conservation strategy based on increased interaction and coordination of local, regional, national, and continental organizations is the only viable alternative. This organization form can bring great weight to bear but keep goals and messages grounded. In any event nation-states remain primary actors in globalization. Global businesses remain headquartered in particular nations and rely on national power. Mobilizing to pressure national governments to change their behavior, business behavior, and international regime behavior is still the primary path of influence.

Observers have argued that a global green culture is emerging (Daniel Deudney, 1993: 300) and an element of this is the recognition of interconnection making the distant local. Certainly this culture does not now include most of the human population so is not a basis for mobilizing most groups. The developing content of this emergent culture is contested. Deudney believes it could replace nationalism among the more cosmopolitan, influence nationalism by making it more eco-friendly, or be captured by nationalism in support of sovereign prerogatives to destroy biodiversity. It could also be captured by nationalism to serve the interests of evolving global business arrangements. One thing is clear: the mobility of capital and its insatiable appetite for "resources" will cause it to seek those resources where the opposition is weakest. Conservationists must be able to mobilize to shore up the weak spots or much will be lost.

Some cognitive (almost) universals

Brain organization and perceptual processes predispose most people to interpret events in particular ways. Variation exists in these predispositions but not to the degree, for example, it does with notions of the sacred. Events are seen by most people as purposive (Deborah Keleman, 1999: 278–9, 289). Humans routinely project their own intentionality onto all sorts of events and processes, from the weather and evolution to the failure of machines. Imputing purposiveness and intentionality to natural events may not reflect reality but it is probably so ubiquitous a trait because it engenders responses that are usually adaptive. Sacred propositions and cosmological axioms impute purpose to the universe or divinities. Language defaults to intent. It is difficult even for scientists to find words that convey function without implying purpose, so genes are "wise" or "selfish." Nature doesn't strike back at humans for their misdeeds, like some angry god, although a grizzly bear might attack someone stealing its huckleberries. Yet, something akin to what the US Central Intelligence Agency terms blowback exists in the world: human actions affecting Nature in turn affect humans. Killing all the predators causes herbivore populations to explode. Herbicide and pesticide runoff kills fish that people like to catch and eat. The use of purposive language in messages if not literally accurate is metaphorically accurate; it is also required if listeners are going to respond.

Humans are also predisposed to explain events in personal rather than structural terms (Murray Edelman, 1988, 1964). Thus, Napoleon's victories and defeats are seen to reflect his personal ability, not the relative wealth and institutional strength of the various governments involved, their ability to support and field armies, and the technology available to each. Likewise, the ongoing failure of a series of leaders to protect species is explained by their lack of caring, lack of knowledge, or bad advice, rather than pressure from powerful constituencies whose interests conflict with conservation goals. Individuals do make a difference because they do make choices, but their choices are constrained. Neither Clinton nor the second Bush is a conservationist. Their differing conservation records are mostly explainable by the coalitions that put and kept them in power. In the case of Bush this included extractive industries and "wise use" groups; Clinton's coalition included conservationists and they achieved some victories but not at a cost to more powerful groups in the coalition that favored economic growth, for example, Clinton gave the store away to gain NAFTA approval. The national monuments Clinton declared at the end of his presidency were about establishing a legacy, not conservation. Had roadless area protection been a Clinton passion it would not have come at the very end of his second term.

It often takes formal education to develop the capacity to see structural factors. Millennia of living as hunter-gatherers in small societies wherein everyone was known would not have selected for structural insight either biologically or culturally. Contemporaneous efforts to obscure structure and discourage critical thinking also blunt the ability to think in structural terms. The human capacity for insight and to play hunches is also oriented to the personal (Gigerenzer, 2007); structural analysis of mass societies consisting of tens and hundreds of millions of people and a vast array of interlocking and conflicting institutions is a conscious process.

Mobilizing most audiences does not require turning them into strategists, but it is often important that they understand why institutional change is necessary rather than just a change of leadership or policy. Even in these situations it is often necessary to personify institutions with a leader's face in order to arouse the passion necessary for mobilization.

In piercing the veil of the purposive and personal it is important to understand that both predispositions make people feel safer. If the many lethal forces loose in the world are impersonal and random it is much more frightening than if they have purposes that can be placated through ritual or other actions. Bereft of these "opportunities" to bargain with powerful forces many people feel powerless and are inclined to passivity.

It is not surprising, then, that when the risks of political involvement are significant, many people need to believe larger forces are on their side and that victory is inevitable (Georges Sorel (1961 [1907]: 26–56). It is important that such beliefs do not interfere too much with concrete and practical analysis of the present and near term, or disaster may result. It is also common for those working for major change and taking risks to exaggerate what's achievable and to discount the enormous odds they are up against (Raymond DeYoung, 2000: 520; Evans, 2001: 121–2). At the very least a strong sense of efficacy is important to mobilization according to several studies (Stephen Kaplan, 2000: 505; Raymond DeYoung, 2000: 517; Elisabeth Kals and Heidi Ittner, 2003: 146). People are more responsive, for example, to appeals that they can make a difference, are competent, and can solve problems, than to appeals asking them to sacrifice for Nature. There are exceptions, of course. Active supporters of the El Salvadoran insurgency were moved by outrage at routinized atrocities and believed that in order to maintain their humanity they had to act although no victory was in sight and the risks were great (Elisabeth Wood, 2001: 268–72).

The better messages take all of these predispositions into account – along with group-specific attributes – the more likely they will resonate. Leaders and strategists must be careful not to let themselves get caught up in the distortions these predispositions lend themselves to. Indeed, those successful movement leaders who have also been theoreticians readily abandoned their theories in the face of a nonconforming reality (Murray Edelman, 1988: 52). This is not to say those who are delusional cannot take and hold power; it happens too frequently, but they do not realize their aims.

Messages are more effective when they create situations in which characters are rewarded for favored beliefs and receive disapproval for contrary beliefs (as with behavior, noted above) (Bator, 1989: 536). One message which followed these rules was nonetheless a failure because it showed the majority of people rejecting the favored belief, that is, they weren't responsive to the disapproval (Robert Caldini, 2001: 281).

Messages need to take into account that many targets of mobilization will look for excuses not to act. Opponents of conservation know this. Business consultant Philip Lesley advises clients that there is no need for a clear victory in the public debate (Robert Cox, 2006: 344–5). It is enough to create doubt in the public mind because this short-circuits their motivation. It is also much easier to raise doubts that prove an opponent wrong. Those who rely on this approach know they cannot stop the eventual emergence of truth, but as with climate change, they can delay action for decades.

Messages and the whole person

Throughout this discussion of messages it has been impossible to entirely separate the realms of need, emotion, and the cognitive. Need-states trigger emotions and are subject to interpretation by cognitive brain centers. The power of sacred propositions proximately rests on emotional investment in them and the innate need for propriety. The intertwining of needs, emotions, and the cognitive in individual and culture is best captured by the notion of identity. When an individual asks Who am I or a group asks Who are we, the answer they give constitutes the conscious aspects of identity. Identity is not monolithic but full of ambiguities and contradictions. Some inconsistencies are striking as when a conquered people adopt the divinities and sacred of their conquerors while also keeping their old divinities in order to hedge their bets (Elizabeth Barber and Paul Barber, 2004: 62–3). More often ambiguities and inconsistencies are difficult to discern.

That identity is not a seamless whole does not diminish its central role in mobilization. Mobilization taps into existing identities and refashions them by asserting and deploying old and new aspects in action (Charles Tilly, 1999: 262–9).

Conservation mobilization requires touching and especially deepening people's sense of connection with the natural world, making it a much more salient part of their identity; this includes providing a compelling explanation of purpose.

The relationship of messages in time

If people are going to answer calls to action to address an issue or solve a problem they need to recognize that there is a problem and they must care about it. Some critics believe conservationists talk too much about what's wrong and not enough about how to fix it (William McGuire, 1989: 64), but few people are aware of the extinction crisis or understand its implications. Awareness must be increased through messages before people become receptive to calls to action. One message cannot simultaneously do both in most instances. Public relations practitioners think that the great success of the designated driver program was due to the years of groundwork done by Mothers Against Drunk Driving (MADD) (Jay Winsten and William DeJong, 2001: 293). People understood drunk driving was a serious problem and were ready to hear about practical solutions.

In contrast to drunk driving, few people in the developed world where millions are enamored of their cars and trucks, recognize that roads are an enormous biological disaster. Vehicles directly kill countless animals annually, provide access to poachers, allow movement of exotic species and pollution, fragment habitat, destabilize soils and hillsides, and increase the sediment load of streams. Before groups can be mobilized to support dramatic changes in road policy it must be clear to them that roads are a serious problem. Mobilizing people in support of large-scale conservation represents a similar but greater challenge. Before people become receptive to calls to action in support of large-scale conservation solutions they must see that the threats to wildlife and ecosystems exist at a large scale, usually the result of large-scale human activity. They must also see large landscapes as something tangible and meaningful, not just as a hodgepodge of smaller regions. Support does not necessarily hinge on people or groups identifying with a large region as a whole; it is enough that they see that to protect the parts they care about they must protect the whole.

Even on an issue where there is much support for conservation – over 49% of Americans want more forestland protected as wilderness – a poor understanding of the issue can be an obstacle to action. Over half of Americans surveyed don't know how much forest is protected and one-third overestimate how much is protected (Campaign for America's Wilderness, 2003: 8, 23). If people knew how little was actually protected as wilderness it could serve as a basis for higher levels of mobilization in support of wilderness legislation and the 30% that think there's enough protected might no longer agree with that.

Most mobilization, then, requires at least a two-step approach. Conservationists must first sensitize groups to the existence of a problem and evoke caring. Proposing solutions to unrecognized problems or providing answers to questions people aren't asking is not very effective.

Much else rides on the initial messages raising awareness of a problem. As noted repeatedly above, how the question is stated or framed is crucial to how people see the problem and how they respond. Defining or framing the issue or problem makes all the difference. Thus, when proponents of California's "Big Green" initiative framed it as a vote about whether citizens and the state should pursue sustainability there was broad support. When opponents of the measure, mostly business interests, joined the election fight they successfully reframed the issue as whether big government would dictate how people ought to live their lives (Ronald Libby, 1999: 89–127). Other factors also played a role in the measure's defeat after initial strong support, but this was a major factor. Problem definition messages require the same attention – if not more – than messages about proposed solutions. Message competition and the need to achieve saturation levels of repetition are much more important at this stage.

Messages that frame a problem and those that propose solutions need not be physically separate messages. But framing and calls to action in support of a solution are distinct tasks and in most mobilization efforts some staging in necessary. When time is limited – as when additional resources need to be mobilized to meet a legislative deadline or raise the ante on a boycott before a board meeting – there is often no choice but to combine them or to simultaneously run messages that address each aspect.

Because the targets of mobilization are at levels of awareness and understanding of an issue, and may for other reasons take shorter or longer periods of time to gain sufficient awareness and understanding to act, consideration should be given to a separate temporal sequence of messages for each. This is particularly critical if the goal is to have a wide range of groups prepared to act simultaneously by a specific deadline.

A common mantra in outreach is that repetition is important. It is. Day after day as commuters head into work they need to hear the message, until they start to hum the "conservation music" in their heads throughout the day. For complicated issues – even once distilled down to basics – this presents a challenge. Messages addressing the various aspects of the issue will need to be sequenced and each message repeated for a time.

Adaptive messages

The importance of keeping on message is often stressed as critical to success, but this is true only if the message resonates. As anyone who has stood before an audience knows, communication is not a one-way street. A presenter's ability to key off the response of an audience, to make adjustments, and to take risks are also important. Mass media don't offer the immediately accessible feedback that public speaking or more personal interaction does. But there is just as great a need to know and to make adjustments in message and presentation in all media. Changes in mass media messages should not be dictated by anecdote; adjustments should be thoughtful, although there is a role for intuition and hunches here, just as there is for the speaker responding to a crowd.

Audience response is not the only factor that calls for message adjustment. Opponents' messages and actions can change the context of a mobilization effort. So can events and social processes, large and small, such as the rise and fall in unemployment or prices, the outbreak of war, the outcome of an election, shifting demographics that reach a transformational point, a hiker attacked by a feral property rights activist or an old and sick mountain lion. During the struggle against Apartheid in South Africa several steps taken by the regime called for message adjustment (Anthony Marx, 1992). When the state sought to divide people by imposing strict racial categories the antiapartheid movement messages shifted to stress what Blacks, Asians, and mixed "race" people shared in common. When the state felt flush enough to offer material concessions to Black unions hoping to create a labor aristocracy that would moderate the labor movement, the unions used the openness to mobilize new members in many sectors of the economy. Long-term demographic shifts from countryside to cities, factories and mines created new organizing opportunities that required new messages. Messages must be adjusted, often rapidly, in response to such changes.

13 Message as story and symbol

In the United States alone over 40,000 new ads make their appearance each year and they constitute only part of the communications noise that includes TV and radio programming, E-mail and web, journalism, book and magazine publishing, movies, concerts, and much else. Rising above this incessant noisiness is rarely achieved on the basis of message alone. Nor is money simply the answer; it can buy media time, the media itself, and the talent that is important in political and commercial marketing, but it is no guarantee of success. In authoritarian and democratic regimes story and the elements that make it effective are central to the outcome of struggles for power and over policy (Murray Edelman, 1988: 90; Renee Bator, 2000: 531–5; William McGuire, 2001: 27). In US elections an unappealing story propagated with millions of dollars does no good; a good story with the same money – or even a little less – can carry the day (Evan Cornog, 2004: 115–6, 5). Stories are the primary vehicles for messages. A compelling story credibly propagated matters more than personal fitness for office or a candidate's positions on issues. The historical record demonstrates, in Cornog's view, that a good story trumps the truth every time.

Story is central to the success and failure of social movements, including efforts to undermine them (Polletta, 2006). Those who challenge the powerful tell stories of wrongs righted, vengeance, and triumph. The powerful tell stories to their minions that the path to success in an individual one and that punishment awaits those who engage in collective efforts to change policy. More simply put, a "list of issues does not stand up to a story with villains and heroes" (Polletta, 2006: viii).

The importance of story rests on its central role in people's lives (see Chapter 4). People understand the world and themselves primarily through stories, not scientific treatises or philosophical texts. Groups express their shared understanding of the world in story. Shared stories bind groups, societies, and civilizations giving pleasure and voice to interests, peoplehood, and values (Joanna Overing, 1997: 2, 12; George Schopflin, 1997: 20; Susan Grant, 1997: 88–98; Smith, 2003: 59–64). Stories weave together systematized and tested knowledge, beliefs, values, and meanings into a worldview that gives coherence to events and purpose. Stories bring together the disparate and contradictory aspects of the world and personality, fashioning functional wholes at social, group, and individual level (Joseph Campbell, 1959: 149–50, 467, 1988: 37–67; Smith, 1993: 12). The sacred is primarily conveyed through story, as are cosmological axioms and the many rules and understandings that govern daily life. Apart from the innate pleasure story gives and its capacity to relieve anxiety by making sense of things, story binds people because it provides role models to them, making it easier to find themselves in a story and gain guidance from it. They have a script.

Story is not merely shaped by our experience but shapes our experience and memory of experience (Elizabeth Barber and Paul Barber, 2004: 91–3; Carol Tavris and Elliot Aronson, 2007: 76–7). Stories can successfully break through stereotypes, as when a jury, following a judge's instructions, considers the life experience of a battered woman who stands accused of killing her abuser (Polletta, 2006: 125–6, 169). Dominant stories, of course, reinforce existing patterns of thought and experience, but the less powerful have their stories of resistance.

Stories are not the only means by which important cultural information is conveyed, nor are stories always the best or appropriate means to achieve mobilization or influence policy. In German society stories are generally given less credibility than "rational analysis," whereas in the US stories are generally accorded high credibility and taken to reflect reality (Francesca Polletta, 2006: 137, 107). In some formal US settings, however, story is considered inappropriate. Despite cultural differences story's role is pervasive in all cultures. If given a choice most will watch a film over reading a textbook because stories offer greater pleasure.

Because critical faculties are relaxed during pleasant experiences stories are ideal vehicles for misleading people – hence the equation of the term myth with falseness or telling a story with telling a lie. Factors other than pleasure are at work in the capacity of story to mislead. People are disposed to ignore the disconnect between reality and a good story if reality (e.g., in the form of political candidates) has disappointed them too often (Ira Chernus, 2004). Story also entices people by enabling them to live vicariously through the powerful and famous (Murray Edelman, 1964, 1988).

Stories may be used in a variety of ways in mobilization. Political pronouncements, most journalism, and even scientific argument rely on or are contextualized by reference to stories. Scientific articles on the decline of biodiversity often reference the larger causes of habitat loss by referencing the story of human conquest. Stories are used within other forms of communication to illustrate a point, or as a way to convey examples or connect with an audience. Public speakers first talk about their own story as a way of evoking audience identification with them before launching into the substance of their talk. When a political leader gives a speech the setting (before a flag or monument) becomes part of the story by adding to the status of the presenter.

Many types of communication rely on the basic structure of story: a clear beginning (problem), middle (working it through), and end (resolution). Journalism relies heavily on the story form with significant consequences (Polletta, 2006: 181–2).

If a news story begins with a cougar sighting near a house the cougar appears to be the problem. If the story begins with human incursion into previously wild areas populated by cougars then it's easier to see that humans initiated the problem. Political and economic leaders play into the media's tendency to present a problem only if they can end the story with a solution or reassurance, for example, we've learned from this and it won't happen again (Murray Edelman, 1964: 16–8, 57–8). Reality is usually different: the problem has not been solved, will not go away, or has not been addressed in the way that most audiences would prefer. Journalism relies heavily on another feature of narrative: the use of familiar and evocative stock characters such as small town workers, independent ranchers, environmentalist do-gooders, or extremists (Robert Cox, 2006: 188).

Politics as competing stories

Political campaigns including mobilization do not so much pit story against other forms – Carter preached and Reagan told stories, conservation scientists lecture and opponents spin yarns – as story against story. In one story conservationist Davids struggle to defend Nature against brutal Goliaths; in the competing story they are well-to-do elitists seeking to impose limits on freedom and deprive people of their livelihoods. Political leaders paint themselves as modest heroes in a compelling narrative and their opponents as unappealing characters (W Bennett, 1980: 168). Those battling to increase logging of ancient forests cast themselves as caring and sensible public servants desirous of protecting people from fire. Conservationists cast the Healthy Forest Initiative as a subterfuge perpetrated by unreformed timber interests and their hirelings in government to skirt existing laws prohibiting logging in roadless areas.

Many factors contribute to which story prevails with a group and shapes its decisions about mobilization. The ability to saturate an audience with a message helps, especially when the other side cannot respond in kind. US tobacco companies spent $40 million on an ad campaign to defeat legislation taxing them by casting the bill as creating more big government; the opposition could not remotely match their spending. Messenger and channel are important. When the mass media are an important channel much depends on which story they adopt to frame the issue. When the media plays lapdog rather than watchdog conservationists have a tougher time. When the Associated Press spread a false story that lynx researchers had fabricated evidence, they refused to issue a correction, stating it was old news and would be too difficult to correct (Michael Williams, 2002: 31). Because of the pivotal role of the media in reaching many audiences, and the difficulty in going around the media – as some have advised and done (Richard Viguerie and David Franke, 2004) – they must be considered a critical target audience when evaluating whether a story is compelling enough to carry the day. If the media don't pick up a story and at least treat it sympathetically it will not reach the intended audiences with the needed effect.

Several factors make one story more likely to be well received by an audience than another:

- its fit with the audience's mythology and repertoire of other stories;
- the context provided by current events;
- genre appropriateness;
- simplicity;
- appeal of characters and plot;
- its capacity to strongly evoke important emotions through shock, drama, vividness, and symbols.

A story that fits

When the American landscape painter Charlie Russell was introduced as an artistic "pioneer" before the Great Falls Montana Booster Club he took offense. He was not a pioneer, he said (Charlie Russell, 1923). Pioneers are people who come to a wild place and poison the water, cut down the trees, drive off wildlife, put up barbed wire fences, and call it progress. His blunt words were not well received because his message ran contrary to booster ideology and it challenged the dominant story of the brave pioneer who brought civilization to a savage land. Russell offered neither an alternative story nor did he try to situate his statements in the context of the mythology accepted by his audience. His message was easy to dismiss on these counts. In contrast, from the 1790s on US abolitionists, and later the US civil rights movement, cast many claims in the language and story of the US Declaration of Independence with its promise of equality. Framing their claims did not work magic but it helped to build both movements. Could Charlie Russell have done something similar? How? Keep reading.

Fit and myth

More than message content must resonate with a target audiences' emotional makeup and worldview. The message package – primarily story – must also resonate. Successful stories, like successful messages, must make use of much that is familiar and understood to an audience. When they run contrary to existing views and feelings they need to justify themselves with appeals to the sacred, cosmological, and associated emotions. If to convey their mobilization message, conservationists can use existing stories or familiar themes and characters in new or reworked stories, so much the better. If they cannot, then new stories must be explicitly nested in a group's higher order stories, that is, those stories that convey the sacred and cosmological. Unless messages are conveyed in stories, which are part of the existing repertoire of lower order stories or nested within the context of existing myths, they will be ignored or rejected. Stories in conflict with a group's mythology will be seen as threatening or not understood.

Myth is the term anthropologists give to narratives about the sacred and cosmological. (Recall that the sacred legitimizes, but it is from legitimized cosmo- logical axioms that lower order rules are derived and explained.) Myths are the overarching narratives explaining the origins and order of the world and a people, the purpose of life, what must be defended at the cost of one's life, and much else. The narrative aspect of myth is important because it is this which links the present to the beginning time, anchors purposes and meaning in the order of the universe, and so provides a compelling understanding of the world that mere statements cannot provide (Percy Cohen, 1969; Roy Rappaport, 1999). Philosophical and ana- lytical treatises have always played a poor second to myth even as they rely on mythic structures and principles enshrined in myth. This is partly so because story as story elicits devotion, not just the meaning it conveys. It is story, not just its informational content, that is internalized, becomes familiar and the object of intense emotional investment.

In his study of US presidential campaigns Columbia Journalism Dean Evan Cornog (2004: 51) finds that the candidate who best makes use of the characters, themes, story lines, morals, characters, and overarching patterns of meaning embodied in myth, or stories sanctified by myth wins the election. The combination of familiar structures, plots, purposes, and characters into coherent wholes triggers positive emotions and simultaneously rationalizes them.

As with the sacred and cosmological, the myths or master narratives that convey them change slowly and change is usually imperceptible to believers. Personal and social crisis can produce rapid and visible change (conversion, loss of faith), but change is otherwise generational or associated with the life cycle. This is why polit- ical practitioners and marketers aim to mobilize people with variations on existing stories that fit within existing mythology. When they cannot they make appeals directly to the audience's myths. How conservationists might influence the evolu- tion of myth is discussed below. First, back to Charlie Russell and how he might have conveyed his message in a manner less likely to trigger quick rejection by boosters.

Russell had two options. He could have framed his statements within the tradition of Thoreau, Emerson, Teddy Roosevelt, and those who protected Yosemite and Yellowstone. Although not the dominant cultural tendency, framing his statements in terms of the story of the importance of wilderness to the American character and the grandeur of its landscape to American pride would have made it more difficult to dismiss them outright. To an audience other than boosters this approach could have struck responsive chords.

Russell's second option would have been to frame his views within the most sweeping American myth: that Americans are a chosen people with a special mission to create a civilization that sets a standard for the world to emulate (Richard Hughes, 2003). This myth has several variations dominant among different groups at different times, but American exceptionalism remains the centerpiece. The version most amenable to nesting his message in is the earliest, promulgated by the Puritans themselves. Although Puritan views of wilderness and other aspects of their culture were far from Nature friendly their notion of right living included caring for others and laboring and suffering as well as rejoicing with them. They were to bring the world along by setting an example of right living, not through conquest (which appeared in later variations of the myth). Russell and conservationists today could find good ground to stand on in this version of the myth, stripped of its religiosity and narcissism, and generalized to the entire community of living things. Indeed, some religious groups in the late 20th century have come to exactly this place. Similar to abolitionist appeals to the US Declaration of Independence, Russell's statements couched in this story would not have mobilized millions overnight. But his statements would have resonated with those less narrowly self-interested in the domination of Nature. The idea that much of humanity has lost its way by destroying Nature and that Americans by their good fortune live with a relatively undiminished natural heritage and can choose a more enlightened path has a familiar structure. It speaks to specialness and showing the way by reembracing the lost mission of caretaking.

The downside of any appeal to national exceptionalism is that it reinforces an imperial mentality that easily and quickly deteriorates into conquest when the means for conquest exist and patience grows thin waiting for others to see the light of one's example.

There are elements common to all mythological traditions that offer conservationists a means of nesting their stories. Myths invariably concern themselves with how to fix things that have gone very wrong, such as how to stop destruction and begin healing (Mircea Eliade, 1991: 5). Although many groups do not see the conversion of the natural world as destructive in itself, there exists increasing recognition that something is wrong with the human relationship with the Earth, if only because of the consequences for humans. Conservation stories that link to this perception and mythic notion of healing will move many people. It also falls to myth to explain death (G. Kirk, 1970: 79). The approach to Nature of the conquistador necessitates the denial of death. This puts humans in the same place as the slaveholder, wherein unease and fear of retribution permeates the psyche. Livelihood based on extinguishing whole species and ecosystems makes people aliens in the land, never quite at home; hence the itch of the pioneer. Conservation can offer a more satisfying understanding of death by reconnecting it to the larger patterns of life and death in the natural world. This variation explaining death's meaning will not erode dominant meanings overnight, but setting out an alternative that addresses a mythic concern, especially if it fits with other aspects of a mythology, can be attractive if it provides solace, a better understanding, and is more attuned to changing sensibilities.

Indeed, myths are like the sacred concepts they convey – usually sufficiently general that different interpretations are possible and one aspect can be emphasized over another as being especially relevant to an issue with new salience, such as conservation. This allows stories that convey lower order rules for behavior, like the rules themselves, to evolve or be replaced without calling myths themselves into question. Groups and societies, however, are not equally open to changed or new stories even if compatible with dominant myths. Indeed, differences of interpretation *within* a mythological tradition can inspire more rancor and nastiness than between traditions. The heretic is more severely condemned than the infidel.

Myth, society, and group

That the same myth may give rise to divergent interpretations is historically obvious; clashes within traditions are ubiquitous. Understanding the relationship between mythological traditions, cultures, and subcultures enables conservationists to craft their stories more effectively. Although all groups within a tradition share the same overarching myth and many lower order stories, cultures, and subcultures within the tradition differ in mythological interpretations and emphases. These differences define in part define the identity of (sub)cultures and are more important in mobilization than the commonalities. Thus Christians share a dominant story (the New Testament) but there are countless sects and the distinctive attributes of each are usually prominent in the group's mind. The story of the evolution of democracy in the west admits to similar variations of interpretation.

Geographic mobility and mass media are undercutting many cultural and subcultural distinctions, but in response groups often cling to their separateness all the more. Groups still experience life differently and give meaning to that experience by drawing upon distinctive interpretations of myths that are shared with others.

Most cultural groups have institutions – if only informal – that are devoted to maintaining the integrity of the group's stories. Group leaders keep group stories alive (and evolving) by using them in their capacity as leaders. Leaders are valuable informants about their group's stories and other information useful to those seeking to appeal to the group via story. Whether leaders will act as informants is another matter. They will have their own agenda, an interest in withholding information, and perhaps in being a gatekeeper.

If a group has a name it is likely that their (sub)culture, including stories, has been studied by anthropologists, marketers, or others, meaning that information exists that is useful in crafting messages and stories. Group leaders or cultural guardians may also have documented the group's stories. Some of this information will be published; other information will be held closely as proprietary (see Chapter 10).

Parties to a political conflict express their differing interpretations of events through different stories crafted to appeal to groups they are seeking to mobilize. These stories may come from the group being mobilized, they may be stock stories shared by the larger culture, or they may be new stories created by those doing the mobilizing. John Kerry's (2004) presidential campaign portrayed him as a war hero – someone who had faced adversity, made tough decisions, and came through it with a strong character. It seemed a good choice because many American groups have supported war heroes for president, for example, Grant, Eisenhower, and Kennedy (Evan Cornog, 2004: 25). Kerry was running against a president that had just started a war, but had also avoided combat service in an earlier war. Kerry's war record was called into question by supporters of the president with competing stories. These stories portraying Kerry as a fraud were not necessarily believed but they cast enough doubt that Kerry's story as hero was ineffectual with many. The lack of a decisive response hurt Kerry but more decisive was the Bush campaign's successful casting of their man as a steady, god-fearing statesman and cowboy avenger fearlessly battling evil (Ira Chernus, 2004). Kerry had trouble settling on a single story while the Bush campaign kept its focus on one story.

Competing stories are common in most conservation conflicts. In many parts of the world where predators remain or ought to be repatriated they are the subject of many negative stories, portraying them as cruel, destructive, and avaricious – the enemies of safety, order, progress, and wealth. These stories are indigenous to many rural regions and reflect real if exaggerated fears about threats to livestock. Some stories are holdovers from an earlier period or reflect broader cultural values, for example, taming Nature. Anticonservation forces can easily tap into these and need not create new stories. Conservationists will not generally find usable stories directly on point in the repertoire of groups in these areas. But conservationists will find other stories with characters, plot lines, and parallel themes that can be useful. A story about injustice suffered by the group at the hands of the more powerful – a common theme among rural audiences – may be reworked to apply to predators, making them another group persecuted by the powerful. Because rural groups often see predators as competitors it may be useful to rely on reworked labor movement stories that show workers they are not competitors for jobs but people with a common purpose. In the Pacific Northwest one nongovernmental organization (NGO) used story frames about pride (this is a special place and grizzlies are part of what makes it special; they are important to the region's ecology), about self-sufficiency (we know how to live with bears), and that affirm safety (there are easy, practical steps that make living with bears safe) (Chris Morgan et al., 2004; GBOP, 2007). Story propagation relied on local messengers who could add their own effective gloss.

Precise targeting is sometimes not possible because the means of communication or available resources do not allow it. In those instances stories or aspects of stories more widely shared should be relied upon even though they are usually not the most salient with individual groups.

Stories and change

Stories, like the rules and views they convey, change for similar reasons and in similar ways. Myths change imperceptibly absent crisis. Lower order stories change when the information they convey no longer works well and alternatives exist; they change in response to changes in cultural sensibilities – the information may remain good but the vehicle has become unappealing; and stories change as a result of human playfulness. Conservationists can utilize all of these mechanisms to create appealing stories to carry their messages.

People are receptive to new stories when old ones don't explain the world well or no longer offer good guidance. But there is a caveat. Groups may continue to keep stories alive if they like them because of their other qualities; for example, if they are dramatically appealing or offer reassurance because they have long been a part of their lives. Stories that seek to replace it in order to bring a new view of things must be very aesthetically appealing and they must point to the failure of the existing story to provide good guidance. Existing stories may also evolve to incorporate new understandings and rules under the pressure of alternative stories (Karen Johnson-Cartee and Gary Copeland, 2004: 151–2).

For a century in the United States forest fires were viewed by professional tree farmers (e.g., foresters), the timber industry, and rural residents as something to be prevented or suppressed at any cost. Stories told of fire's destructiveness, lost jobs and property, damage to wildlife, and landscape ruin. As a consequence of fire suppression heavy fuel loads built up; combined with other forestry practices such as clear-cutting and planting even-age monocultures, catastrophic fires resulted, killing most trees and sterilizing soils. In regions that have experienced such fires, new and competing stories (conservationist and industry) have emerged as they typically do out of crisis. The struggle continues over which stories will be adopted but the old stories lack credibility. Because fire is generally confined to certain regions it has not created a society-wide crisis that lends itself to adoption of new stories across the society or to change major stories, for example, those having to do with human management capabilities. Climate change will likely create openings for new major stories and conservationists need to be prepared. Even myths and the assumptions they embody are subject to substantial transformation or replacement in a profound crisis.

Crisis, as with lesser change, trigger denial, reaction, reliance on authoritarian leaders, and simplistic ideology in the more rigid (sub)cultures and individuals. Even in more democratic cultures positive change is only likely when attractive, alternative stories are available. The conservation record is not great in this respect. As indicated, inroads have been made in changing fire suppression practice in some parts of the world, but no conservation equivalent of the Smokey Bear campaigns of another era have been created that could bring about the mass mobilization required to pressure agencies and other decision makers to make significant changes. The insider strategy has had limited results. Some smaller NGOs, as part of their forest protection work aimed at gaining support for forest integrity, have created fire programs that help communities protect themselves from fire by thinning and removing fuel loads near settlements.

Changing sensibilities cause old stories to lose appeal and become subject to displacement or evolution. To craft effective stories conservationists need to be attuned to changing trends in what makes characters, plots, dialogue, and other aspects of story appealing to an audience. Observing how electronic media commercials and programming change and which ones are successful provides hints regarding what is working with particular groups. Programming and ads can also drive changes. Both have contributed to making stories faster paced. Many groups now require visuals before they will pay attention – words alone require a concentration they can't summon. Much popular programming reflects ripples on the surface rather than deeper current, but the point is that reviving Smokey to do a new kind of fire duty would require a different bear.

New stories work their way into a group's repertoire as the result of human playfulness and interest in the new. Groups differ in their playfulness and in how much they innovate or are open to innovation. Play in the realm of story (and art generally) mostly involves reconfiguring familiar elements in new ways. These innovations reflect new ways of seeing things by the innovator; when adopted by others innovations change the way they see things (Homer Barnett, 1953). Whether an innovation catches on depends on its usefulness, elegance, aesthetics, and other factors. Minor innovations in stories occur constantly. Major changes to stories, especially higher order stories like myth, tend to originate among cultural workers (Murray Edelman, 1995: 69). Artists, including novelists, filmmakers, painters, and musicians, generate new patterns for organizing experience which become templates adopted in other realms, such as politics and marketing (Murray Edelman, 1995: 39, 52). VanGogh's painting style changed perception throughout society as did the first photo of the blue and cloud-swirled Earth against the darkness of space.

The consequences of innovation are difficult to predict: which groups will embrace a new or changed story, how deeply will they integrate it into their identity, how rapidly will it diffuse, how it will be applied? Goddfrey Reggio's film, *Koyaanisqatsi* (1983) (the Hopi word for "life out of balance"), was a scathing and highly innovative visual critique of modern consumption and inequality. Film critics observed that Reggio's ability to make the substantively ugly appear photogenically beautiful – for example, napalm scorching the Earth – undermined his message (Michael Dempsey, 1989). Indeed, his techniques were quickly adopted by commercial television advertisers. Those entities that perceive changes in story substance and technique earliest and apply those lessons to their stories will be more effective.

An important factor in the acceptance of new or reworked stories is how well they integrate the familiar and the new. Stories must generally begin with familiar characters in familiar situations and once identification is made the story can take the audience to new situations or new outcomes. Most groups will not easily or strongly identify with stories or characters centered on habitat protection or the intrinsic value of predators, for example. But if a story includes these concerns, and is about themes that a group can easily and strongly identify with because they are moving and important, the audience can come to better understand habitat and predator protection. As her mother was dying of cancer Terry Tempest Williams (1991) found some solace at a wildlife refuge. The presence of a wild place and its seeming continuity in the face of loss and accompanying pain helped her keep her balance. When the rising waters of Great Salt Lake threatened the refuge – *her* refuge from her mother's impending loss – the comfort that its continuity had given her was suddenly lost. She came to find a greater continuity in being part of the larger rhythms birth, life and death that included all of Nature, including herself and her mother. Her story has touched many thousands of people who had no profound experience with or connection to the natural world. Through her book they achieved some familiarity with and understanding of the importance of Nature in human life. Many came to understand that the loss of places and other living creatures is a loss like that of anyone close and dear, and even more terrible because it forecloses the opportunity for a connection that has great healing power. Able to find themselves in her personal and universal story of loss and renewal, they discovered the value of connection with Nature.

Stories such as *Refuge* not only link conservation to those things most meaningful to humans such as birth, death, coming together, separation, change, alienation, and identity. They make clear that conservation *is* about these things. It is about individual and collective decisions that result in life and death for individual creatures and whole species. It is about the inevitability of death and even extinction, and about the tragedy and monstrosity of unnecessary death and extinction.

In crafting new stories or reworking with an eye or ear to familiarity it is important to note that stories tend to leave out information widely known and shared within the group (Elizabeth Barber and Paul Barber, 2004: 221). It is an economizing measure. Thus, inclusion of too many obvious details can make the story appear awkward or the work of outsiders – even unfamiliar – lessening its influence.

Characters and roles

A story's influence on behavior depends in large part on an audience's ability to find themselves in the story by means of one or more characters. Myths and the stories derived from them – folktales, proverbs, morality plays of all sorts – contain a variety of characters and roles. It is largely through the characters' actions that purposes, understandings, and rules are expressed. Built into the characters and the roles the characters play are codes of conduct, social expectations, and guidelines for problem solving and achieving fulfillment (Doug McAdams, 1993: 240, 247–8). Typically a character plays many roles – lover, warrior, healer, caregiver, sage, traveler, survivor – but emphasizes one over the others (Doug McAdams, 1993: 122–4). The repertoire of characters available influences both adult behavior and individuals during growth and identity formation, when a personal story is adopted (Doug McAdams, 1993: 102–10, 129–30, 166). Because available stories have a life of their own, the individual cannot alter them at will when adopting one, but must reach an accommodation. Thus aspects of a personal story may clash with other elements of personality – a reminder that personality is not monolithic. Stock characters do change and they are added and deleted from a culture's repertoire; it is also true that in different periods some stock characters and roles are much more prominent. Mobilizing stories enjoy greater success when they tap into identity by means of the salient characters in myth and personal stories. Redefinition of roles and offering new, conservation role models (the redefinition of identity in mobilization) need to be incremental; familiar attributes need to remain.

Mobilizing stories should link the roles embraced by people to the tasks they are asked to perform as a mobilized part of the conservation movement, for example, join us and become a warrior for conservation, care for Nature or others in the movement, put your desire to understand the world to work for conservation. This is as important as a story's aesthetic attributes and provision of purpose. People find themselves in story by way of familiar circumstances, plots, themes and meaning, and by way of familiar and valued characters. Audience identification or sympathy with a character's motivation is especially important in initial mobilization (Polletta, 2006: 51).

Stories crafted for mobilization should also keep in mind that groups are no more monolithic than personalities. Members of groups based on class and ethnic attributes share age and gender attributes with those outside of their group (Adams, 1993: 102, 110, 202–35). Thus, regardless of class, stories aimed at adolescents should take account of their emerging sense of history, ideological development, and decisions about livelihood. Stories will be stronger with class-based elements, but the development issues are the same. People in their 40s are starting to think about their legacy and passing on what they know to the next generation. This role can be enlisted in appeals. Indeed, appeals to leave a conservation legacy can be very influential when conveyed with a compelling story.

Manipulation

It is common in politics everywhere for participants to utilize the stories of a group in ways that are designed to mislead them into supporting or acquiescing in policies they would not otherwise accept. Thus many entities claim that their policies are informed by the best science when these same entities have suppressed or distorted scientific findings. Such misrepresentation is not confined to policies affecting the natural world. Claims can be quite absurd, as when it is stated that mergers in an industry with only a handful of companies will increase competition, but absurdities repeated by an authoritative figure have the same qualities as the emperor's invisible clothes. The appropriation of a group's stories and their use in manipulating the group can be quite effective, but it is not a good strategy for conservationists. Conservation success depends on the ability to refashion policy over the long haul and sustain those changes against the political and economic inertia of the last several millennia. Reliance on misrepresentation, like reliance on propaganda, effectively undermines not just the conservation movement's integrity but also those aspects of the social fabric on which positive change depends: a grasp of reality, the capacity for genuine communication, and well-grounded social trust. Chronic misrepresentation breeds arrogance in perpetrators and dullness, not healthy skepticism, in those who absorb it.

Context

To result in mobilization stories must fit in another way – they must be attuned to the times (war, economic ups and downs). When economic times are good, for example, people tend to be more optimistic and stories that stress taking advantage of opportunities will garner a better response than those which focus on obstacles. During a downturn stories that stress overcoming significant obstacles, provide practical advice on getting through the crisis, and that include some humor along with the drama – especially humor that deprecates opponents or obstacles – will work better. Inspiration is never enough by itself. When conservation leaders or opposition leaders are the subject of mobilization stories their characters and roles must be convincing and timely; only some roles may be credibly worn by a particular leader (Evan Cornog, 2004: 51).

Different venues call for different types of stories. What works in a courtroom is not usually well suited to a legislative hearing; and when courtroom or legislature is fictional as in a film, still different types of stories are appropriate. Stories in a speech attuned to the faithful will not work in a speech to a rally that includes the merely interested and skeptics. Presenter and audience will share values and knowledge in the former case and less so in the latter. Stories presented in support of a policy proposal made to business groups will require a different tone than those included in the opening remarks at a press conference. Thirty-second public-service announcements must rely on the use of icons and other powerful symbols to imply major parts of the story.

Genre appropriateness

If every political context has a limited set of suitable story types, it is also true that stories must conform to the expectations audiences possess for each genre without becoming too predictable. Genre attributes include distinctive story development, types of characters employed, and denouements; they also include typical audience responses. To violate genre rules can make a story ineffective by breaking the spell and inviting disappointment. Myth is usually dramatic; humor and irony have places in drama, but not cuteness or satire. Humorous stories may deal with serious subjects, but if they turn frightening or genuinely dark audiences may become disoriented and withdraw emotionally. When done with skill, however, breaking genre conventions can deliver consciousness inducing shocks or more gently nudge people to insight and action. Edward R Murrow broke with television documentary's rules with *Harvest of Shame* (1960) and his challenge to Senator Joe McCarthy (*See It Now,* 1954). Both programs atypically and fearlessly attacked powerful interests and had enormous mobilizing effects. Chris Freddi's (1983) fictional tales about several animals begin in a manner that leads readers to expect sentimental, anthropomorphic portraits; instead he delivers realism.

Political and other leaders must conform to certain expectations that are genre-like. Some apply within a culture across the board. In the United States the private lives of candidates and office holders – especially their sexual peccadillos – play a major role in shaping their fate. In other countries the private lives of politicos are regarded as irrelevant or are carefully guarded. Candidates are also expected to express their individual styles (within an acceptable range of variation) suitable to the character and role they lay claim to. Were George W Bush to speak as articulately as John F Kennedy it would be out of character, violate expectations, and cause people to doubt they know the person and likely question their support.

Simplicity

The sine qua non of *political* story is its capacity for appealing simplification. Effective political stories are like mainstream Hollywood cinema – superficial, technique-dominated, and stereotypical (Evan Cornog, 2004: 61). Just as Hollywood films poorly capture the complexity and nuance of the great novels they are sometimes adapted from, political stories rarely capture the complexities of life and issues. Most people are content with a simple view and find reassurance in simplicity (Evan Cornog, 2004: 48).

Groups differ in their depth of political understanding and the time they can devote to politics, but most are not interested in the complexity of issues, including conservation. Some conservation NGOs grasp this, and it is reflected in communications that focus on, for example, charismatic species. Other groups do not understand or reject simplification as dumbing-down and a violation of their standards. Scientists in particular have difficulty abandoning complexity, nuance, and uncertainty in communicating with nonscientists. This attitude results in abandonment of the field to those who deliberately misrepresent the world and it ignores the truth observed by social scientists and political practitioners about how people gather information and use it.

There are limits to the information people can effectively manage, whether as providers or receivers (Elizabeth Barber and Paul Barber, 2004: 154). Even in societies where much is written and documented for reference much communication remains oral and economy is important. The norm is simplification in all cultures and involves encoding reality in stories according to each culture's rules. Simplification need not be manipulative; it can be like making a map, which must leave out detail unimportant to the map's purpose (to guide vehicle travelers, geologists, etc.).

The film *Flock of Dodos* (2006) made by PhD marine biologist-turned-filmmaker Randy Olson takes the complex battle between knowledge and belief in the United States and distills it to its basic elements through characters who are sympathetically portrayed regardless of their views on evolution and by treating the controversy with a bit of humor in places. He neither oversimplifies the antagonists' arguments nor demonizes and this has contributed to the film's influence. Olson also understands that documentaries are not exempt from the need to tell a story. It is ironic that storytelling in longer formats such as novels or film often simplifies less than the news or other formats aimed at closely representing reality.

Conservation communication *as story* remains mostly underdeveloped by the conservation movement. There is a large body of Nature writing not directly associated with conservation NGOs or campaigns. "Nature" painting abounds but its subject is pastoral and bucolic, not about the wild. Conservation as the subject of novels, short stories, films, music, plays, and similar art forms is very limited – poetry and photography being exceptions. These forms are largely ignored by the movement even though they collectively and routinely transform cultures over time.

It is not just the politically unsophisticated who seek simplicity. Reporters looking for a story hook understand the value of simplicity and respond to those who offer them a hand, especially if the theme is timely and well thought out (Tim Ahern, 2002: 45–7). The electronic media in particular, given ratings battles and the limited time available to tell a story, seek simplicity in story lines and in stock characters which serve as useful shorthand (Evan Cornog, 2004: 48, 249). This is as true for news as it is for entertainment programming (James Combs and Dan Nimmo, 1993: 78). US agency scientists, chafing under the Bush administration's political distortion, delay, or deep-sixing of their findings, turned to media leaks only to find that many of the issues they wanted to air were regarded as too complicated by reporters. More stories would have been picked up had they been predigested.

Simplicity is not only important in mass communication. Small group and one-on-one interaction rely on story, face time constraints, and benefit from formulations that are concise. Conservationists who recoil at the need for simplicity should keep in mind that conservation's success depends on long-term relationships with many groups. Over the course of those relationships it is possible to tell a more complex story. Not everything need be communicated on the first date.

Simplicity is not without its dangers. Like a road map it must not be confused with the territory it represents. It's a tool – a means of making information understandable within the constraints of time and the human capacity for understanding.

Character and plot

Effective stories emphasize characters and their journeys. A character who sets out on a quest is more interesting to most audiences. Interest is enhanced when the character shares traits with the audience or confronts similar circumstances. Stories that better mirror and tap into salient aspects of an audience's attributes tap into a primary interest of all people – themselves. Stories in which characters' work through internal states that mirror the audience's can help them work through doubts about action and its consequences, making mobilization more likely.

Emphasizing characters in a story makes the story more effective in other ways. When the message emerges from the characters' actions it comes across as less preachy and generates less resistance (Murray Edelman, 1995: 49–50). It is easier to identify with a prominent character(s) than with a mass of people or with institutions. Kings, presidents, and prime ministers know this and use it in their struggles with congresses, parliaments, and other institutions. Stories about characters tend to trump stories that focus on the issues. Because issues are often complex and people don't want to take the time to figure things out it is much easier to look for guidance from a character who appears to know what they are doing. Characters themselves are usually simplified to make their different roles more obvious and the plot clearer.

In the end it is usually a character in the form of a leader that personifies a movement or an NGO, such as David Brower or Martin Luther King Jr. Myths that record the journey or transformation of a single hero – who is nonetheless recognized as a stand in for a whole people or group – reflect this human predisposition (Elizabeth Barber and Paul Barber, 2004: 250). Other characters are also stand-ins for whole groups. This process involves a kind of stereotyping. Although stereotyping causes much grievous mischief it is so prevalent because it is an economical means of creating and manipulating categories. Such shorthand is necessary to living in a complex world.

Because it is largely through plot that a narrative organizes and gives coherence to reality – an element that lies at the heart of story's appeal – plot is important to the success of story. Well-crafted stories help people make sense of the world. Unimportant things are left out, important things have center place, and narrative imparts directionality. The most engaging plots unfold around a character's transformation: the birth and death of a person or a whole people, journeys of maturation and redemption, passage through crisis caused by the loss of loved ones or of meaning. In myth and many less grand stories the main protagonist is often ordinary and so is "everyman." The purpose of stories, after all, is to offer guidance to everyman. Stories aimed at mobilizing most groups will be effective when they take this approach – when the plot features a journey in which ordinary characters rise to meet extraordinary situations and do well. Some people gravitate to stories of the great but they generally imagine the great to have their attributes and successful stories about the great usually portray the great in terms of their intended audience.

The theme of transformation that gives stories purpose is also a central aspect of conservation and this is a happy coincidence. Conservationists need to transform policy, institutional behavior, and more generally the human relationship with the natural world. Although many people are sympathetic to the aims of conservation and have a modicum of empathy for other creatures, they are wary and even frightened of transformation whether it be personal – becoming politically active, changing how much they consume – or societal. They are fearful of where change will lead and of losing the familiar. By entwining larger purpose with attention to personal concerns, stories can assuage fear and free people to act and find fulfillment in action.

Plots should demonstrate that the transformations required by conservation are not about sacrifice and deprivation, but about reconnecting and reestablishing rewarding emotional ties with the natural world, with others, and with parts of one's self. As in *Simplemente Maria*, rewards such as social approval and achieving goals must accrue to those who do the right thing and who are willing to undertake transformation.

Drama and vividness

Several scientists applauded the film *Day After Tomorrow* for raising concern about climate change among many different audiences. They were disturbed, however, by the "implausibility" of events portrayed in the film (Andrew Balmford et al., 2004: 1713). Couldn't Hollywood combine truth with appeal? Probably, but the literal truth would reduce the chances for achieving blockbuster status, which US film production companies seek. This film demonstrates a point anthropologists have long made – that the accuracy of a worldview is not always important to achieving the desired (i.e., adaptive) behavior (Roy Rappaport, 1974, 1969). Many cringe at this (not just scientists), pointing out that inaccurate views lead to (or rationalize) slaughtering tigers to make virility potions, to politically oppressive policies, or to the reduction of adults to frightened children at the mercy of their divine creations. There is merit in this argument, but it is overstated. Although no studies have been done to measure audience assessment of this film's believability, most audiences understand that filmmakers take significant license with reality. People are able to enter into a story (as they do ritual), be moved by it, and still understand that it is a story, and that story relies on conventions that necessarily distort reality. Lacking the habit of critical reflection, however, people remain susceptible to the vicissitudes of the next good story, which may move them in a direction that serves, for example, greenhouse gas profiteers.

There are other practical dangers attendant on conflating "poetic license" with reality or belief with tested knowledge. Stories can create distorted expectations that go unrealized and undercut support for desired behavior. A steady diet of overstimulation, unreality, and misinformation can reinforce people's already underdeveloped critical abilities. Enhanced critical capacity is good for conservation, but purists seeking scientific accuracy in fictionalized stories are impractical. It is better to be accurate, all else considered, so that people absorb knowledge rather than nonsense, but without a compelling story there is likely to be no communication.

A compelling story depends on several elements. It must be dramatic (W Bennett, 1980: 168; Murray Edelman, 1988: 90). Contrast a river described in terms of flow volume, chemistry, and similar metrics, and the river running narratives of Edward Abbey (1982) or John Powell (1961 [1975]). Descriptions of wolf weights, tooth wear, and scat analysis by weight and volume of prey species are less likely to stir the passion needed for mobilization than the stories Barry Lopez (1978) recounts in *Of Wolves and Men*. Drama involves a conflict the resolution of which matters to the audience because their sense of justice is invoked, they find the characters sympathetic, or they are curious about how the story will unfold or its outcome.

Stories are more dramatically appealing and appear less contrived when the conflict emerges from the characters. Conservation has advantages in this respect because there is much in the ordinary lives of creatures that is dramatic when well rendered in story. Robinson Jeffers (1965 [1925]: 37) captured this when he wrote that "... all the arts lose virtue/Against the essential reality/Of creatures going about their business among the equally/Earnest elements of Nature." Most audiences are not so attuned to Nature that they easily see this drama, especially given the degree to which modern media overstimulates with virtual action. It must be cultivated. But there is also much drama in the human side of conservation and this can serve for most audiences. Filmmakers know that there must be a major character drawn from the audience they intend to reach for the film to work. Thus, in *Dances with Wolves* (1990), a film ostensibly about Native Americans, a White character plays a major role.

Stories are more engaging and emotionally evocative when characters and situations are vividly drawn (Renee Bator, 2000: 531–5; William McGuire, 2001: 27). Vividness is important because it helps to raise a story above the noise and it contributes to message recall. Recall is important because people are disinclined or can't act on a message immediately. They need to hear the message repeatedly for it to sink in, receive a series of messages so they can connect the dots, or need time to evaluate the message before acting. Humans are visual animals and visual images gain attention and influence more easily, giving advantage to television, film, photography in magazines and newspapers, posters, maps, billboards, and murals. The written or spoken word and music can also evoke images, sometime more powerful because they rely on audience imagination. Rhyme and meter can greatly enhance words' ability to evoke vivid images (Elizabeth Barber and Paul Barber, 2004: 155), as do catchy slogans. When aspects of a story are well tailored to a group's experience highly useful they are more likely to conjure or retain a more vivid image (Renee Bator, 2000: 533). If an individual has experience losing a nearby open space to development a story seeking support for forest protection will be more effective stressing images of clear-cuts than stream degradation.

Symbols

Whether messages are cast as stories or not, and regardless of form or medium, their content is composed of symbols. Even aspects of messages which are not wholly symbolic – touching someone in the course of a conversation – have symbolic aspects. Body language probably comes closest to direct communication, involving smells, touch, hormonal reactions, but it too is entwined with symbolic overlays of interpretation, intent, and other aspects of meaning. Symbols that strongly resonate add great power to a story; symbols that confuse or repel will cause a story to fail.

The centrality of symbols is evident from a look at how small variations can produce significantly divergent responses. Naming and renaming is particularly potent (Margaret Keck and Kathryn Sikkink, 1998: 225). If prevailing in a policy conflict depends on framing the issue, success in framing the issue can depend on a name (Alex Williams, 2004: ST1). An "estate tax" doesn't sound so bad to most Americans; a "death tax" sounds like government has run amok and will prevent parents from passing on their assets to their children even though the tax applies to less than 2% of Americans. "Tax relief" sounds good. Isn't relief always good? A "tax cut" may or may not be a good thing, depending whose taxes are being cut. To be a "religious extremist" is not a good thing; but to be a member of the "religious right" – well, what's wrong with being right? Being right is almost as good as being warm, especially if you live in a cold climate. So "global warming" sounds to many much less ominous than "global climate change," or "climate instability." Williams agrees with others that naming, like framing, often depends on repetition (as well as a catchy name) and this makes money important to the outcome. He also notes that use of symbols must avoid confirming to people what they suspect – that symbols were chosen in an effort to manipulate them.

Maslin Fairbank, Maullin and Associates (2004) have advised conservationists to be more careful in their word choice. After 30-plus years of industry attacks many groups in the United States are mistrustful of anyone to whom the label "environmentalist" is applied, but "conservationist" is still okay. "Wildlife protection" evokes support, mostly because people want to see wildlife and because their sense of justice calls them to see animals as voiceless and needing looking after. But the term "endangered" can evoke a contentious reaction. The list goes on.

The power of words to mobilize people (and to mask and reveal reality) are evident in the slogans that are so much a part of politics and marketing campaigns. "Stay the Course," "No Child Left Behind," "Solidarity," "Black Power," "A Chicken in Every Pot . . . ," "It's the Real Thing," "Healthy Forest," and "Happy Meal." Slogans are symbols that help to create and hold groups together (Daniel Rodgers, 1987: 5). They achieve this, as other symbols do, by calling forth entire stories, including myths. Although stories are composed of many symbols, the right symbol can stand in for a story. Symbols that by themselves evoke a story are often very dense, that is, they represent complex phenomenon with extraordinary economy, as when a flag represents a country or a lone animal the great wild. They can also represent ideas, places real and imagined, living things, and processes with great economy, and alternately with great precision or ambiguity.

Although symbols such as individual words and slogans usually cannot compare for effectiveness with multimedia presentations, much depends on context. Slogans shouted during a demonstration forge group identity in a way seeing them written does not. On the other hand the British Conservative party slogan "Labour is Not Working" shown against the photo of a long queue at the unemployment office evoked a more energetic response than the words alone did (Dominic Wring, 1996). Some representations attain the status of icons – widely recognized and powerfully evocative symbols such as the Statue of Liberty, the hammer and sickle, or the golden arches.

Because of their cultural authority, symbols, like stories, are contested. NGOs seeking to influence decisions lay claim to their ownership and attempt to favorably define their meaning. Symbols are always somewhat ambiguous and contestation over meaning can exacerbate ambiguity (Daniel Rodgers, 1987: 10-1). To one group the dam at Hetch Hetchy in Yosemite is a symbol of progress – public control of urban water supplies ensuring public safety and health. This is how they sought to present it to other groups whose support was wanted. To conservationists the dam is a symbol of human foolishness and myopia and a call to ensure such a transgression never occurs again. A symbol's ambiguity is not without limits. The dam at Hetch Hetchy is a highly unlikely symbol for the plight of the urban underemployed.

The battle over the control of symbols and their meanings is sharpened by the limited number of politically potent symbols available at any time and by other factors affecting the vitality of symbols: cultural drift, overuse, misuse, and the waxing or waning of resonance. Some symbols, like the Nazi Swastika (a symbol other cultures have used), have significant baggage, making them potent but of limited utility.

Sometimes a symbol is so troublesome for one side in a political struggle that they seek to ban it from discourse. George Orwell (1968 [1946]: 127–39) observed that by abolishing or prohibiting certain symbols – for example, an image of the devastation mountaintop mining causes to streams and rivers or mass arrival of coffins of the war dead – it becomes difficult or impossible to convey certain messages.

There are additional considerations in selecting symbols to represent an NGO, to frame an issue, to represent a position on an issue, or use in story composition.

- Symbols' arbitrary and flexible qualities do not just enable ambiguity; some simultaneously convey contradictory meanings. An imposing public building evokes both national pride and the intimidating power of the state. The Grand Canyon conveys both a sense of awe and aesthetic pleasure, and a sense of human smallness and even fear. An image of the Canyon that emphasizes the wrong meaning can undercut the intended message of a story.

- Although symbols may be highly abstract (consider the letters of alphabet), they can evoke passionate emotions and action as well as convey information.

- A symbol can be compounded from many other symbols, as in a logo that combines words and other objects, a parable, an anthem, or a story character. Creating compound symbols that resonate and take advantage of synergies among their elements is high art and can be very effective. But sometimes the simplest symbol can be more powerful – the cross or the red star.

- Story themes and meanings are created by means of the combination of symbols such as words, musical notes and chords, and images to create characters, setting, and context. Symbols may also stand in for a story but usually accrete over a long period of time.

- The use of very dense symbols – symbols that represent very complex phenomenon such as a people, Nature, or a government – requires a careful reading of the intended audience and what they bring to the interaction. Emotional predispositions, history, and cognitive structures always make a difference but with dense symbols the range and intensity of reaction is greater.

- The ambiguity of symbols can be deployed to minimize differences within a coalition or alliance, not just to win a framing debate with opponents. Those who support wilderness with carnivores may join forces with urban park advocates who are not anxious for the presence of even small carnivores, under the banner of open space to further mutual goals.

- Symbols simplify as a map does. The necessary simplification of reality that a map presents may not deliberately mislead, but some maps are made to mislead just as words or images in a newscast or political speech may intend to mislead. They may do so very directly or by reliance on ambiguous generalities.

- Conservationists typically rely on charismatic species that represent the wild, freedom, natural heritage, or land and waterscapes that evoke epiphanies or other memories of being emotionally engaged with Nature. These symbols only go so far, especially in competition with symbols opponents use to speak to basic security needs and primary emotions. Conservationists should not hesitate to employ powerful secular and religious symbols not usually associated with conservation: monuments to ideals and dead leaders, historic places, flags and other emblems, technological icons (often as negative examples), and temples of various gods.

Story's limits

The centrality of story to mobilization cannot be overly stressed. Stories, however, have many weaknesses that those relying on them need to keep in mind. They always provide an incomplete and simplified, though not necessarily untruthful, rendering of reality. They impute purposes that may not be present, attribute meaning and intent to the "forces of Nature," and almost invariably they overstate the ability of human actors to realize their intentions (Charles Tilly, 2002: 28). Conservation (and other) leaders live by story as well, but the aspects of reality that story makes visible, attractive, and dramatic do not present the full picture leaders need to adequately lead. Stories cannot replace the scientific paper that describes the habitat needed by grizzlies or salmon for survival, the organizational memorandum that sets out campaign strategy, or well-grounded and informed intuition. Nor can story replace the theoretical structures needed to guide political action in complex human societies.

14 Mobilization and action

Collective action is the means by which policy is influenced (lobbying, protest) and other programmatic goals are achieved (fence pulling, exotic species removal). Collective action is also a method of mobilization. Events such as demonstrations, rallies, and celebrations convey messages more powerfully to observers than do radio spots or direct mail, creating a more lasting impression and making mobilization more likely (Ted Brader, 2006: 108). Some people participate in collective action prior to their commitment to the cause and this participation fosters commitment and further action (David Kertzer, 1988: 136; Doug McAdam et al., 2001: 55; Willet Kempton and Dorothy Holland, 2003: 321).

Properly staged demonstrations, tree sitting, theater, rallies, and lobby days are dramatic, vivid, and exciting ways of making demands on decision makers and of calling others to action. This is so whether audiences personally observe action or see media reports, although direct experience is more influential. Collective action releases a contagious energy that attracts observers (Joao Stedile, 2004: 33) and makes those who do act for reasons other than political commitment more likely to act again by increasing their commitment.

Action is an economical form of communication offering opportunities not otherwise available to nongovernmental organizations (NGOs) with limited resources. Its economy arises from its greater impact on direct observers and greater likelihood of gaining media attention. In addition to conveying substantive messages, action conveys messages about the strength and commitment of those taking action and the demand to be taken seriously (Charles Tilly, 2004: 3–5). The combined messages of action are particularly important when conservationists are marginalized. Even when conservation is an accepted political player, protest action is a reminder to others that proponents take their cause seriously. Potential recruits, opponents, and decision makers are influenced by the commitment action expresses. Action thus makes it more likely that opponents will bargain and that decision makers will take account of this commitment, making either concessions or repression more likely. Staging a major action can itself be portrayed as an achievement. Many people are attracted to winning causes (Doug McAdam, 1997 [1983]: 399). They may root for the underdog, but they do not like losers or the pathetic (Francesca Polletta, 2006: 134).

Action and innovation

Action's capacity to convey a resonant mobilizing message depends on projecting a sense of efficacy (success is coming, we're making it happen) and excitement, as well as achieving real progress. These in turn depend upon constant tactical innovation. (More about tactical innovation in Chapters 18 and 19.) Innovation is an element in making action attractive to media and observers. Sit-ins, road blockades, protesting in animal costumes, and unfurling banners from unlikely places are all examples of actions that have garnered significant attention and resulted in new recruits to conservation. Over time, however, tactics lose their political effectiveness and drama. When something isn't new it isn't news. Opponents figure out how to counter tactics so the desired effect is undercut (e.g., people sitting down in front of bulldozers are no longer brave and committed but a nuisance). They learn to work around tactics by avoiding the confrontation that makes for attractive drama and media coverage or that makes authorities look heavy-handed.

Innovation includes using older forms in new ways or settings. Zuni efforts to stop a major coal mine, discussed above, relied on a centuries old form of communication – sending runners with messages. In a contemporary setting the runners attracted much positive media coverage.

Actions organized around existing events and holidays get more media attention (Francesca Polletta, 2004: 163, 166). Native Americans organized protests on Columbus Day – normally a day commemorating the European discovery of the Americas – to press for redress of grievances that have their roots with the coming of Europeans to the New World. The media coverage contributed to mobilization of churches, students, and many others. Chinese students demonstrated in Tiananmen Square for greater democracy on the occasion of a party reformer's funeral to point out that the government was hypocritically honoring the reformer while stifling reforms (Francesca Polletta, 2004: 158). The politicization of funerals was not new in China, but organizing a grass-roots protest around it was. It did get the attention of decision makers who responded with armed repression.

Campaigning to establish a holiday or celebration to commemorate the struggle of a movement or to honor its leaders and achievements can be an effective mobilization approach. Efforts to establish a Martin Luther King holiday gave new energy to the civil rights movement in conservative and unsympathetic times. Although many opposed the holiday, it was difficult to argue against honoring a nonviolent leader who pressed for such "radical" reforms as an end to Jim Crow and the right to vote.

The King holiday continues to be a source of political contention, providing opportunities for mobilization (Francesca Polletta, 2004: 153, 170; Dennis 2004). Activists seeking to honor King and what he stood for by mobilizing people to become activist like him have good foils in those who use the holiday as a platform to call for individual community service or who see it as another day off work to go shopping. Earth Day is similarly contentious (Francesca Polletta, 2004: 16–8). Not a legal holiday it is nonetheless celebrated throughout the world by hundreds of millions of people. Many NGOs use the day's focus on conservation to mobilize people in support of major change and to note major achievements. Similar to some King holiday celebrants, there are those who call for greater individual efforts such as increasing recycling (likened to bailing out the Titanic with a teaspoon (Hawken, 1993: 5, xi–xvi). For others Earth Day is an opportunity to greenwash, deflect people from activism and reform, and encourage current ecologically destructive behaviors while feeling good. The contention can heighten message effectiveness by drawing more attention.

Innovation is not without risks as Doug McAdam et al. (2001: 316) have observed. If audiences can't make sense of new tactics the resources invested by conservationists are wasted. Even supporters may find new tactics inappropriate or threatening and turn away. New tactics may simultaneously succeed with important new audiences while alienating some supporters and raising the ire of other groups, leading to countermobilization. Even when new tactics succeed in throwing opponents (including decision makers) off balance, leading them to stumble or act stupidly in the eyes of important others, the costs to a movement can be high. If authorities regard the tactics as threatening or simply as outside of what they will tolerate, they may impose serious physical damage that discourages new active support (Charles Tilly, 2006: 186). Successful tactical innovation requires both a sensitivity to the political landscape and a willingness to take risks and bear the costs that come with meaningful political action.

How action mobilizes

Not everyone wants to be where the action is, but many do. Action excites and feeds feelings of efficacy. One of the reasons for the success of the McGruff crime prevention program – as compared with similar programs that failed – was the many opportunities for action it provided to participants (Garrett O'Keefe and Kathleen Reid-Martinez, 2001: 274). When mobilizing messages rely on channels such as word of mouth or the media the positive effects of the message quickly dissipate if they are not soon associated with action (more on this in Chapter 18).

The desire for belonging and purpose that so often contributes to mobilization is only meaningful to some if the promise of action is also present. Action provides for self-expression and the exercise of agency regardless of how well the cause is understood. Taking action releases energy and is pleasurable; pleasure is enhanced and reinforced when cohorts are attracted to action as something to do together

(James Jasper, 1999: 73). All of which brings people back for more. Collective action also enables people to overcome individual timidity that often blocks action.

Action is almost always transformative which is why and how it leads to mobilization. The more physical or otherwise costly the action is, the more transformative it is. Although public-service announcements successfully encourage people to contact an organization and obtain and display a decal, commitment remains weak (Lewis Rambo, 1993: 138–9; Renee Bator, 2000: 537). When people make a public statement by signing onto a newspaper ad a higher level of commitment is generated and these people are more likely to answer subsequent calls to action. Action that goes beyond public statements – such as demonstrations and boycotts – is a more powerful predictor of future committed action. The initial outpouring in support of the bus boycott by Montgomery, Alabama's Black population – by no means mostly composed of civil rights politicos – changed participants (Ronald Aminzade and Doug McAdam, 2001: 33). People became less afraid, more hopeful, and developed a sense of their power. Most stayed with the boycott despite threats and the material and physical hardship of finding other ways to work. This dominant White community was also transformed (McAdam and Sewell, 2001: 111). Faced with over 90% Black participation in the boycott, Whites generally and the segregationist leadership in particular were made to come to terms with a new social force. Segregationist leaders did not have the option of using their most repressive tools because of national media attention.

Action transforms people in a number of other ways that makes future action more likely. People come to understand themselves and the world better, including the cause they are acting on behalf of, and their commitment deepens. Action is one of the primary means by which people come to see, understand, and feel the links between the local, regional, national, and global arenas and to think and act on those larger scales (Doug McAdam et al., 2001: 332). When participants perceive that a larger scale is relevant they act differently: local actions become part of actions taken simultaneously by others across a nation or internationally. Local actions may become more centrally coordinated or people may travel to concentrate their strength in a few places for a massive protest. These large-scale events in themselves can become both transformative for participants and attract others. The March on Washington in 1963, a giant feat of mobilization, increased participants' commitment and made them more likely to continue movement participation. It mobilized many new people because they directly observed it, heard about it from participants, watched it on television, or read stirring accounts. The Seattle anti-World Trade Organization (anti-WTO) demonstrations – described as a master coincidence rather than the result of a master plan – were organized mostly from the bottom up and by loose, electronically linked groups (Naomi Klein, 2004: 222). These differences did not diminish the transformation of participants nor the attraction of new recruits, although some observers have noted that people mobilized through electronic communication only to be attracted to the action and show no penchant for sustained organizing and movement building (Charles Tilly, 2004: 97, 121; Cassen, 2004: 158–9). The lack of organization building is troubling because it is organization that fixes the level of mobilization over time and brings the constant pressure needed for significant policy and social change.

Participants in action typically come to see that the political (and personal) limits they imagined are either not real, or can be transcended. They become bolder. Obviously, if repression is severe enough, some will become frightened and back off, but most often repression generates defiance and a shift in tactics, not quiescence (see Chapter 19).

Action as a mobilizing tool need not always be programmatically meaningful. A demonstration may be intended to achieve nothing other than make participants feel good, strengthen their common identity, and increase emotional investment in the cause and organization. Social events or nonpolitical conservation work such as counting wildlife creates bonds between people and the organization, making future action much more likely, including overtly political action.

Action also leads to commitment and future action for less ennobling reasons. Because people need to feel that their actions are proper and justified these feelings are often strengthened when people encounter information that suggests the actions were not positive or a mistake (Lewis Rambo, 1993: 116; Carol Tavris and Elliot Aronson, 2007: 15–22, 35, 60–1, 76–7, 136, 147–50). Resistance to admitting error given past action is not just proportionately related to the amount of physical and emotional energy invested in the previous action. It is also the product of personality and subculture attributes. Some are more open to new experiences and able to integrate new information and insights, admit error, change their minds to accommodate new knowledge and adjust their behavior accordingly. They can do so because they do not feel that mistakes, regardless of the emotional investment or cost associated with the original behavior, make them bad people. Others express the stereotypical military disposition: having committed troops to battle, lost lives and spent treasure, there can be no turning back; it would be shameful. Thus more lives are squandered.

The psychosocial dynamic that precludes admission of mistakes is important for conservation in two respects. Lack of flexibility and the inability to learn from mistakes can seriously hamper the movement if enough participants possess such traits. On the other hand the opponents of conservation, who are mostly involved in defense of the status quo, can benefit from such traits in participants.

Not all actions that mobilize are directly political action. Many scientists, for instance, are wary of politics out of concern for their credibility or distaste for many of its attributes. Scientists do act in many ways that further conservation goals, however, and these actions have the same effect on them, that is, increasing commitment and likelihood of future actions. In mobilizing scientists via action it is necessary to keep in mind the actions scientists consider appropriate to their identity as scientists. Among actions scientists undertake are reporting the results of research; reporting and interpreting the results for decision makers, advocates, and others in a position to act on them; working closely with decision makers and others to integrate findings into operational activities; advocating for preferred policies; and becoming decision makers themselves (Robert Cox, 2006: 358). Although scientists are only comfortable with the first two of these actions – reporting on findings – broad segments of the public were comfortable with the first three.

Mobilizing people through action seems a tautology. But if we take mobilization to mean action accompanied by a commitment to purpose, then the process becomes clearer. People are moved to purposive political action by being drawn to action for a variety of reasons that lead to their commitment.

15 Overarching tactical concerns

Mobilization tactics (the tools used to realize strategy) have been discussed throughout the earlier chapters 8–14. Deciding which tactics to use in what combination and when is situation-specific. Because of the enormous variety of situations encountered it has not been possible to do more than give some examples and discuss the basic principles. There are some overarching tactical approaches worth highlighting (A Ginsburgh, 1955; Doug McAdam et al., 2001; Karen Johnson-Cartee and Gary Copeland, 2004; John Sellars, 2004: 186; Charles Tilly, 2005: 44) and some damaging tactics used by opponents for which conservationists should be well prepared.

Overarching tactical approaches

Surprise

Innovative action, action in an unexpected venue or at an unexpected time is more likely to gain media attention and excite participants and observers; the excitement attracts recruits. Surprise will also knock the opposition off balance and give the initiative or greater momentum to those using surprise.

Working behind the scenes

Working quietly, below radar, by word of mouth, can be quite effective at mobilizing some audiences while avoiding the attention that generates countermobilization by opponents. Much depends on the target audiences' accessibility to quiet outreach. Coalition building can be accomplished quietly, and lobbying decision makers is often more effective when low-key.

Have others pave the way

Mobilization is almost always a multistage process of defining and framing issues, defining a limited range of good solutions, and advocating for a preferred solution.

When people that have influence with an audience provide introductions for conservationists and also frame the issue, the audience is much more receptive to the solutions proposed by conservationists.

Have a respected independent party pronounce positive judgment

Those considered by the targeted audience to be independent and unbiased can do much to aid mobilization by endorsing the issue definition, the preferred solution, and urging supportive action. Their statements are considered authoritative and reassuring.

Work through opinion leaders when mobilizing mass audiences

Target audiences look to a handful of people to tell them what's important and how to consider or respond to events. These figures may be religious leaders, business people, civic leaders, political leaders (officials or nongovernmental organization [NGO]) or news people. Opinion leaders often speak for, as well as to, a group. Mobilizing opinion leaders has a significant multiplier effect.

Utilize word-of-mouth and personal connections whenever possible

Personal connections are not always more persuasive than other sources but mobilization relies heavily on social networks and the influence of peers. One-on-one communication is critical within elite circles and among NGO leaders. The more politically savvy the messenger is seen to be, the more persuasive the message.

Emphasize who one is, not just the mission

Personality is frequently more important than issues in making political choices. It is also much easier for people to identify with an individual than an institution or abstraction. Emphasizing that conservation is about *people who care about* wildlife and Nature – preferably with personal examples or testimony – is more effective than calls to protect biodiversity or Nature writ large.

Speak for justice

Emphasize that conservationists work for justice – justice for those creatures who have no voice in the councils of state, and to ensure that all humans' right to a healthy biosphere is realized.

Prepare for crises

Crises offer openings; understanding the dynamics of recurring crises (e.g., major economic downturns, political succession, and elite scandals) is critical to making good use of them. Political and natural contingencies do not automatically lead people to insight about the source of problems. When conservationists are well prepared to explain a crisis, speak to people's fears, concerns, and need for understanding, and offer solutions, they can make headway in attaining ecologically sound behavior and policy.

Give the appearance of being everywhere and of having broad support

Many people are only prepared to go with the flow, never against it. When conservationists appear numerous and influential these people are influenced to support the movement or at least refrain from opposing it. Groups differ markedly in their propensity to embrace changes in behavior and policy: some are bold and early joiners of a cause; others wait until they see it's safe or the benefits are clear; still others won't support changes until they are largely in place and they do so to remain in the mainstream (Lewis Rambo, 1993: 95–6).

Do not act arrogantly

In giving the appearance of being everywhere or powerful it is possible to appear arrogant. Arrogance causes support to evaporate. Humility is easiest to maintain and project when conservationists remind themselves that no victory is secure and that the political struggle is not about how history will judge them but about preventing extinction.

Multiple channels are better than one

Single channels do result in mobilization, but not as often or consistently as when multiple channels are used. Even saturation public-service announcements and entertainment programming, which have been shown to move people when done right, are more effective when combined with organized action such as rallies, meetings, and other involvement.

Do the media work for them

Aside from generating newsworthy activity, the likelihood of getting the intended message across effectively is enhanced by providing lead-ins, text, photos, videos, and sound bites. Personal testimony offers the press drama, as do powerful graphics. The more that is provided and the better the quality and fit with the medium's needs, the better the chance the material will be used and that coverage will be favorable. Meeting press deadlines, being available, returning calls, and being fully prepared tend to keep the media coming back for more. Aiming to influence those media that set the tone for others, for example, in the United States the electronic media take their lead from major newspapers, allows for greater influence in the face of limited resources.

Keep the tone of the message appropriate to the audience and circumstances

Messages with the right tone (e.g., heartfelt, don't scold, television and radio are usually more conversational than the printed media) are more likely to arouse the right level and type of audience affect (passion, seriousness, humorousness) needed for mobilization in each case.

Link calls to action to local concerns

Issues and messages should always be linked to what the audience experiences as local and immediate. The links need not be material, for example, halting overfishing will ensure long-term local availability of fish for human consumption, but can be about notions of justice, for example, imposing fishing limits and buying out fishermen is better than continuing to subsidize fishermen to further deplete fish stocks.

Messages should always be concise

People are busy and attention spans are limited so even stories need to be compressed. Better to leave people a bit hungry than overfed.

Remind constituencies of the benefits conservation provides

Pointing out benefits received in the past and that will continue to be received due to their participation in the movement, creates a sense of fulfillment and satisfaction. Benefits include conservation successes such as protecting wild places and wildlife, places where the soul can rejuvenate, and the more personal: being able to associate with other smart and thoughtful people, or having the opportunity to meet an exceptional prospective spouse.

Pressure on decision makers is most effective from multiple sources

To be effective these sources must have good leverage, that is, be part of a decision maker's base of support (interest groups, voters) or be peers whose cooperation is needed or approval desired – no one wants to be shunned at cocktail parties or in the back room.

Tactics of attack and responses

Conservationists are too often unprepared to deal with harmful tactics regularly used against them (Thomas Dye, 2002; Karen Johnson-Cartee and Gary Copeland, 2004: 158–9, 164–71; Rob Walker, 2004; Matthew Fellowes and Patrick Wolf, 2004; Carol Tavris and Elliot Aronson, 2007: 52). Lack of preparation results in ineffective responses, mobilization being undercut, and resources wasted. Some of the more pernicious tactics used by opponents include:

- using paid "journalists" to pose as legitimate journalists and write biased reports that appear to be by disinterested parties;

- spreading disinformation through channels audiences normally rely on, including manipulating the media by providing disinformation to them, using media ownership to control content, pressuring the media to toe the line by threatening to withdraw advertising, obtaining the agreement of media outlets to misrepresent through appeals to personal connections or common interests, coercing the media, or bribing individuals within the media;

- using pro-conservation names for front groups that oppose conservation in order to obscure who is organizing, funding, or providing major support to the group;

- name-calling, such as labeling others as propagandists; it is particularly effective to falsely accuse another of what the name-caller is doing because it makes their response appear defensive; a favored variation on this tactic is for large, profitable enterprises to attack conservationists' character and motives, accusing them of self-enrichment;
- deliberately misusing words to obfuscate facts or intentions;
- using distorted statistics;
- using false analogies and simple-minded polarities;
- using hate speech;
- disingenuously using "secret sources";
- presenting opinion as fact;
- attempting to generate or appealing to irrational fear;
- using neuromarketing and product shills;
- using bribery in the form of "gifts" to sway opinion; small gifts often prove more effective because recipients are convinced that small gifts can't sway them;
- attempting to sow dissension and discord among conservation groups;
- making threats of material or economic injury;
- using legal process such as civil suits or criminal prosecution to discourage conservation action;
- using private and official violence.

Because of the many variables careful consideration needs to be given to responses to these tactics (Karen Johnson-Cartee and Gary Copeland, 2004: 184). Sometimes responding is not worth the resources. Sometimes it is clear an opposition tactic is backfiring and it's best to let it develop without a response. Conservationists are under no obligation to discourage their opponents' self-destructive proclivities. In order to protect their sources of information about opponents' activities conservationists may need to bite their tongue and allow things to play out. At other times it is necessary to respond rapidly, directly, with all the resources an NGO can muster. A rapid and well-thought out response can nip damage in the bud (Eric Dezenhall, 1999). Authoritative third parties can be especially helpful.

Countering disinformation with the facts is not usually very effective because people tend to defer to the initial information. A well-crafted response at saturation level (most conservation NGOs cannot afford this) can work, but it is usually most effective to expose the prevaricating entities' funding, corruption, and hypocrisy in a well-told morality story. When facts are embedded in a story that offers an alternative frame it can undermine the attacker's frame and thereby their factual accusations. When memos were written by the Fox News Channel's business office to the news people telling them what stories to cover and how to cover them and this was framed in a story calling into question Fox's claims of fair and balanced reporting, some eyes were opened. Oil company, tobacco company, and others have suffered from such exposure despite their massive public relations expenditures. Because much of the media are willing to settle for he-said-she-said reporting, it is important to provide independently verified information to the media and challenge them to corroborate it and report it as fact in contrast to claims made by the parties. In the film *Controlling Interest* (1978) as a company bureaucrat expounds how businesses simply can't afford higher wages his hefty salary is flashed on the screen.

Business- and state-conducted repression – dirty tricks, lawsuits, prosecutions, threats of violence and violence – are not as rare, even in democracies, as some middle-class conservationists would like to think. Conservationists, like everyone, tend to project onto others their own mode of thinking and behaving. They usually do not expect, and are surprised by and unprepared for subterfuge, bald-face lies, appeals to ignorance, and official lawbreaking. To indulge the tendency to project their own standards onto to others and ignore others' behavior is neither a testament to their own standards nor smart politics. When powerful interests are threatened they rarely let the rules limit them. Repressive tactics and responses are examined in detail in Chapter 19. For now it's enough to remind conservationists they need to have done nothing illegal or unethical to be subject to these tactics. Building and maintaining good relationships with other NGOs makes it more difficult for provocateurs to successfully sow dissension and mistrust. Familiarity with public and private police tactics, and the tactics of other entities (mercenaries, private security and investigative firms, freelance thugs) that do the heavy lifting for the powerful is very helpful in spotting such tactics early and responding appropriately; a wealth of documentation exists and is easily accessible (see Chapter 19). Developing good security and safety habits is not the same as embracing paranoia and can help avoid the pitfalls of manipulation and entrapment that are common; to ignore the lessons of other movements is foolish. Good relationships with savvy lawyers, media contacts, contacts with sympathetic members of the elite, and influential foreign connections offer some safety and help ensure a rapid, more effective response.

16 Monitoring and evaluation

...success will be measured...by the effectiveness of our approaches to sustaining the diversity of life and the health of ecosystems, and by the respect for the living world we are able to foster within our varied cultures and within the human heart.

Curt Meine et al. (2006: 647)

When asked if the millions of dollars his company spent on advertising worked, a Fortune 500 chief executive officer said he thought about half of it did. But he didn't know which half and so didn't know which half to cut. Because mobilization is a complex effort, involving variants of message, delivery, channels, levels of saturation, source credibility, story themes, and different combinations of these, it is often difficult to know which elements are working well, somewhat, or failing. The final results are usually clear – mobilization targets are being met or not – but investigation and analysis is needed to determine what's contributing to the results and how much. It's not just failure that needs investigation, so does success. Success in a particular case may be contingent on factors difficult to duplicate. Mobilization for organization building or a campaign depends on the ability to understand what is happening and why – on monitoring and evaluation and making appropriate course corrections (Karen Johnson-Cartee and Gary Copeland, 1997; Judith Trent and Robert Friedenberg, 2004; Sandra Beckwith, 2006: 189–96; Arjun Chaudhuri, 2006; Robert Cox, 2006: 243–88, 367–410). Cultivating experienced judgment and grounded intuition – the source of good decisions in the heat of battle – should be an equal priority with analysis. Both are better than either.

Some nongovernmental organizations (NGOs), mostly because of limited resources, resist allocating needed time and energy to monitoring and evaluation, preferring to devote resources to program; but it can be more wasteful to invest in activities that don't produce results. Time is often the limiting factor. Not just because personnel are in short supply, but because politics moves rapidly and doesn't present time-outs for taking stock. Time must be made for this. Richard Margoluis and Nick Salafsky (1998) provide an excellent example of a systematic approach to monitoring and evaluating projects that is far from onerous or resource-intensive. It is tailored to local conservation projects heavily dependent on local cooperation for success rather than on policy choices made by national institutions so there is little discussion of how progress might be measured in larger political arenas.

Monitoring and evaluation admit to several missteps that are worth noting. Among them are confusion in analysis, preoccupation with interim steps, and overemphasis on analysis or judgment.

Risk taking is essential to conservation success and too much stress on close monitoring can breed timidity. It can also breed mobilization *interruptus*. Because mobilization efforts can take time to show results, the impatient are prone to see failure and abandon an approach before it has shown itself a success or failure. Mobilization approaches are dumped on occasion not because they failed to bring in new resources, but because the resources did not translate into policy change. Conflating one type of failure with the other confuses mobilization of resources with their use; adjustments based on this confusion will not be sound. One means of avoiding early abandonment and conflation is through a monitoring and evaluation approach that includes identification of the critical steps on the path to the goal. This allows progress to be assessed and only those factors important to each stage to be taken into account.

In many cases evaluation focuses entirely on the interim steps and fails to keep the goal – mobilization in this case – in the forefront. Measuring the number of hits on a web site, brochures printed and distributed, and a host of other such things is meaningless if they are achieved but fail to produce the desired end result. The much vaunted DARE program, intended to indoctrinate millions of children (at the cost of tens of millions) on the evils of illegal drug use, did not reduce drug use among graduates of the program as compared to those who did not attend (US General Accountability Office, 2003). Millions did attend, films were watched, testimony given, literature distributed. By these measures the program was meeting its interim goals, but not *the goal*. The dairy industry spent millions of dollars on ads to overcome flat milk sales resulting from public concern with cholesterol and animal fat (Matilda Butler, 2001: 310–2). According to polls it succeeded in convincing audiences that low-fat milk was nutritious. The ads were considered attractive and memorable and they changed attitudes – but not behavior. Conservationists experience similar results all the time. Meaningful evaluation must focus on the critical connection between interim steps and the goal of action. The US legislative process provides a good analogy. For a bill to become law it must gain approval at each of many, many steps. Success may be had at all of these steps except one and the bill fails. An evaluation process that only examines the steps in isolation and that fails to keep the goal of final passage in the forefront is not useful.

Stressing the need for – and pitfalls associated with – the monitoring and evaluation of NGO's effectiveness at reaching goals is not intended to overstress the role of analysis. Many of the most politically successful people are not well versed in the social movement literature nor do they rely heavily on formal analysis. Their decisions are based upon their judgment and insight. They invariably rely on a team of advisors and other leaders – a team they have assembled based on their judgment – that represents a range of experiences, skills (including for critical analysis), and connections (Aldon Morris and Suzanne Staggenborg, 2004: 181, 188). Just as a great orator on the stump doesn't contemporaneously evaluate his or her word choices according to the formal rules of grammar and instead relies on what sounds right to his or her ear, so an effective politico avoids minutiae, formulae, and textbooks and thereby avoids the stumbles that come from inundation in detail and rules (Philip Tetlock, 2005: 67–188). Indeed, nonexpert intuition or hunches regularly outperform expert overanalysis with regard to the future (Gerd Gigerenzer, 2007: 34, 85, 91, 109).

Judgment and intuition are not mysterious. They make use of long-tested (by evolution) nonconscious processes for quickly discerning what's important and how to respond. These processes are not just neural, but integrate experience, that is, how a situation and the response to it played out in the past. A critical element of judgment, then, is the comparison of the present with past situations. One reason narcissists so often excel at politics is their long practice of reading situations and people in order to manipulate them to get what they want.

Intuition and judgment should not be overrated either. Many situations call for both. Novel situations benefit from analysis and an examination of assumptions and facts. Fighting the last war again is costly even when it doesn't end in defeat. Accommodating value conflicts requires analysis because judgment often fails to extend across subcultures. The business executive discussed at the beginning of this chapter who didn't know which half of his advertising was effective will probably not be able to find an answer absent gathering and analyzing information.

Achieving higher levels of mobilization among more groups is not the whole answer to the extinction crises. The ebb and flow of political opportunities such as opponent or regime weakness, play a significant role. So does how effectively resources are used once they are mobilized. The biggest and most well-provisioned army will falter with bad generalship, or achieve nothing useful if it takes the wrong territory, doesn't know the terrain, or isn't matched to the opposing force's method of fighting. How successfully resources are used in turn has a profound affect on the ability to maintain high levels of mobilization.

Part 2B
The care and maintenance of the hammer

Political success requires a clear vision and goals on which there is no compromise, persistence, and a hammer. Mobilization creates the hammer. Organization is the hammer.

Organization

Constituting the hammer

Mobilization brings people and other resources together into a pattern of ongoing relationships that is the organization (or coalitions and alliances among organizations). Mobilization creates power and authority, making available energy for politics not previously available (Amatai Etzioni, 1968: 403). In order to exercise the power and authority created by mobilization and use these to achieve goals organizations must do more than aggregate resources – they must give structure to them. Political effectiveness depends on *how* resources are combined and used over time. Organizations break complex and long-range goals into manageable roles and tasks and integrate them in a way that moves the organization toward attainment of those goals piece by piece and step by step. They define the relationships between tasks and between roles and allocate resources to them. They define the meaning and purpose of the various roles and tasks and how they relate to the organization's overall purpose. Organization provides for the regular transmission of clear information between the parts and from the top down and bottom up. Organization keeps the orchestra working on the same score by providing direction and propagating an ideology that provides purpose for the parts and the whole.

By combining the resources of many, planning for their use, and bringing those resources to bear strategically – in the right sequence and at the appropriate moment – organization (or organizations) accomplish what an individual or spontaneous collectives cannot. Most individuals' resources are too puny to make much political difference. Twenty-five dollars or a few hours of labor devoted to politics is like a small wave in the global ocean. But combine a half million $25, and a half million people with a few hours to give, and what can be done changes dramatically – and not merely in an additive way. Organization creates a multiplier.

An organization is a collective, more than the sum of its parts, and thus a distinct organism with an identity. Individuals forget, become tired, move on, and die, while organizations keep political books. Organizations have a collective memory and can learn from their actions and those of others. They can persist in ways that individuals do not although they depend on the persistence of individuals that are part of them.

Organizations such as nongovernmental organizations (NGOs) differ markedly in their formality, specialization of roles, size, leadership, and much else. They differ in how leaders, staff, activists, supporters, and members figure into setting goals and purposes. They differ in their flexibility and ability to respond quickly to opportunities and threats. Large organizations tend to be slower and less adaptable because people are vested as much in their niche within the organization as the organization's purpose. The greater people's role in defining organizational relationships and culture, the greater their investment and commitment. Organizations are more effective when internal interaction encourages trust, openness, critical thinking, and risk taking. Nonetheless organization always possesses hierarchical attributes. People are not equally endowed with vision, prescience, risk-taking predispositions, organizational abilities, or other skills.

Using the hammer

Providing an ongoing structure to the resources mobilized is what permits the sustained *exercise* of political power and authority. Organization, which fixes the power and authority created by mobilization, is the basis for an organization's ability to obtain desired actions from decision makers – those who hold even greater power such as corporate executives, presidents, premieres and prime ministers, and legislators. Power permits an organization to reward and punish, to help put people into office or remove them from it, and to make their lives pleasant or miserable while in power. It is also a step toward capturing governmental power and being able to use the institutions of government to further conservation goals. Generally conservation NGOs have sought to influence government or have a role in government, but not to take power as political parties or revolutionary movements do. Influencing or playing a substantial role in government makes available some of government's power to regulate and repress, tax, spend, and speak with authority. Influencing economic institutions can be just as important and sometimes more important than influencing government.

Organization also permits the effective use of the authority created by mobilization to further conservation goals. The exercise of power is most effective when it is regarded as legitimate. Legitimacy or authority is itself an important source of influence in the political arena. When a cause, an NGO or its goals enjoy legitimacy decision makers' options are constrained. It is more difficult to ignore claims regarded as legitimate. It is easier to mobilize broader support for authoritative goals. Decision makers who support goals with broad legitimacy can lay claim to doing good and doing good because it is the right thing to do. And opponents must work harder in order to find the means of opposing legitimate goals; usually they do so by advancing their position in the rhetoric of legitimate goals, for example, the Healthy Forest Initiative.

An organization's influence is not solely determined by the mobilization of resources. Power and authority, like mobilization, are relative; the power of opposing forces counts. Whether a political system is open or repressive makes a difference in the costs of political action and the opportunities presented to influence policy. Basic communications and transportation infrastructure make a difference; bad roads and a lack of telephone and E-mail add costs to organizational operations. Because no human society is ecologically neutral many conservation NGOs recognize significant social change is needed, not just a change in a few policies. To be politically effective requires adapting to some existing relationships and processes even as change in them is sought. Managing the tension between ways of doing business necessary for policy change and holding onto and living within an emergent vision for a different set of political and social arrangements is difficult.

Sustaining mobilization

Organization depends on sustaining mobilization and is necessary to sustain mobilization. All organizations from empires to corporations to NGOs rely on maintaining mobilization of resources previously mobilized. When an organization cannot sustain mobilization it will be ignored or opponents may decide to hurry along its demise.

Conservation NGOs must attend to three tasks to sustain mobilization: show progress toward programmatic goals, dedicate resources to sustaining mobilization, and build a conservation community. Unless these are successfully undertaken mobilization dissipates (Amatai Etzioni, 1968: 408; Lewis Rambo, 1993: 80, 136).

Organizations are constantly battling entropy. Countless individual, interpersonal, social, and structural factors erode the willingness to engage in collective political action. Outrage wanes. Victories lead to complacency. Personal desires or crises pull people away from politics. People become overwhelmed, fearful, bored, or burned-out (Amatai Etzioni, 1968: 411). The material or emotional support for mobilization is inadequate or community is lacking.

Showing progress toward goals is necessary because it tells participants they are making a difference. Their labor, their money, and the risks they take all count and without them the outcome would be undesirable. There are exceptions to this; people struggle when they believe they have no choice or no more meaningful alternatives regardless of apparent progress. Most NGOs understand they must demonstrate progress to maintain support. In difficult times progress may amount to stopping opponents' effort to inflict increased damage on the natural world. Organizations must be able to credibly explain the ebbs and flows of political opportunity and influence that are part of every system's politics.

Resources need to be directed at sustaining mobilization. The need-based, emotional, and cognitive rewards extended in initial mobilization must be met. People's decision to mobilize must be reinforced by peers and by leaders (Jacques Ellul, 1972: 121). NGOs must provide a pathway for deepening connection, identity, understanding, and commitment to their goals and the organization itself by way of collective action, ritual, and learning.

To sustain mobilization over the long haul and generate the sort of commitment that has led to achievement by other social movements, conservation NGOs must create a strong community. This is what Emma Goldman (1934: 56) was talking about when she said that if being part of an important cause precluded having a good time dancing, there was something wrong. Community provides what Lewis Rambo (1993: 5) calls holding – interaction that nurtures and reinforces an identity that includes cause and organization. But such interaction includes more than that. It is what Black churches provided to their members in the course of the civil rights struggle and what the labor movement provided in many countries during industrialization. Community provides venues for a range of nonpolitical activities from drinking beer and hanging out to recreating and finding a mate. To maintain mobilization large or geographically dispersed organizations must establish local communities that are experienced as part of a larger community as some religious organizations and fraternal organizations do. Organizations often need not create community sites from whole cloth; existing sites (churches and universities), for example, were used by the US civil rights movement.

A small conservation community exists, mostly consisting of activists, volunteers, paid staff and board members of NGOs, but that base is too narrow. Outside of this community most support for conservation is shallow and lack of community is precisely the reason.

In the chapters of this part we will examine in greater detail what organizations must do to sustain mobilization and deepen commitment among those mobilized. A threshold factor is leadership.

Leadership

The US civil rights movement has sustained itself through ups and downs and generally grown more powerful and effective over the long haul. Other movements have faded from the scene after a brief flash even though the circumstances that called them into being did not fade. The US labor movement struggled for decades in the first half of the 20th century, won significant victories, only to have their power blunted by economic trends, coordinated and aggressive repression by state and business, and their own myopic and self-satisfied leadership. At the same time the European labor movement on the whole prospered. Some religious institutions show remarkable longevity, although their role and functions have changed.

The ability of movements and NGOs to sustain themselves and remain effective depends on many factors like political opportunity, but among the most important is leadership. Although amateurs found most movements and NGOs they must become professionals. Meeting the challenges of contentious politics requires committed, full-time leaders with vision, sound strategic sense, who can raise money, manage staff, lead successful campaigns, and handle growing organizational complexity (Doug McAdam, 1997 [1983]: 437; David Meyer and Sidney Tarrow, 1998: 12–5).

Direct action is argued to be the domain of amateurs, but direct action has played a major role in labor, nationalist, or other NGOs led by professionals. Because direct action often involves breaking laws and creating disturbances there can be political advantage in having organizational distinctions between those engaged in direct action and those engaged in other types of activity. Earth First! in its first decade proclaimed itself as a nonorganization with no organizational infrastructure – people were members by subscribing to the journal. Such nice distinctions are lost on police and media who do not require a formal relationship to cast blame and aim repression. Most often taking responsibility for direction action is an important part of pressing claims and organization building. In cases where rule-breaking involves civil disobedience a public affiliation is usually an important part of the act – accountability for civil disobedience is central to its political effects. Direct action is also the province of state professionals, as Central Intelligence Agency mobs and Nazi brownshirts attest.

Leaders' attributes disproportionately affect organizational actions and culture for better and for worse. Some smaller organizations can maintain informal and fluid leadership styles, but institutionalized leadership emerges when organizations attain a certain size – an orchestra needs a conductor, a string quartet does not.

The founders of movements and NGOs are innovators and in most spheres of life leadership entails innovation and originality (Murray Edelman, 1988: 51–2). At the highest levels of political leadership in society real innovation and creativity are rare, largely because those obsessed with getting and keeping power must avoid offending their benefactors. Leaders, of course, strive to maintain an image as innovators.

To what degree is this lack of creativity typical of NGO leaders who are not founders? Social movement researchers have found that nongovernmental leaders, as their organizations become more successful and powerful, become so vested in their position that it becomes more important than their organization's purpose (Robert Michels, 1962 [1915]; Francis Piven and Richard Cloward, 1977). Others argue this process only applies to centralized organizations and most conservation NGOs are not centralized (David Meyer and Sidney Tarrow, 1998; Christopher Rootes, 2004). Certainly many modern conservation leaders closely connected to wealth and power seem timid in their demands for fear of endangering those connections, although they claim they are just realists about what can be achieved. The image of boldness and creativity can sustain support for an NGO provided supporters receive no contrary information and remain vague on goals. To the degree this happens the movement is wasting resources.

Students of leadership have identified other important effects of leadership styles that flow from the personality of leaders. Max Weber (1947: 324–41, 358–86) identified three types of leaders, two of which concern us: charismatic and bureaucratic-legal. The former is visionary, inspirational, entrepreneurial, and creative. They neither rely on nor stress formal organizational structures or rules and are often poor at organizational tasks and intolerant of those who regard such tasks as important. The latter type of leader is a manager, concerned primarily with establishing clear goals and reaching them by means of an organization with clearly defined roles and staffed with those who do the job. Managers, however, often lack vision and the ability to generate significant enthusiasm and energy.

A similar distinction is made between leaders who are people-oriented and those who are task-oriented (Ronald Aminzade et al., 2001: 129–30). Most leaders possess some combination of traits although most have attributes that fall into one category. Because political success depends on both sorts of attributes – able administration and the ability to inspire and motivate – NGOs that are able to build leadership groups that integrate both types of attributes while avoiding factionalism and power struggles are more effective (Ronald Aminzade et al., 2001: 133, 136, 141). Some leaders, such as King and Chavez, demonstrate great insight in recruiting lieutenants who can make up for what they lack (Aldon Morris and Suzanne Staggenborg, 2004: 188).

Charismatic traits are important in maintaining mobilization because people require ongoing inspiration and emotional stimulation. Leaders must be able to arouse in supporters, staff, and others feelings that they are good, worthy, and important by virtue of being associated with the organization or cause (David Kertzer, 1988: 108). Leaders are a mainstay of holding (Shierry Nicholsen, 2002: 196), enabling people to feel safe (enough) in the face of external threats and the psychological challenges associated with political action. If tension turns into anxiety it is debilitating. Those with charismatic traits usually possess the insight that makes them good at holding, but the insight rarely extends to themselves. Leaders with charismatic or people-oriented traits can prevent organizations from becoming rigid and losing sight of the vision. However they are often restless, constantly generate new ideas they expect the organization to adopt, and demand personal loyalty from their lieutenants, all of which can generate turmoil and uncertainty.

Administrative traits are important if resources are to be used effectively and with minimal wasted energy, that is, organizations need to be organized. Organizations with more than a few staff rely for effectiveness on a division of labor in which roles are defined and integrated, and internal communications are clear and regular. This is the essence of bureaucracy as an organizational form; although the term is used as a pejorative, bureaucracy is simply the organizational form for all but the smallest groups. It is an effective, if not usually efficient, form of organization, and leaders who understand bureaucracy – who know how to hold the pieces together and make them work, recruit the right staff, and set up functional processes – are an essential part of the organizational mix. These task-oriented people are central to organization building, making and seizing opportunities, and carrying out long-term strategies and campaigns (Ronald Aminzade et al., 2001: 153; Aldon Morris and Suzanne Staggenborg, 2004: 188).

Administrative leaders and bureaucracy have built-in limitations, however. They tend to resist innovation even when helpful, because it threatens their control of their niche (Douglas Torgerson and Robert Paehlke, 2006: 6; Douglas Torgerson, 2006: 18, 16). Bureaucracy relies on expert and technical knowledge, which leaves much important information out. This makes administrators unable to take holistic approaches (which include much outside the control of an organization) and deal with wider and longer term issues and effects.

The traits associated with different leadership types are best looked at as being in perpetual tension. Rather than trying to resolve that tension in favor of one type or the other, or by trying to fix the right balance for all time, it is most useful to see the tension as something that calls for ongoing adjustment on the basis of an organization's changing internal strengths and weaknesses and the political environment.

Leaders also tend to be either self-effacing or self-aggrandizing (Ronald Aminzade et al., 2001: 131–2). One need only think of Nelson Mandela and Mao

Tse-tung. Both types are concerned with their historical role, both are probably motivated by a big dose of narcissism (they must live up to a grand image and have that image mirrored by others in order to feel good), but self-effacing leaders are able to see the movement as something bigger than themselves and to put it first. They are willing to adjust their role as political necessity dictates. They share power and understand the need for institutions that will survive them. This is not true of self-aggrandizing leaders, who see themselves as the movement and any criticism of themselves or the movement as a personal attack. Their power and authority is their ultimate concern; because they conflate themselves and the movement, whatever reinforces their power and authority is good for the movement or organization. Bruce Mazlish (1990), Wilhelm Reich (1946), Theodore Adorno et al. (1950), and many others have observed that such leaders can be enormously innovative and destructive when they possess great charisma and attract a large following that is unable to see their faults. They are also adept at obscuring their self-aggrandizing behavior from followers and at manipulating their organizations. That such leaders have brought destruction on organizations, movements, and entire societies is not solely attributable to them. Social structures and circumstances make such leaders possible as does the desperation of large numbers of people who look for a leader to deliver them from misery. In egalitarian societies humans are pack animals and look to individuals to lead only in those endeavors in which they excel. In hierarchical societies humans are herd animals. Unlike other herd animals this quality of humans is a pathological condition, so it is not surprising herds follow pathological leaders.

17 Organization and identity

To maintain mobilization organizations must foster participants' identification with them, with the movement, and its purposes. To achieve this organizations must engage supporters in ritual and action, and demonstrate success. Multiple organizations are necessary to provide different types of participation.

Identity refers to the cognitive, emotional, and need-related aspects of the self, whether conscious or not. Developmentally the achievement of identity integrates the aspects of personality, enables an individual to feel grounded in themselves and the world, and answer basic ontological questions. Who am I? What's important to me? What do I want? What do I need? What feels right? It includes impulses, the expression of identity in behavior, and the internal responses to behavior. Although identity integrates aspects of personality it is not a seamless integration; tension among emotions, needs, values, and actions always exists. The self is also defined in terms of relationships with others.

Initial mobilization depends on at least a tentative identification with a group or cause, although other factors (cohort bonding, the attraction of action) may be dominant. Sustaining mobilization depends on explicit and deepening identification such that participants regard and feel the organization and cause is an important part of who they are. Identification entails emotional attachment to the cause, organization, and those involved in it, making the cause part of one's own story and fitting one's own story within the movement ideology and mythology.

Nurturing identification with conservation requires social venues in which people can interact on matters important to the cause and organization and to meet personal needs. A nongovernmental organization (NGO) need not provide all social venues directly. It can rely on other sympathetic institutions to do this, but the NGO must ensure it is identified with the venue. Overreliance on other institutions can result in divided loyalties.

Needs and sustaining mobilization

NGO supporters must be satisfied that they belong, are effective, and have a worthy purpose; they must feel their safety is being looked after; they must find pleasure; they must feel committed to an organization and cause and be able to say to themselves "this is where I belong," and they must feel supported by the organization (Susan Clayton, 2000: 304–5) and that all of this will continue. Caring must be demonstrated, not just proclaimed. Which doesn't mean it is a one-way street; movement supporters should be involved in NGO's efforts to address these issues, not just leaders and senior staff. Among most conservation supporters deep commitment is lacking and it is not central to their lives (Deborah Guber, 2003: 5, 32–52).

Once an individual, cohort, or group is mobilized a major step has been taken and the presence in the organization of friends who have joined is one factor in keeping them mobilized. But many other groups and purposes are competing for their attention and the presence of friends begs the question, which simply becomes: how to maintain the cohort's mobilization? There are a number of specific actions NGOs should take to address belonging, safety and livelihood, efficacy, pleasure, and purpose needs.

- Leaders should communicate regularly with staff, supporters, and partners about the NGO's vision and mission and why it is important. It should never be assumed that the importance is understood and in any event it needs to be continuously reinforced. The more fully supporters understand mission and vision the more likely they will become important personal priorities.

- Interaction among supporters and others involved with an NGO should be structured to make plain and reinforce the worthiness and importance of the NGO's vision and mission. Many movements have accomplished this through a variety of collective events, including those involving communication on organizational purposes and goals and opportunities for public displays of commitment by supporters (e.g. Doug McAdam et al., 2001: 168; Elisabeth Wood, 2001: 267–8). Collective action proclaiming the NGO's purpose to a larger audience also reinforces the purpose and makes participants feel worthy through association with the purpose.

    The importance of purpose to identity is made plain by the amount of greenwashing that occurs. Businesses and others would not be touting their products and services as green if it didn't help sell them. Business knows people want to think of themselves as green. And businesses know they are really selling identity and not just a product (Robert Cox, 2006: 374–84). If not challenged, as Cox points out, people may think they need not reduce consumption to realize conservation goals. They may also be misled into thinking green consumption instead of collective political action is enough.

- Collective action and other mass events deepen feelings of belonging by evoking and reinforcing comradery. Gatherings that include singing and passionate speeches energize people and provide a sense of power and solidarity (Emile Durkheim, 1995 [1912]). Song has been extraordinarily important in sustaining the collective identity of the US civil rights movement. Songs that lend themselves to shared performance – those easy to sing, that rhyme, and are stirring – are best at creating belonging (Jeff Goodwin and Steven Pfaff, 2001: 291).

- Mass action and events foster close personal relationships that provide encouragement, raise the costs of dropping out, and enable supporters to resist outside pressure to quit or become an informer (Jeff Goodwin and Steven Pfaff, 2001: 287–8).

- Public demonstrations permit a group to define itself in relation to others – those who are not members and those who are opponents. This definition reinforces being part of a group or cause. Polarization often imparts a sense of "no going back" – a more available and less dramatic version of burning the ships (Doug McAdam et al., 2001: 322). Excessive polarization can also lead to a group's sense of insularity and superiority. Some US antiwar activists in the 1960s wrote off most Americans as too dumb or self-interested to join their cause and effectively isolated themselves – not a good strategy in most instances.

- More intimate bonds (sexual, familial), or the prospect of finding those in movement venues or social networks, bind people to organizations and groups. They can also distract from political work and cause people to leave an NGO if they go bad.

- Movement venues for social interaction around nonpolitical interests (e.g., astronomy, travel, gardening, childrearing, solving personal problems or dealing with losses) deepen attachments to others in the movement and to the venues and networks that "hold" such relationships.

- Political action involves psychological and social risks as well as political ones. Creating an organizational environment that addresses risks so that people feel safe is important. Addressing psychological and social risks requires a setting in which people can express their fears or uncertainties and receive reassurance from peers and leaders about the worth of the cause and the importance of action. NGOs must also be prepared for repression by taking security precautions and by preparing supporters and staff so they do not feel powerless in the face of repression (see Chapter 19). Repression can strengthen identity with cause and NGO, but it can also destroy organizations.

- Some social movements (e.g., antiapartheid, US labor, most revolutionary struggles) have provided many material benefits to members and supporters including shelter, protection from violence, food, jobs, and health care and these have sustained mobilization (Elisabeth Wood, 2001: 267–8; Doug McAdam et al., 2001: 168). In poorer countries some conservation NGOs provide important material benefits such as jobs or access to education. As globalization unravels already threadbare safety nets this will become more of an issue and difficult to address given limited resources. Although conservation NGOs lack the power to reform economic structures they must give thought to economic alternatives that are ecologically friendly. Because of the growing impact on Nature of expanding human consumption and population, conservationists cannot much longer ignore the false promises of sustainable development and the delusion that no hard choices need to be made. Real, ecologically benign livelihoods are needed, and conservation must address their creation. Offering high-interest and high-fee credit cards from banks of questionable reputation is not a meaningful material benefit.

- In stressing the near-term benefits of adopting conservation policies on human well-being, such as clean water and air, safety needs are affirmed. Clean water and air have high salience. The US animal rights movement provides consumer and health information on factory-farmed foods as well as companion-animal health insurance and conservationists should look at this model. In some parts of the world NGOs are operating businesses that bring in revenue while meeting material needs of supporters and the larger society, for example, restaurants, birth control clinics, insurance services, and telephone services (Michael Shuman and Merrian Fuller, 2005).

- Climate change is a growing source of fear and anxiety and warming will continue even if emissions were markedly curtailed soon. As conservationists attend to maintaining habitat permeability, enabling species to move in response to climate, one result will be the maintenance of ecological resilience and services – a material benefit to humans, not just supporters. (Broad benefits generally provide less incentive to remain mobilized than benefits accruing only to supporters.) Climate also provides the opportunity to raise the salience of the notion that humans are not smart enough to manage Nature (David Ehrenfeld, 1979) and that humanity's best chance lies with keeping enough Nature intact to keep things going. This is the basis for a strong appeal to purpose: it is the task of conservationists to give natural systems the room to operate.

- Conservation NGOs can do a much better job of providing supporters with consistent networking opportunities that are so important to finding jobs, services, and much else.

- Wilderness and other species are essential to human emotional health. Protecting the wild means protecting this lifeline. As the world appears poised to slip into another dark age of fundamentalist inspired slaughter (Mark Juergensmeyer, 2008) this lifeline is all the more important to the health of the rest of humanity.

- People find a sense of efficacy and take pleasure in being part of an organization that achieves things. Sustaining mobilization requires that NGOs go beyond touting accomplishments – they must link them to the *work* of members and supporters. Without this linkage – and some basis for it in reality – commitment does not deepen.

- Much pleasure is taken from the many relationships and actions discussed just above. Acknowledging this in internal communication helps keep it in people's minds.

Emotion and sustaining mobilization

The conservation movement's support among most self-identified supporters is weak (Deborah Guber, 2003: 36, 51). People have no deep emotional attachment to conservation goals or NGOs (Ronald Aminzade and Doug McAdam, 2001: 37–8). This must change if mobilization is to be sustained and NGOs are to become more effective. When supporters become emotionally invested in an organization, a cause, and others involved, they are more likely to remain committed and take positive steps to stay involved and avoid ostracism.

What can NGOs do to strengthen emotional attachments and thereby identification with conservation? Many things. Most of the actions noted in the previous section of this chapter not only satisfy needs but in doing so strengthen emotional attachment to the objects that meet those needs or give pleasure. Successful NGOs recognize that among their tasks are acculturating supporters to a particular emotional orientation toward goals, mission, participation, and opponents; they must also nurture emotional change, such as deepening commitment and overcoming fear of taking risks. Organizations communicate an emotional tone via their activities and messages; this needs to be done deliberately rather than haphazardly. People are attracted to different organizational moods and an inconsistent mood can undercut mobilization by causing confusion and disappointment. This is one reason a movement needs different types of organizations. Some people want to be part of feisty organizations involved in the contentious politics of pushing what's possible; others want to be part of NGOs that are businesslike and focused on achieving specific, incremental goals such as land purchases or rule-changes. When NGOs harmonize the orientations of those mobilized with the actions the organization requires of them participants feel emotionally at home. At home does not mean always comfortable; pushing people is necessary because many are not inclined by habit to political action.

Not all those with the same emotional orientation toward collective action, for example, comfort with contentious politics, have the same emotional expectations of an NGO. Individual, cohort, and group expectations vary (often nonconsciously) depending upon level of commitment, past political experience, and other factors. Some people act from fullness – they love the natural world and must express that love in action. Others act more from emptiness – they need love and recognition and often require excessive attention from peers and leaders. Levels of political sophistication vary. Many lack the political experience to put evitable frustrations in context. These differences affect the dynamic of supporters' emotional attachments and must be tended to. If individuals or groups are to be written off because they require too much investment it should be an explicit decision.

- Leaders at all levels need to model desired emotional expression (and justify it ideologically) in order to set an example and to give permission to the hesitant to feel and express emotion. Many NGOs in the US women's movement focused on giving women permission to feel and express anger and to act on it politically thereby increasing levels of mobilization (Ronald Aminzade and Doug McAdam, 2001: 39).

- Public displays of emotion maintain and strengthen identification with the cause and NGO (Goodwin et al., 2001: 23, 291). Shared emotional expression reinforces emotional responses such as anger at injustice, pride in taking a stand, and happiness with accomplishments. "Keeping together in time" as when marching, chanting, or singing, bonds people to each other, to the purpose for which they are assembled, reinforces determination, and helps overcome fear (William McNeill, 1995; Eugene d'Aquila and Charles Laughlin, 1975; Jeff Goodwin and Steven Pfaff, 2001: 293–4). The expression of emotion – the flow of energy and interactivity – is usually experienced as pleasurable and often cathartic. At a minimum shared emotional expression legitimates the emotions expressed.

- Events or other attention focused on movement or NGO achievements evokes and increases satisfaction with being associated with an NGO and its purpose. This is particularly so when the communication *credibly* references participants' role in the achievement. To know that one has protected a place feels good; to be recognized and applauded for it, even better. Both increase identification and nurture mobilization. Flattery is okay, but when people believe praise is fatuous the source loses credibility.

- Settings other than narrowly political ones (e.g., rallies and meetings) should be provided under the auspices of NGOs for people to engage each other, form friendships, and otherwise find in each other emotional satisfaction. Opportunities for satisfaction also mean opportunities for disappointment. Particularly in smaller organizations, personal attachments that become disappointing can be disruptive and easily spill over into organizational matters (Jeff Goodwin and Steven Pfaff, 2001: 288). When political differences are entwined with intense personal feelings or cliquishness organizational politics can get as nasty as familial or religious disputes.

- Emotional expectations that are part and parcel of the leadership-led relationship produce positive and negative emotional responses. Mature leaders can help head off or diffuse problems by avoiding too much focus on themselves. The more active movement participants are, the more they contribute to development of strategy and to executing it, the more likely they are to look for rewards (and deal better with disappointments) in their own work than in looking exclusively to leaders.

- Regular reminders to supporters that they are important people doing important work fosters pride and accomplishment. There is pleasure in causing discomfort to the comfortable who afflict the Earth.

- Getting people out to the places they love or in contact with the species they care about is a means of expressing their love and reinforcing it. Shared expression adds to the reinforcement. This is the great value of the US Sierra Club's outdoor programs.

- Engaging supporters in political action aimed at realizing NGO goals enables them to learn by doing (demystifying politics) and overcome any fear of politics. To succeed in this there should be regular opportunities for people to discuss their experiences and how they feel about them. In these settings people receive the support they need when their naivete is challenged, for example, when a political fight is lost because opponents didn't play by high school textbook rules for civic engagement. To prevent energy lost from disillusionment, another model of politics – grounded in realism and the notion people can still make a difference – must be offered.

- Pathways that accommodate emotional development must be provided so that as supporters grow and change, roles are available commensurate with their development – roles that continue to nurture them and allow satisfying expression of emotion. People change emotionally for a variety of reasons. Political experience changes them. As they progress through the life cycle emotional focus shifts – young adults want to change the world right away and those in their fifties are looking to pass on experience and values.

- Emotional development includes burn-out and what psychiatrist Robert Coles (1964) terms soul weariness. In the face of ongoing, intense, and risky political action, for those who experience everything intensely, or when infighting seems endless and progress nonexistent, the chances of exhaustion increase markedly. NGO efforts to increase awareness of the symptoms among activists and providing for mandatory "rest and relaxation" can avoid the loss of important people to burn-out. Leaders should model appropriate behavior;

if a leader can get 'buy' on three hours sleep a night and thrive, keep it a secret.

- NGOs should be sensitive to changes in the political environment that require emotional reorientation and take active steps to bring supporters and others along. For example, organizations that have successfully pursued insider strategies may find that with a change in governmental leadership or economic climate more confrontation is inevitable. A shift to hardball politics requires not only ideological justifications for the emotional shift but explanations that make people comfortable and enthusiastic about the shift. Many of the actions described in this section – for example, events, good and continuous communication from leaders, modeling the emotional shift (think of the transitional characters in *Simplemente Maria*) – are the tools for guiding this change.

- Repression and other attacks on NGOs can enhance identification by evoking solidarity and defiance (Elisabeth Wood, 2001: 267–81). The US civil rights movement was generally strengthened by attacks, though it increased fear in many participants. Addressing security concerns, as noted in the discussion on needs, can ensure fear does not become debilitating and diminish mobilization. Repression also sharpens group boundaries, which not only increases identification directly but also heightens effect and thereby identification (Ronald Aminzade and Doug McAdam, 2001: 43–4).

- Fanning anger toward opponents and those who destroy Nature contributes to enhanced identification and sustained mobilization if releases for the energy are provided, for example, a demonstration at a business's headquarters. Leader and peer approval for feelings of frustration with those who stand idly by, or who claim to care but do not act, also reinforces identification. The existence of free riders can sap the will of those doing the work, but free riders can be an opportunity to make those who do the work feel special and empowered.

Transnational NGOs, coalitions, and networks present special challenges. Cultural differences include differences in emotional predispositions. Forging a truly transnational identity is not just difficult, but may never be satisfactorily attained short of cultural homogenization (Sidney Tarrow, 2005: 7). Culturally associated emotional differences do not create insurmountable problems but cannot be ignored. Thus, conservationists working in western North America across the United States–Canadian border must cope with the Americans' greater fear of predators and Canadians' greater political politeness and desire for more inclusive politics.

Ideology, mythology, and sustaining mobilization

Without emotional attachment to some aspect of conservation it cannot become part of identity, but attachment is not enough. People must explain the world to themselves including their place in it, give it all purpose and meaning, and find those explanations, purposes, and meanings mirrored by others. So in order to sustain concerted political action, more than passion must be sustained – ideological justification for the passion is necessary (James Combs and Dan Nimmo, 1993: 145; Ronald Aminzade and Doug McAdam, 2001: 40). Ideology and mythology are required. All movements have an ideology (and variants) that describe what is, what ought to be, and *how to get from is to ought*. There is no political organization without an ideology, although it may not be well worked out or comprehensive. The authority of leaders in large part rests on their providing a functional and appealing ideology that sets out an organization's purpose and program.

Ideology is generally confined to the political realm and the political is not compelling for many; explicitly nesting ideology in mythology increases its salience and makes it more effective at sustaining mobilization by providing a strong connection to the sacred and larger purposes. The activist's reward is closeness to the sacred (Roger Gould, 2003: 250–1). Although ideology, when functional, provides and links higher *political* purpose with practical, day-to-day guidance, it generally does not offer the broader meanings and purposes mythology does.

Ideally it is possible to nest an ideology in more than one mythology so that some supporters do not feel excluded. So an NGO ideology that values and pursues the protection and recovery of native species should be accommodated in appropriate anthropocentric and theocentric mythologies, not just ecocentrism. Nesting conservation ideologies in a variety of sacreds is not difficult and many NGOs have done do. Obviously not all ideologies can be tied to all mythologies and a movement more easily embraces many mythologies than a single organization (see Chapter 21 on the importance of a range of NGOs).

Ideology and myth have the ability to unite people around common programmatic goals despite other differences. This is made possible in part because ideology and myth are somewhat ambiguous and accommodate differing interpretations. For this reason they are also a source of faction. Many disputes over ideology and myth (or sacred propositions) are genuine and have practical significance. But too often the commitment of supporters to one ideology or another is exacerbated or exploited by leaders seeking political advantage. Myth and ideology can become a "hired gun" to support a range of meanings and purposes (Wendy Doniger, 1996: 119). Thus, ideological (or mythological) meanings are a means for leaders to exercise power, a means for lieutenants to exercise patronage, and are only taken at face value by the followers who are mobilized on that basis. Which is not to say either lieutenants or leaders lack an ideology. Karl Rove is an agnostic and apologist for big business; the religiously conservative legions he helped to mobilize for Bush were none the wiser. There is no greater transgression of integrity that a leader can make than to use mythology and ideology for narrow political purposes. To use mythology in this way distorts its adaptive role and ultimately empties it of meaning.

Despite the problems of faction and manipulation that inhere in ideology and myth they are both vital to sustaining mobilization because participants, including leaders, need them. They are indispensable tools for guiding action. Supporters vary in what they require to sustain mobilization. Some want a comprehensive ideology; doctrine is important to them and they rely on its categories to think and reason, at times overriding their emotional predispositions. Others are more experientially based in their decision making – ideology is a necessary but malleable supplemental guide. Differing needs for certainty and clarity cause some to cathect ideology more or less strongly and to be more or less able to accommodate new information and learn. Some want to shape ideology, others are content to consume it and look to others for guidance. Ideological sophistication usually increases with greater political involvement and decision-making authority. Many charismatic or people-oriented leaders are less focused on working through problems within an ideological framework (leaving justifications to deputies) and rely on their hunches, insight, and judgment. *Successful* movement leaders – even those who have produced wheelbarrows of doctrinal works – have not let attachment to or authorship of a worldview to cloud their perception of reality when reality takes a turn unexpected by their worldview.

The tendency to faction and ideological conflict is moderated when conservationists keep their eyes on the prize and maintain some humility. There are many paths to realizing the vision, and given uncertainties, there is a need for diversity. Each organization must have a focus and this is one reason movements need many organizations. This is not to say NGOs can avoid criticism and conflict over movement direction. Fearless debate and discussion are essential to identifying and correcting mistakes. What can be avoided with regular communication is criticism based on misunderstandings.

The problem of excessive criticism by the ideologically pure is matched by problems generated from the use of ambiguous ideological principles and mythological propositions to hold together a coalition or alliance or to build a very diverse organization. Papering over some differences is part of social lubrication, but obscuring serious differences – for example, that liberty meant one thing in the North and another in the South as Abraham Lincoln (1953–1955 [1864]) noted – can cause huge problems down the road. From the standpoint of sustaining mobilization an ideology is dysfunctional if it is too vague to provide meaningful guidance for action or a framework to reinforce belonging and commitment. There is no algorithm for resolving this tension – it is a matter for judgment about how best to achieve organizational unity around a clear direction while avoiding the pitfalls of overgenerality or overnarrowness.

The population issue within the US conservation movement is a good example of difficulty of dealing with an ideological matter that has important mobilization consequences. Although views may reasonably differ over specifics, one must be blind to ignore the trend of the last 10,000 years – growth in human population and consumption has meant the destruction of ecosystems and species and the trend is accelerating. Although grass-roots NGOs recognize the problem they have little capacity to act other than to point out the need for appropriate policies. Larger conservation organizations have tried to avoid the issue or deny that population is a problem in order to avoid conflict or offending those of their supporters who cannot admit human population is a problem because it runs contrary to their views that limiting population is somehow racist, antifeminist, antireligious, or impinges on the freedom of women or families to make their own reproductive decisions (Roy Beck and Leon Kolankiewicz, 2000). NGO leaders are no doubt correct that challenging or debating these articles of faith would result in demobilizing parts of their membership and take a huge amount of time, weakening the organization in the near term. They weigh their loss of influence on immediately pressing issues against an uncertain ability to affect population policy and its consequences down the road. If leaders admitted their predicament and kept the question open and on the agenda (although policy decisions are postponed) it would be a step forward. But many who have sought to avoid the debate have justified their stance by accepting that population growth either doesn't matter or can't be stopped. This abandonment of strategic responsibility is potentially disastrous for biodiversity.

Some larger US conservation NGOs maintain their membership by avoiding any political controversy or asking little of their members (Ronald Shaiko, 1999; Neil Carter, 2007: 144–54), and usually achieve very little because of that (Deborah Guber, 2003: 6). Groups that ask more of their members and do engage in contentious politics are smaller and have fewer resources and often achieve significant results. Certainly they push the political envelop and consciousness, ultimately making it culturally safer for larger organizations to address important issues.

What steps can NGOs take in the ideological and mythology realm to sustain mobilization?

- The tools important to initial mobilization remain important to sustaining mobilization: messages and stories that are compelling, that explain, that guide, that offer people attractive personal stories and roles, that provide purpose, and that are delivered by those with credibility. Once people are mobilized NGOs have direct channels of communication with them, but these channels must still compete with others offering pleasure, guidance, and information. As with initial mobilization, multiple channels are better than one and presentation counts. Newsletters, E-mail, interactive web sites, magazines, events and gatherings all have a role. Electronic interaction is better than none, but personal interaction at events and other gatherings where supporters can meet each other and interact with NGO staff and leaders creates more lasting impressions and better sustains mobilization.

- The content of communication should address action desired from supporters, explain its importance and why acting is urgent. Calls to action should be placed in an ideological context and general communications with supporters and others in the organization should aim to increase understanding of the organization's ideology. Those new to an NGO or movement require more exposure, as do people during times of strain when they are apt to question their commitment. The less internalized an ideology the more exposure supporters need to it. Presentation of ideology must not simply explain what's going on and the path forward but how supporters fit in.

- Acculturation to ideology should be formally organized and involve discussion. It should provide a path over time to deeper ideological understanding. As commitment deepens – and to reinforce deepened commitment – a more comprehensive understanding is needed of the problems the NGO sees as primary and its role in solving them. As individuals take on more responsibility, especially decision-making responsibility, their engagement in the organization's ideological development as well as their own becomes more important.

Because NGOs recruit people at different levels of ideological sophistication it is important that acculturation efforts start where people are at or they will become lost and lose interest.

- There are some across-the-board lessons acculturation to ideology should emphasize: what conservation is up against and the need for basic societal change in order to meet conservation goals; that politics is not about reasoning but about the relative resources of opposing parties, using resources intelligently and forcefully, and taking advantage of available opportunities. Experience does not teach so well on its own as when combined with a clear intellectual and ideological framework. Experience is also a more fruitful teacher when people are able to observe how others work and see what succeeds and fails. Ideological development and accompanying emotional changes do not transform personality in fundamental ways – turning the anthropocentric into the ecocentric – but it makes the stakes clearer and supporters come to see more clearly the need for thorough-going societal change and the purpose of long-term goals.

- Defining conservation goals in terms of principles of justice must be a central aspect of acculturation. Although the sense of justice is emotionally based, explicit codification of justice principles and their relationship to conservation goals (What is a just relationship with Nature?) and the proper means of achieving them, build identification with NGO programs. Through an understanding of principles, supporters become participants. Like abolitionism, conservation offers the moral high ground – standing for the web of life and against the brutal stupidity of accumulation. To sustain mobilization requires ongoing affirmation of the legitimacy of these principles in the minds of supporters and other important constituencies (Doug McAdam et al., 2001: 319).

- A good example of the importance of imbuing and affirming principles of justice can be seen in the fight over regulation as a tool to achieve conservation goals. Because regulation has been effective at reigning in those whose actions degrade natural systems (Barnaby Feder, 2004: C13), and because it is difficult for them to attack conservation goals on their merits, opponents have attacked regulation per se. They have argued convincingly to many that state regulation is elitist and therefore unjust – a way of shoving things down people's throats (Robert Durant et al., 2004: 496–7). Conservationists must make clear to supporters and others sympathetic to conservation goals that regulation is a just (and vital) tool in dealing with powerful entities. Moreover, regulation needs popular political muscle behind it to ensure regulation is not deflected by its targets through the capture of legislatures, regulatory agencies, or other means.

- Whenever possible principles of justice and other aspects of ideology should be linked to tradition – that is, long-standing principles. If traditions on point are lacking they can be invented – a phenomenon common in politics (Hobsbawn and Ranger, 1984: 1, 8–9). Conservatives have made good use of this technique, and not because conservatism is intrinsically about tradition (it isn't). Progressives have generally rejected appealing to tradition, invented or otherwise, because it is regarded as an antirational justification for a principle. Nonetheless invented traditions are hugely important in state, institution, and movement building (Hobsbawm and Ranger, 1984: 267, 283; Amitai Etzioni, 2004; Olav Hammer and James Lewis, 2007). A major legitimizing US holiday, Thanksgiving, does not date from the 17th century as many Americans believe, but was created in the mid-19th century by state proclamations; a justification given in the proclamations was the celebration's lineage dating to the first 17th century pilgrims (Diana Muir, 2004: 209).

- Communication and reinforcement of principles of justice, mission, vision, and other aspects of ideology and mythology are economically and forcefully conveyed through symbols and icons. These symbols or icons often represent values and feelings that are part of identity and this tradition-like aspect to them adds to their effectiveness. If not already cathected they become so through their use in communication and association with what the NGO stands for. Symbols appeal to the cliquish aspect of identity – something marketers and branders well understand.

- To sustain most supporters conservation must fashion a broader social agenda. It cannot confine itself to trying to influence others' vision of a good society so that they are compatible with conservation. Conservation certainly cannot rely upon others to come up with a ecologically sound vision of human society. It is just as certain a broader social agenda will be divisive within conservation, unattractive to some potential recruits, and a sticking point with allies. But what is the alternative? To abandon the field and hope for the best? Without a broader social vision that addresses the broad range of problems a society faces, most people will turn elsewhere and conservation will remain a subsidiary concern. A conservation-fashioned vision for society can ensure that the social relations necessary for reversing ecological degradation and maintaining its integrity are central and more importantly that all other elements of the vision are compatible. To address division caused by the need to settle on a broader vision, NGOs will need to allocate time to the internal debates necessary to achieve understanding and unity around that vision, and then to promulgate it just as others do, even as they pursue coalitions and alliances around priority conservation issues.

- Ultimately conservation success depends on fundamental ideological and mythological shifts among politically important constituencies. Keith Hart (1999: xvi) observed that a new mythology is needed "founded on post modern science grounded in ecology rather than astronomy. Human society would be conceived, then, as inside rather than outside the planet." It would call on humans to live as citizens of the Earth rather than its conquerors (Aldo Leopold, 1987 [1949]). This change will mostly occur generationally and the movement as a whole must attend much more to the broad socialization and acculturation process. But that is another book. It is also important that current leaders identify the next generation of leaders and work to ensure the core of this new mythology in them. Fundamental mythological change also results from profound personal and social crisis. When existing explanations and purposes fail people become open to alternatives, but alternatives must exist and be attractive (Anthony Wallace, 1956, 1970). Conservationists need a mythology at the ready and must know how to communicate it effectively. What does this have to do with sustaining mobilization in the present?

It means devoting existing resources to longer-term mythological development and propagation, but it also means recognizing that some people come to the conservation movement seeking better explanations of the world and purposes consonant with a vibrant Nature (Lewis Rambo, 1993; Dillon and Wink, 2007, 119–36). Existing ideologies and mythologies aren't working for them but they are not able to construct something new on their own. Although historically the emergence of new mythologies is popularly associated with a prophet they are invariably products of their times and of many people. By identifying and encouraging those seeking a new mythology, NGOs sustain mobilization based on an ecological understanding of the world – a task essential to the long-term effectiveness of conservation.

Conservationists have already started on this mythological task. New stories tell about who conservationists are, where they come from, why they must dedicate themselves to conservation, and how to achieve their vision. The Earth First! of the 1980 s was probably the most successful example of culture creation in North American conservation since John Muir's voice was heard. The forms of action Earth First! pioneered only partly explain its effect on the larger movement; its ideological and mythological innovations embodied in story, song, and dance generated an influence far beyond its member numbers and meager budget (Martha Lee, 1995: 49–50).

Identity, strategy, and sustaining mobilization

Although the opponents and sources of problems conservation confronts are the same ones many other human groups confront it does not mean there are common solutions. The Green parties of Europe proclaim common causes for the range of primary problems they address, including biodiversity loss. But the policy solutions they pursue are neither focused on biodiversity nor have much effect on it (Neil Carter, 2007: 48). Although ecological responsibility – one of four primary Green goals – is given ideological priority and includes recognition of limits to growth, it is not seriously pursued. The most successful Green political party, the German Die Grunen, is focused on unemployment, citizenship laws, and energy taxes and policy (Neil Carter, 2007: 123–4). Conservation is not getting lost in the pursuit of power or in political haggling over coalition formation but is a weak force to begin with. Its place is even weaker in more mainstream parties. Although EU conservationists are piecing together what's left of relatively natural areas and having some success in protecting areas where predators remain, the fate of a brown bear that entered a southern German forest is telling: it was shot. Meanwhile the English debate with much gravity whether the ferocious beaver ought to be restored to Dartmoor and Exmoor.

Lacking strategies based on a broader social vision that puts conservation interests paramount, it is not possible for the conservation movement to be clear on where its interests lie with regard to other issues and where its interests diverge with allies. No other social vision gives primacy to conservation. Conservatives place "the market," profits, and the self-aggrandizing ego first. Progressive's see development as the solution to many global problems and are willing to balance the books on the back of Nature rather than constrain the overall human footprint. Whether it is Teddy Roosevelt's decision to defer to the Hetch-Hetchy dam in Yosemite (*Battle for Wilderness,* 1989), the constraining influence of the wealthy on the creation of the Redwoods protected area (George Gonzalez, 2001), or the actions of colonizers whose concern for wildlife was limited by their insatiable demand for land, resources, and cheap labor (Peder Anker, 2002), the outcome is the same for other species. Conservationists will forever be unprepared when allies' interests diverge and their own supporters sympathize with diverging interests. They cannot effectively address those divergent interests without a conservation alternative. This is how self-proclaimed conservationists get sucked up in confused notions that conservation must support development even if it damages biodiversity.

Identity and sustaining conservation over the next decade

Conservation mobilization is under siege from many sources. However, burgeoning population and increased consumption in societies with skyrocketing growth rates. Traditional opponents – profiteers and power seekers – continue to mount ever more sophisticated attacks on the natural world, fatuously justified by a concern with the environment and with the needs of the poor. Leaders of some large conservation NGOs – out of a desire to be taken seriously by the powerful, out of timidity or political naivete, to prove their humanitarian credentials, or simply because they don't care enough about wild places and creatures – are willing to place conservation as an offering on the alter of development to gain some dispensations.

To beat back these challenges and move forward – to sustain conservation mobi-lization and keep conservation at the service of the wild – the conservation move-ment must address the following tasks:

- transforming supporters from check writers and fellow travelers into people for whom conservation is central, from people who say "I like this organization and what it does," to people who say "This cause/NGO/movement is part of who I am.";

- creating new tactics that gain attention and keep opponents off guard (Doug McAdam et al., 2001: 319–20, 332);

- establishing and nurturing a conservation community that deepens commitment to necessary political action by meeting supporters' needs, strengthening their emotional attachment to conservation, and enhancing their understanding of politics, conservation goals and the means of reaching them, and the direction social transformation must take to make societies compatible with biodiversity and wild lands protection. In authoritarian and poorer countries safety and livelihood needs must be a major strategic focus;

- establishing programs and organizational pathways that encourage supporters' growth and development emotionally, organizationally, politically, and ideologically, for example, from the recognition of specific complaints to an understanding of their systemic causes and ultimately to the need for fundamental social change and the capacity to take it on;

- establishing programs that socialize and acculturate the next generation and nurture future leaders;

- encouraging development of ideologies that encompass a conservation compatible vision of human society that is concrete enough to organize around and negotiate with potential allies, and broadly attractive enough to gain support from those for whom conservation is not central;

- encouraging development of conservation-based mythologies that provide satisfying answers to questions of ultimate purpose and more specifically answer how humans can live with the natural world;

- strengthening conservation NGOs and the movement as a whole so they are more attractive allies and are in a better position to bargain.

18 Organization, action, and ritual

Action and sustaining mobilization

When supporters of a movement or nongovernmental organization (NGO) join in collective action it increases their identity with the cause and the organization and thereby sustains mobilization. Among the actions discussed in Chapter 17 were rallies, demonstrations, and other public events directly political and otherwise. These activities are important for achieving programmatic goals but may be undertaken solely to initiate and sustain mobilization. The role of action in sustaining mobilization is worth discussing further because it is important to the organization strengthening needed to bring about good conservation policy. Many NGOs lack both the structure to involve people more deeply and the desire to do so – perhaps because leaders feel politics is best left to them as professionals. If much of the US civil rights movement or the African National Congress had adopted this view segregation might still be with us and Nelson Mandela still in jail.

Stronger organizations are a key to conservation's ability to apply greatly increased pressure on decision makers and gain policy changes. Stronger means more supporters, more deeply committed supporters, and the ability to attract more powerful allies. Action contributes to bringing about all of these. It does so in part, as noted in Chapter 17, by increasing commitment to both cause and organization. When people act they embrace the ideological and emotional basis for action more deeply, they develop bonds with others based on shared experience and pleasure, and they feel a stronger sense of determination and power (Lewis Rambo, 1993: 136; David Kertzer, 1988: 119; Ronald Aminzade and Doug McAdam, 2001: 37–44; Carol Tavris and Elliot Aronson, 2007: 32–5). These all sustain mobilization and ensure that supporters are available for political action in the future.

Action operates to sustain mobilization in other ways. Humans are creatures of habit. Generally, levels of mobilization are low in modern societies (Amatai Etzioni, 1968: 400–1); minimal political participation such as voting is the norm. Voting on the basis of conservation issues, actively campaigning for candidates, and lobbying or protest are not the norm even for supporters of conservation NGOs. By organizing action that involves supporters on a routine basis, habits change and action becomes the norm. People are then more likely to act in the future. Action is contagious, drawing others in with the release of energy and sustaining those who act with a chainlike reaction (Amatai Etzioni, 1968: 402–3; Joao Stedile, 2004: 67).

Action sustains mobilization because it is efficacious. When habits change and supporters become used to taking action in significant numbers progress is made on achieving goals; even structural change results from increased mobilization (Amatai Etzioni, 1968: 415–7). Achieving goals rewards action, diminishes a sense of limits, encourages participants to reach for more (provided they have accepted long-term NGO goals as theirs), and makes future action more likely (Ted Brader, 2006: 107). Absent immediate progress toward goals, action imparts a sense of efficacy and involvement that sustains mobilization. Indeed, opportunities for meaningful political action are not always available. Elections are sporadic and campaigns call for alternate periods of intense action and waiting. Rallies, protests, annual conferences that combine inspiration, networking, workshops and play, concerts, hikes, weekend bird counts, and restoration activities can help keep fatigue and entropy at bay. These actions deserve ongoing investment. Conservationists will have to create most of the venues because unlike other movements there are few other institutions like universities or churches that easily lend themselves to this role.

NGOs differ in the types of activities (restoration, lobbying, litigating, and protesting) that further their work and they attract supporters on that basis. In each instance it is important that pathways be created for political development, for example, from participating, to planning, to leading a demonstration. Not all will take the same path or go the same distance. But NGOs must become real organizations that accommodate active involvement. Without that even low levels of mobilization are difficult to sustain. Much has been accomplished by organizations of check writers supporting professional staff. But not nearly enough, and turnover is high.

Stressing action attracts action junkies – people who enjoy the adrenalin but show no propensity for the less dramatic work of politics (Cassen, 2004: 158–9). Although protest may bring down a regime it will not be replaced by anything very different unless there are organizations with a vision and the capability to usher in change and stay with it.

Joint action by NGOs at the leadership and mass level is important in sustaining coalitions and alliances. It strengthens relationships and alignment around a goal, campaign, or cause (Frank Dobbin, 2001: 78). Action can strengthen trust and allow for more efficient and effective use of resources, which in turn reinforces a sense of efficacy. Organizational (and individual) interaction is multiplicative rather than additive. For joint actions to be successful a shared framework or common goals must be clearly articulated and agreed to by coalition partner or allies so that expectations are clear. Differing understandings of expectations are a major source of conflict and mistrust – much more frequent than those caused by dissembling or deliberate failure to deliver on promises.

Sustaining transnational coalitions and alliances through joint action – or even coordinated action within a single political entity – presents additional challenges

(Sidney Tarrow, 2005: 7). Because political repertoires are sociopolitically specific NGOs can find it difficult to agree on a joint or coordinated action. Because of differences in political systems or property regimes, actions that produce results in one country can generate serious blowback in another, or just be duds. The proliferation of international institutions and treaty regimes does provide new venues for common approaches to joint action (Sidney Tarrow, 2005: 26–8; Linda Malone and Scott Pasternack, 2006), but often the most effective means of influencing a treaty secretariat, for example, remains via influencing more powerful signatory states and thus their distinctive politics.

Because coalitions and alliances consume resources NGO leaders must calculate the costs and benefits of involvement. If coalitions and alliances are not producing results or movement toward results or saving resources then they cannot be justified. Joint action makes it more likely that they will achieve results, but these must still be weighed against the cost.

Sustaining coalitions and alliances through action need not rely entirely on political action. As with individual NGOs, social interaction at events, hikes, campouts, concerts, and victory celebrations or holiday celebrations creates incentives to remain involved. Social networks are extended at these venues and offer unanticipated opportunities down the road. Activists from North America and the Russian Far East met at an evening party given as part of a conference in Kiev; their connection became important in developing a grass-roots strategy for landscape level protection in Kamchatka.

Ritual and sustaining mobilization

All the actions discussed above – demonstrations, rallies, meetings, and litigating – have ritual aspects. The ritual aspects of action and ritual per se – an invariant sequence of behaviors encoded with meaning the performance of which signifies acceptance of the meaning encoded – deserve more detailed review. The richer a movement or organizational culture is in rituals, stories, songs, and heroes, the fuller and more pleasurable participants find their experience and the more the movement becomes part of their identity. Rituals and other aspects of movement culture engender joy, attachment, enthusiasm, pride, and other emotions. Ritual is particularly effective at evoking emotion and no organization can sustain mobilization without it. Unionists at the Lenin Shipyard in Gdansk – the birthplace and nursery of solidarity – employed ritual extensively. Singing was a regular feature of gatherings and formal meetings (Colin Barker, 2001: 187). New members were always announced, and much of the extensive interaction followed stylized patterns.

Performance of a ritual indicates acceptance of the meaning and purpose of the ritual and obligates participants to abide by their acceptance. It reinforces group values, sentiments, and adherence to group standards of morality (David Kertzer, 1988: 62, 28). Ritual transforms the cultural (that which is created and changeable) into the natural: the purposes, meanings, and orders are seen not as created by humans but to be in the fabric of Nature (Roy Rappaport, 1999: 166–7, 346, 405; Jeff Goodwin et al., 2001: 18–9). Ritual makes the obligatory desirable (David Kertzer, 1988: 40). By evoking strong emotions in a highly structured setting ritual disciplines emotion and channels its expression so that outside of the ritual participants' emotions reinforce their obligations. As a concrete series of physical acts ritual creates physical states – increased bodily awareness, shared focus, and mutual awareness of focus, feelings of solidarity – that result in participant's strongly cathecting that to which they have obligated themselves (Eugene D'Aquila and Charles Laughlin, 1975, 1979). Mere verbal statements – even public ones – rarely generate the same level of emotion and commitment because they lack the physicality and drama of ritual, as well as the feelings of efficacy and power ritual can impart (Terence Turner, 1977: 61; David Kertzer, 1988: 132).

Ritual may reinforce belief, but it does not require that people renounce doubt. Ritual constitutes public acceptance of the obligation to adhere to group purposes while allowing individuals autonomy in their beliefs. It thereby furthers two essential aspects of culture: the replication of uniformity (binding people to a common purpose) and the organization of diversity (integrating differing beliefs, attitudes, subgroups, and personalities; Anthony Wallace, 1970: 23–4, 123–9). In short, ritual enables collective action without a fully shared belief system.

Ritual's invariant aspects take on the authority of tradition (Hobsbawm and Ranger, 1984: 279–80). Both tradition and ritual may be recent, however, despite claims to the contrary; both hide their own invention.

Rituals are usually associated with a particular setting (e.g., a solemn place), with calendrical cycles (annual holidays or celebrations), or are triggered by events, such as elections, a political victory, or in the face of sagging fortunes, waning discipline, or corrosion of purpose. Those rituals most concerned with commitment to a group and its purpose need to be repeated more frequently for those newly mobilized, in the face of serious challenges to purpose, or when action is needed (Roy Rappaport, 1999: 323–5; Amatai Etzioni, 2004: 10–1). Rituals need to be more frequent in a hostile political environment and absent good news about progress toward goals. Rituals can raise spirits and reconfirm the importance and justness of goals.

For all of these reasons ritual sustains mobilization and is thus central to organization and movement building. It sanctifies institutions and their purposes and fosters attachment to group, cohorts, and purpose. Myth, ritual, and organization "produce one another" (Bruce Lincoln, 1996: 166).

Opportunities for mobilization-sustaining ritual

When rituals are performed before nongroup members they are a means of outreach and also define participants as a group. Ritual's primary role in societies and groups is internal – to invest group members in a common purpose. The conservation movement and NGOs have rituals that are consciously used to maintain mobilization. But the movement does not use ritual to its potential for mobilization maintenance. In particular ritual is not used enough to transmit conservation mythology: the overarching purposes and meaning of conservation, why and when it came into being, why it is important, and a vision for the future. Earth Day serves to do this to a limited degree with large rallies and high-profile speakers looking backward and forward, but no comprehensive story has emerged and such a story is needed to sustain people. Failure to develop and inculcate a mythology is partly a failure of imagination and a failure to recognize the need for it. Fear of contention over what the mythology should be and how it might conflict with supporters' existing mythologies are other factors. Contention cannot be avoided and is part of the process of creating a new story. Moreover, contention need not be destructive; for those who are not fundamentalists, adoption of a conservation mythology can be done piecemeal and combined with elements of existing mythologies. Indeed, it is through recombination of elements (with some new ones) that mythologies typically evolve and are transformed, while continuing to distinguish one group from another.

It often makes sense to use the rituals of others and either seek to co-opt them or use them to define a counterritual that propagates new aspects of myth; both can be used to reinforce group identity. Christmas has been reengineered from a pagan carnival ritual celebrating the winter solstice into a Christian version of carnival; from that into other forms of religious celebration, and finally into a secular shopping holiday (Stephen Nissenbaum, 1996). Native American activists and sympathizers have used the celebration of Columbus Day to critique the candy-coated version of the coming of Europeans to the Americas and their impact on the New World's people, other species, and ecology (Francesca Polletta, 2004: 166, 163). It has become a quintessential "teachable moment," perhaps in part because protests and similar events tied to holidays get more coverage. Such efforts are not always successful as the battle over the meaning of the US July 4 celebration indicates: it has not be possible for critics of current US imperialism to drive home their point that the once revolutionary United States is now a counterrevolutionary power.

Most ritual is not about mythological or sacred purposes and meaning but about more concrete matters. Rituals celebrating and raising awareness of the solstices and the equinoxes is a way of highlighting the importance of the Earth's rhythms to humans; they facilitate a deeper understanding and connection with

Nature by enabling people to see through the electric lights and air conditioning systems that dominate their lives. More intensely experiencing the seasons and the way human rhythms are linked to them draws the bonds more tightly and fosters recommitment to healing the Earth and healing injuries to self from ignoring physiological rhythms.

Celebrating the birthdays of conservation leaders and anniversaries of conservation victories is another opportunity to more deeply anchor conservation values and commitment to immediate goals. Rituals that mark victories while also drawing attention to negative trends that call for greater effort reinforce commitment. Marking negative events can help sustain mobilization if a clear way out is shown. Most US conservationists were too timid to use the occasion of US population passing the 300 million mark to point out its devastating effect on ecosystems and other species. It was left to population activists to do so and call for heroic efforts to change direction. Peak oil offers an opportunity to tell a new version of an old story (resource depletion), to celebrate the coming end of an enormously destructive era, and to gain commitment via ritual to a different course of action.

Most rituals need to be tailored to particular groups of the faithful. The many opportunities that offer themselves can be attuned to groups' differing experiences. Rituals typically offer participants some opportunity for individual expression and this allows differences in experience to find expression, creating greater investment in NGO purposes and goals. Thus, a participant might tell their own story of how they became involved in a deeper way with conservation and relate it to the larger narrative of the group. Through these rituals personal narrative, organizational narrative, and myth are interwoven.

Rituals that recognize various types of contributors to the cause – activists, volunteers, donors, leaders, others – are a means of reinforcing commitment emphasizing forms of participation much needed at a particular time. Many NGOs and foundations do this but much more is needed. Joining the desire for recognition with the enjoyment of ritual is a powerful combination. Awards need not require resources much beyond the time involved and should be seen not as a burden or an extra but as an investment in gaining new levels of commitment and action from supporters. It's useful and appropriate that honors be hierarchical – including large cash awards that provide people with the ability to act outside of organizational constraints and indulge their creativity – but it is also necessary to routinely acknowledge all types and levels of contribution.

Conferences and meetings are venues for award rituals but they offer many more opportunities for mobilization-sustaining ritual. Processes for deciding on leaders or policies, networking opportunities, films, plenary talks, and workshops to name just a few, have ritual aspects that can be emphasized. Even business meetings at which budgets are presented or bylaws approved offer opportunities for structured participation that reinforces organizational purposes. When people rise to speak and identify themselves and their place of residence a sense of place may be affirmed or the commitment of participants to travel to attend highlighted. Ritual can highlight the role of supporters in governance.

Other opportunities for ritual exist when organizations attend to the transitions in people's lives related to the organization and otherwise. Induction rituals are very important; they clearly mark membership in the group and create a divide between membership and nonmembership and thereby foster movement toward a new identity that includes cause and organization (David Kertzer, 1988: 17; Lewis Rambo, 1993: 114). Ritual marking or designating other transformations increases commitment and identification (Randall Collins, 2001: 29); these can include celebrating promotions or success at a new task, or launching or winding up a major campaign. Rituals marking deeper involvement, past victories, and honoring participants' devotion or accomplishments also increase commitment and identification. Often boundaries between states of reality are ambiguous and a ritual clears up the ambiguity and maps reality – a campaign starts now, ends now, a person is a member.

Opportunities to link day-to-day life with higher purposes via ritual are important in maintaining enthusiasm with being part of a cause. Audubon Society spring and fall bird counts, Oregon Natural Desert Association fence-pulls, and similar recurrent activities provide opportunities to proclaim commitment to a grander purpose (healing the land). With repetition activities take on the aura of tradition and their cumulative effect can be celebrated.

NGOs can also provide templates for individual or cohort rituals that through their recurrent quality reinforce identity. If, whenever conservationists eat in a restaurant they ask about the source of the food – is it wild, organic, factory farmed, endangered? – their identity is affirmed. So does consulting crib sheets on acceptable fish to buy for dinner and making a habit of patronizing and praising vendors whose practices aid conservation.

Rituals are important links between the local and the national or global, helping to overcome feelings of isolation and providing a sense of momentum. Simultaneously occurring rituals at dispersed sites become a single ritual. Those on the front lines of an action – lobbying in a national capital or protesting a global meeting – are bolstered knowing that support actions are occurring elsewhere; those undertaking support action feel part of the front line action. Even an individual repeatedly deciding to stick with a boycott feels more deeply committed to conservation and more strongly linked with others.

Creating mobilization-sustaining ritual

Conservationists have created ritual and continue to do so, but not always deliberately. Ritual often emerges without intent, coalescing over time as ongoing interactions become stylized and reliant on specific symbols (David Kertzer, 1988: 92). When Del Norte County, California created the Aleutian Goose Festival after discovering that the subspecies thought nearly extinct was actually camping out by the thousands just off the coast, it was looking for a conservation friendly economic boost. But woven throughout the multiday festival's events that draw in hundreds if not a few thousand from across rural Oregon and northern California, rituals have emerged that acknowledge the festival as a conservation success and a great alternative to resource extraction and building prisons. The festival provides opportunities for the conservation community to participate in a variety of events that reinforce belonging and identification with the movement and with wildlife directly. Those in the community who benefit economically see conservation in a new light.

The celebration of Cinco de Mayo in the United States (it is a Mexican Holiday celebrating a military victory) involved a more deliberate effort to create ritual. The day was rarely celebrated in the United States before the rise of the Chicano movement; the movement made it the occasion for celebrating their heritage and instilling solidarity (Amatai Etzioni, 2004: 18). The movement soon lost control of it to commerce as it spread to mainstream groups and communities, becoming another day to drink and eat to excess. Kwaanza suffered a similar fate (Anna Wilde, 2004: 120–30).

Earth First! of the 1980s deliberately created many rituals: annual rendezvous in the woods, songs, ways of dressing, and celebrations of Earth rhythms and movement history (Martha Lee, 1995). The Abalone Alliance used dance and song following staff meetings in order to reconnect with each other at a deep level and repair the tensions caused by disagreements during the meetings (James Jasper, 1998: 192). The US civil rights movement, New Left, labor movement, Solidarity Union (Poland), and countless other movements relied on song, chants, and other recurrent collective actions that reinforce membership and sustain mobilization. One comic commentator on the New Left's affinity for drugs noted that the pleasure taken in the ritual associated with drug use rivaled the effects of the drugs.

The Darwin Awards – given informally by a multitude of observers and admirers of human intelligence – bring at the very least a smile and vague sense of shared perception about the inability to escape the rules of the natural world. They affirm for conservationists the wisdom of Nature and identification with the movement. Formalizing the award or creating a series of awards that honor particular forms of human stupidity aimed at the natural world or self would provide not only a movement affirming ritual but add some much needed humor to very dark times.

New ritual, like new myths, are not made from whole cloth, but fashioned from existing (and a few new) elements recombined to convey new meanings. Sometimes old rituals are co-opted – new meanings and content are given to existing forms

(Lucero, 2004: 525; David Kertzer, 1988: 42). Fundamentalist Iranian clerics intent on overthrowing the shah were able to mobilize devout but nonfundamentalist Muslims by appealing to them to join in the traditional ritual of 40 days of public mourning for fallen martyrs. Thus fundamentalist dissidents murdered by the shah triggered massive demonstrations that included mostly nonfundamentalists. When these were harshly repressed demonstrators took this to be evidence of the shah's anti-Muslim feelings, increasing support for the shah's overthrow (Francesca Polletta, 2004: 160).

Ritual involves the manipulation of symbols – a powerful shorthand (see Chapter 13) – and these are important elements in the creation of ritual. New rituals typically rely on the preexisting power that certain symbols have with ritual's participants. By incorporating symbols such as flags, natural and human-made monuments, icons of nationhood or the people, a ritual gains something of their ability to evoke attachment and identification and puts it to a new purpose (Sally Moore, 1977: 21; David Kertzer, 1988: 121–2, 161). Demonstrations and rallies usually take place at highly symbolic places, rather than at the centers of power (David Kertzer, 1988: 120). Public shrines associated with the demonstrators' aspirations and values are often chosen (e.g., the Jefferson Memorial or site of the Bastille) because their demands are most easily linked to these places and their ideals, thereby legitimating demonstrators' claims. Although seats of power (e.g., 10 Downing Street) carry their own symbolism and are sites of demonstrations, they present greater obstacles to effective communication and claim legitimation.

Conservationists can also sustain mobilization through participation in broader community rituals. This also contributes to establishing a positive NGO presence with others. Several conservation groups located in rural areas offer scholarships to graduating high school seniors, making themselves part of an important right of passage for students and community. Awards are usually given to someone in the natural sciences but there is no political litmus test. Conservationists take pride in being identified with contributions to the community; and in the case of scholarships made at graduation, opportunities are presented to highlight alternative livelihoods.

Just as NGOs designate staff to carry out a campaign they should designate staff to develop organizational rituals. Although rituals already permeate campaign activities, meetings and rallies, paying explicit attention to them can greatly enhance their effectiveness at sustaining mobilization. Although cultural creativity has top-down aspects it is more passionately embraced when it harnesses supporters' creativity. People rely and thrive on ritual from the most mundane – talking over coffee to keep personal bonds alive – to celebrating high purposes, great political victories, and investitures of national leadership.

19 Organization, efficacy, and repression

Efficacy

When people feel they are making a difference it energizes them. It braces them in the face of adversity and makes both sacrifice and success meaningful. Maintaining feelings of efficacy is neither a matter of instilling a psychic state unrelated to actual progress nor is it reducible to actual progress. They run on parallel tracks and the relationship can be complex.

Sustaining mobilization and ungrounded hope

Most people tend to feel effective although this predisposition can be socialized out of them by chronic mistreatment. But all is not made easy for conservationists by the majority's predisposition to feeling effective. This predisposition is associated with a tendency to see the world somewhat less accurately than those with darker perspectives (Joseph Forgas and Michelle Cromer, 2004). So those who are most prone to remaining mobilized because of this predisposition are also those least likely to see how dire the situation is for the natural world and the necessity for uncompromising political action. It is argued that feeling highly efficacious and overestimating one's chance of success because the world is seen less accurately is adaptive (Dylan Evans, 2000: 178–9). This can be so if failure rarely has serious costs and the few successes that do occur are beneficial and outweigh the costs of the more frequent failures. (Failures that result from overestimating the chances for success may also lay the ground for later success because they change the political landscape. Slave rebellions undertaken with high and ungrounded hopes but that were unwinnable nonetheless contributed to setting the stage for eventual emancipation.)

What was once adaptive may no longer be so. There is less room for error in modern politics because the consequences can be disastrous (Wright, 2004). The ecological impact of the errors inherent in humanity's long-standing trial and error approach to adaptation with its wasted effort and missed opportunities is no longer good enough in the face of species extinctions, climate change, and other portents of biological meltdown. A rosy view of things and ungrounded hope have become dangerous. After all "(h)ope elects the politician with the biggest empty promise; . . . Hope, like greed, fuels the engine of capitalism." (Wright, 2004: 123). Hope misleads. It ". . . is only man's mistrust of the clear foresight of his mind. Hope suggests that any conclusion unfavorable to us *must* be an error of the mind. And yet the facts are clear and pitiless . . ." (Paul Valery, 1919).

Conservation nongovernmental organizations (NGOs) seeking to sustain mobilization are thus faced with a conundrum. Keeping people mobilized depends in part on appealing to an inbuilt sense of efficacy which is closely associated with ungrounded hope and a brighter view of the world than it merits. At the same time NGO decisions need to be grounded in reality to get things right politically. If NGOs speak too bluntly about their view of reality on which their strategy is based it is possible to demobilize supporters. Part of the answer lies with leaders' awareness of the problem.

Of course too much reliance on leaders is always dicey because they are often driven by narcissism and other impulses which are just as problematic as reality distorting hope – all interfere with good analysis and sound intuition. Yet there is little choice. If leaders and strategists do not temper their own hope and instead succumb to the uplifting stories they promulgate to maintain the spirits of supporters, the results can be catastrophic for the natural world. And even if leaders can operate without reality distorting hope while allowing supporters to indulge it, this can come back to constrain leaders – they may not be able to bring their supporters along with them to take the steps necessary to win a fight. It's not clear what Franklin Roosevelt's fears were when he told Americans on his coming to power that the only thing they had to fear was fear itself. It is clear that leaders' management of these issues requires some fine balancing.

Organizations can help keep themselves and leaders on track by ensuring that leadership circles are open to a range of views from within the organization and from outsiders that have useful perspectives. The institutional insanity that is all too common in powerful hierarchies where leaders live in a bubble can also infect NGOs. Openness is a powerful antidote to reality distortion whether from ungrounded hope or reified thinking. Leadership circles need a culture of critical thinking and differing personality types that bring a variety of perspectives to decisions (Aldon Morris and Suzanne Staggenborg, 2004: 181–2). Overlapping responsibilities among managers is another means of creating the internal tension that pushes differing views to the fore. These steps together can help NGOs avoid their leaders closing ranks around a distorted view of reality and best manage the issue of ungrounded hope and efficacy.

Other circumstances should be considered in managing the hope–efficacy conundrum. Groups of supporters vary in their level of political understanding and sophistication and the risks they are exposed to. The greater the political sophistication the less delusional a group is likely to be; the greater the physical risk the more likely people need to believe that history or the gods are on their side and victory inevitable, although this need does not necessarily (but often does) impinge on accurately reading the political landscape in the present. The presence and intensity of the demobilization forces at work also varies, and ranges from the incessant distraction typical of developed societies to quite deliberate efforts to discourage political support for conservation. The more intense these forces are the more corrosive of mobilization they are and the more energy needs to go into sustaining feelings of efficacy.

Demobilization pressures and responses

Erosion of efficacy and resulting demobilization is caused by factors in addition to risk, distraction and the work of opponents directly aimed at making conservation supporters feel, hopeless or ineffective. The significant length of time conservation victories can take discourages many (Michael Gunter, 2004: 24, 174; Dieter Rucht, 1999: 209). Policy change can take years or decades; seeing the results of its application even longer, especially in the face of continued resistance by some sectors of society. Many current problems are the result of bad decisions made in the distant past and good decisions made now may take an equally long time to show results. But it's hard to know for sure.

Thus, although measures of real conservation progress are vital, they are inadequate to sustain mobilization. Efficacy-reinforcing milestones are an important adjunct and contribute to maintaining a sense of momentum among supporters in the face of the long term or disappointment. Milestones may be derived from real measures but presented in the manner of community fundraising campaigns that mark increments of money raised, that is, the increments themselves mean nothing in relation to achieving the goals for which the funds are being raised. The broadcast of favorable news coverage or a documentary can be celebrated as an accomplishment even though it may have no measurable effect on achieving policy change. Finding the positive in the midst of setbacks can help sustain efficacy, such as when conservation's opponents win by portraying themselves as conservationists thereby acknowledging the dominance of conservation values. Efficacy-based milestones can also emphasize actions within the control of conservationists: restoration activities, wildlife counts, symbolic demonstrations of commitment. Dramatizing the many disasters conservation's opponents so readily produce – collapsing fisheries, flooding and landslides due to deforestation, callousness toward the communities that depend on them – contributes to maintaining mobilization so long as events are used to explain what needs to be fixed and how it can be done.

Although efficacy-reinforcing milestones contribute to sustaining mobilization over the longer haul, real progress must be demonstrated and conveyed to supporters. Supporters' role in securing progress needs to be clearly and credibly identified, explained, and stressed in communication to them and celebrated in ritual. Feelings of efficacy cannot be sustained completely apart from real progress.

Achieving real progress depends on identifying opportunities and taking advantage of them, constant tactical innovation, exploiting the weaknesses among opponents while hiding one's own (Doug McAdam, 1997: 397, 399), and making use of crises and shifts in power relationships (Ken Conca and Ronnie Lipschutz 1993: 11; Ken Conca 1993: 318-9; Ronnie Lipschutz and Ken Conca 1993: 332). Conservation success depends on agility – the ability to shift geographical and issue focus, from outsider to insider approaches or vice versa, or a new mix of the two – or to shift venues from the local to national or national to international. Success hinges on resources: the right people and strategy, understanding the levers of policy change, and getting the timing right.

It is easy to understand how failure or delay can damage feelings of effectiveness. Success, while not damaging those feelings, can have the same ultimate result (demobilization), however. Success does this by draining the fervor of members who interpret a victory in battle, albeit usually a major one, as victory in the war (Bert Klandermans et al., 1998: 174). The US women's movement lost steam with the passage in 1920 of the 19th amendment granting women suffrage in federal elections. For many the struggle was over. That many Americans believe much more land is protected than actually is, may be what's causing them to think conservation NGOs are pushing too hard by continuing to ask for more (Campaign for America's Wilderness, 2003: 23). When an NGO appears to supporters as too powerful or to be overreaching and misusing its power, those supporters withdraw.

Repression

Of the many forces external to the conservation movement that affect maintaining mobilization the single most important is repression, be it by government or private actors. Virtually all social progress – the end of slavery, democratization, suffrage for women and those without property, and economic, health, and safety reforms – have been accomplished in the face of serious and sometimes lethal resistance from the powerful and often from those who simply fear change. Although those resisting change usually have the benefit of the rules (they make them) they seldom hesitate to violate rules that get in their way. Even-handed application of the rules in political conflict is the exception rather than the norm. Repression *can* destroy NGOs and set back movements. It can also strengthen them, reinforcing determination and vitality. It always shapes them. Repression occurs in democracies and nondemocratic regimes. NGOs that are prepared for it, understand what to expect, and have a strategy for dealing with it, are more likely to be strengthened by it rather than crippled.

Some conservationists hear the word repression and tell themselves that their group does not engage in any activity remotely illegal or otherwise likely to trigger repression. Certainly those groups with significant elite involvement or support have less to be concerned about. But those NGOs throughout the world seriously committed to conservation are experiencing many types of repression; one need not be engaged in illegal activity to be a target. Further, all of the types of repression discussed below occur in developed democracies, not just under authoritarian regimes.

The following are the most important types of repression experienced by the conservation and other social movements (Murray Edelman, 1988; Frank Donner, 1990; Ward Churchill and Jim Vander Wall, 1990, 1990b; Susan Zakin, 1993; Stuart Ewen, 1996; Laurence Zuckerman, 2000; Bud Schultz and Ruth Schultz, 2001; Gerald Markowitz and David Rosner, 2002; Madelaine Drohan, 2003; Marilynn Johnson, 2003; Athan Theoharis, 1978, 2004; David Cunningham, 2004; David Helvarg, 1994, 2004; Karen Johnson-Cartee and Gary Copeland, 2004; Jennifer Harbury, 2005; Michael Isikoff, 2005; Myra Ferree, 2005; John Tierney, 2006; Robert Cox, 2006; Jules Boykoff, 2007; Jim Dwyer, 2006, 2007; Adam Liptak, 2006a, 2006b, 2008).

- *Direct violence aimed at NGOs and their members.* This includes killing, torture, beating, and jailing, destruction of property (the homes, offices, and records of NGOs and activists); these acts are sometimes under cover of law. The message is simple: quit or be hurt. Because violence has costs, most purveyors first try making credible threats. Nominal democracies such as the United States and European countries have used direct violence against many movements although conservationists have not generally been targeted with lethal violence. Nonetheless France's terrorist attack on a Greenpeace vessel resulted in a death. Industry-backed "wise use" groups have threatened lethal violence and used nonlethal violence. (If one considers the death caused by purveying toxics and destroying habitat to be violence then violence is business as usual.) First-world corporations and first-world security agencies have together and in cooperation with the domestic security forces of other countries, used lethal violence against conservationists in third-world countries.

- *Tolerating and encouraging and violence by third parties.* Officials in the United States, for example, long tolerated and even encouraged Ku Klux Klan violence against the civil rights movement. Over the last 30 years US federal, state, and local governments have tolerated violence against grass-roots conservationists and in some cases arrested conservationists when they were the victims of violence. Two Earth First! activists injured by a bomb planted in their car were arrested for possession of the bomb. Charges were soon dropped, but there was no apology or rigorous attempt to identify the real perpetrators. Activists have been the victims of arson, beating, and other acts under democratic and authoritarian regimes. Amazonian, Nigerian, and other conservationists have been murdered with impunity by interests opposed to their work.

- *Movement supporters are materially deprived.* Businesses do not just fire and blacklist labor activists, but others they frown on, including conservationists. Businesses and governments have sought to destroy reputations, block publications by intimidating publishers, stop research by denying or withdrawing funding. Rachel Carson is a well-publicized case, but similar actions continue. In 2007 a public health scientist at the University of North Carolina – Chapel Hill – who was researching the effects of large-scale hog farms on rural health was threatened with litigation by the North Carolina Pork Council to obtain his research files and files on informants. The Pork Council contacted state legislators, university officials, and the National Institute of Environmental Health Science in an effort to discourage funding and sully his reputation. The Federal Bureau of Investigation and those acting on behalf of business interests have sent anonymous letters to employers accusing researchers or activists of personal and political transgressions, suggesting that should such transgressions be made public it would reflect badly on the employer.

- *Laws banning or severely restricting NGOs or critical public expression, prosecutions, and other harassment under cover of law.* In authoritarian regimes laws limiting expression are common. In more democratic regimes laws protecting expression are ignored and laws originally passed to control organized crime and terrorism are used to attack NGOs. In countries where sabotage and similar activities have long been part of politics, governments have always reacted – sometimes harshly – but now sabotage is labeled as terrorism and punished as such. "Independent" courts play their role by punishing politically motivated acts more severely than the same acts absent a political motivation. Legislatures pass laws that violate otherwise protected political rights knowing it can take years for the courts to strike them down; in the mean time they are used to quell dissent rather than terrorism, for example, the USA Patriot Act. Some US media outlets that are far from being fierce watchdogs for democracy have been threatened with prosecutions under the Espionage Act of 1917 for revealing government wrongdoing. A transnational pattern is apparent in this category of repressive action: critics are branded dissenters, dissenters are branded subversives and finally as traitors or serious criminals. The former Russian President, Vladimir Putin, issued edicts constraining the activity of foreign NGOs in support of Russian NGOs as an illegal interference with Russian sovereignty.

- *Police and prosecutorial harassment.* During the Republican National Convention in New York City in 2004 the police arrested nonviolent protestors and held them at special detention sites thereby keeping them off the streets for the remaining days of the convention. Normally those arrested for such offenses are given summons and let go; *if* held in custody they go before a judge within 24 hours and have the charges against them reviewed and bail, if any, set; usually they are released on recognizance. The New York police made many shows of force during the convention aimed at intimidation. Conservationists were among those protesting in New York. US police agencies have targeted activists for repeated arrests in an effort to intimidate, entangle them in the legal system, and to drain their funds until they can no longer make bail. Grand juries in the United States, created as a safeguard against prosecutorial misbehavior, are used to harass and jail activists.

- *Illegal surveillance and searches.* Unlawful surveillance, including wiretaps, eavesdropping, and break-ins of houses and offices are frequently used against movements. Sometimes these activities are intended to intimidate as well as to gather information. Police let targets know they are being followed, listened to, and have been subjected to invasions of privacy. As anyone who has ever been burglarized or robbed knows, these sorts of actions are deeply unsettling and are made more troubling when targets know those undertaking them have the force of the state behind them. Personal information gathered from these activities is used to pressure and manipulate the targets.

- *Infiltration and disruption of groups.* Police, business or contract infiltrators and agents do not just spy – they disrupt, sow mistrust within and between groups, instigate illegal activity, and plant "evidence" of wrongdoing including plans to commit illegal activity (conspiracy is the first refuge of political prosecutions). Infiltrators are often the most vocal proponents of illegal activity, and in many cases illegal activity would not occur with the funding and other material infiltrators bring. Such was the case with the Arizona 5 in the early 1990s. A favorite tactic of infiltrators is to accuse others of being police or industry spies – so-called badjacketing. Disruptive efforts may be elaborate and include the fabrication of letters and other documents purported to be authored by movement figures disparaging each other. They may inject division into meetings. At one large national social reform conference police agents put forward a motion that Black delegates should have half of the votes at the meeting because of their long history of mistreatment. A debate of several hours ensued, preventing other issues from being addressed, and causing divisiveness. It would not be surprising to discover that *some* of the political correctness on migration and population issues within conservation NGOs has been instigated or encouraged by infiltrators. Agents have even created their own NGOs, issuing statements and newsletters containing disinformation, muddying debates, and making divisive accusations.

- *Campaigns of demonization and media manipulation.* Demonization aims to isolate a group and foreclose meaningful evaluation of their claims by linking them directly or by analogy with groups already demonized. In the United States, for example, labor activists during World War I were falsely accused of having ties to the German enemy, in the 1950s and 1960s civil rights activists and civil liberties proponents of having ties with communists. Conservationists in various parts of the world have been accused of being communists, imperialists, foreigners, and members of pariah ethnic groups to name a few. Media manipulation includes business and government planting stories or cooperating secretly with friendly news media to publish stories containing falsehoods and disinformation; leaking false stories anonymously and then confirming them officially or citing the story as confirmation of the official position or the truth; secretly placing reporters or others on the payroll of government or business so their commentary appears as independent; staging nominally newsworthy events featuring false testimony and giving credibility to dubious sources; funding the publication of books or producing films containing disinformation; strong-arming the media with threats of prosecution, tax audits, or exposure of individual reporter's foibles; denying reporters who are "not on the right side" access to sources; and strong-arming publishers to fire or reassign critical reporters or news anchors.

- *Media instigated distortion of information about and stigmatization of a movement.* The media often need no prompting by governments or other powerful opponents of social movements to treat them with bias to undermine their work. A major tactic is to ignore movements by not reporting on their actions or the problems they seek to address and solve. Nonelite views, if included, are disparaged. When movement activities are covered their claims are trivialized, their strength downplayed, and their accomplishments ignored. Movements or NGOs are portrayed as being only against things, uninformed, composed of social fringe elements, being antipeople, or as threats to order and potentially violent. Scapegoating is quite common as when loss of timber jobs is attributed to conservation policies rather than automation. The media also report negative rumors as fact. Media reliance on he-said-she-said stories often includes deliberate disregard for the veracity of those cited, for example, oil company "experts" on climate change. Bias may be ideologically driven, involve kowtowing to advertisers, or result from the many business interests of media owners.

- *Civil lawsuits to intimidate and drain resources.* Strategic lawsuits against public participation (SLAPP) are used to silence opponents of development projects by claiming millions of dollars in alleged damages. SLAPP plaintiffs cannot prevail legally in most countries absent corruption, but courts do not always dismiss them immediately and their real purpose – to intimidate and drain resources – is realized. Recently industrial hog farmers in the US south have used racketeering laws to try and silence critics, in one case claiming that publicly quoting the author Upton Sinclair, an early 20th century critic of the meatpacking industry, constitutes illicit activity.

- *Instilling uncertainty and ambiguity about what triggers repression.* By creating uncertainty about what actions may precipitate the wrath of the state, business, or their surrogates, NGOs back away from anything they think might be close to the line. This creates the desired "chilling effect" on political action. Targeting the most effective or militant NGOs with harsh repression can cause other groups to distance themselves. More effective groups can then be isolated and destroyed and other groups limited to ineffectual activities.

Effects of repression

Some observers argue that on the whole repression dampens challenges to current policy by raising the costs (Charles Tilly, 2005: 218). But there are circumstances in which repression increases protest, for example, when it is aimed at the physical survival of a group, when it triggers the intervention of powerful allies on the side of challengers, or when the repressive apparatus hesitates. Because conservation, more than some other movements (e.g., labor and ethnic minorities), espouses widely shared values and enjoys substantial middle-class and some upper-class support, repression is checked to a degree. Repression increased collective action of the antiapartheid resistance in South Africa (Jack Goldstone and Charles Tilly, 2001: 181), as did repression against the New Left in the 1960s/1970s in the United States (Gilda Zwerman and Patricia Steinhoff, 2005: 87). Action taken in response to repression often takes new forms, especially when repression is severe; NGOs go underground and engage in guerrilla warfare. Not all are radicalized, however; some are scared into quiescence (Elisabeth Wood, 2001).

Other factors shaping the effects of repression on movements include the attributes of targeted NGOs, the type of repression, its timing and its focus. When trust within and among NGOs is high it makes them better able to withstand attacks (Sidney Tarrow, 2005: 165). When repression occurs early in the protest cycle and when it is heavy it can destroy or cripple NGOs by undermining solidarity and raising the costs of protest (Gilda Zwerman and Patricia Steinhoff, 2005: 86; Goldstone and Tilly, 2001: 190, 181–2). Despite the auspicious circumstances in which Chinese protestors found themselves – international media attention, significant participation, and local military units loath to attack them – soldiers brought in from outside Beijing brutally clamped down and broke the gathering wave of protest in Tiananmen in 1989. Mild repression in the early phases can increase protest, drawing in allies who see little risk. If in response to this increased activity repression becomes heavier, protest can be dampened. There is also evidence that if the hubs of an organization or network are attacked it can quickly disrupt operations (Buchanan, 2002: 131–2). Recovery depends on the resilience of individuals and NGOs.

The interactivity of repression

One aspect of preparing for and dealing with repression is an understanding by NGOs of what in their own behavior can be altered to short-circuit repressive actions aimed at them. Some repression results from uncertainty by its perpetrators about what targeted NGOs are up to; action is taken assuming the worst (Goldstone and Tilly, 2001: 187; Charles Tilly, 2006: 130–1). Reducing uncertainty, without compromising progress toward goals, *might* reduce repression. Because movement effectiveness depends on tactical innovation which creates uncertainty – the very thing which keeps opponents off balance – this presents a thorny problem for NGOs. Keeping opponents off balance but not so much so as to trigger the stupid brutality of the Chinese at Tiananmen or the United States against the American Indian Movement or Central American peasants, demands walking a fine line.

Decisions by states and others about repressive action depend on how they perceive the challenge a movement poses (Charles Tilly, 2006: 48). When powerful actors perceive challenges to be focused on an area where they are weakest they feel more threatened and are more likely to respond with repression (Vince Boudreau, 2005: 34–51). Thus, in the United States where the state is powerful and well entrenched, but suffers from legitimacy problems, repression has been more severe against NGOs whose values were most threatening to dominant values (Jack Goldstone, 2003: 14). It is difficult to ignore, however, that repression has been harshest against non-White movements and NGOs, for example, the official murder of members of the Black Panther Party or the American Indian Movement (Jules Boykoff, 2007). In countries where state power is weak in some regions, for example, frontier areas, areas rich in resources, or areas dominated by ethnic minorities, challenges in those regions will be dealt with most harshly. Conservationists thus face serious danger when they advocate for biodiversity in such areas (Michael Klare, 2001: 190–3, 202–11). Often the domestic players are surrogates for more powerful states (or corporations) in competition with each other over resources or commerce. Conflicts can become especially brutal when governments and corporations employ mercenaries or corporations field their own "security forces" (Michael Klare, 2001: 195–7; Madelaine Drohan, 2003; Jeremy Scahill, 2007; Peter Singer, 2008). To the degree NGOs can diminish the perception of the threat they pose with compromising the struggle to reach their goals, decision makers may hold repression in abeyance.

Generally harsher forms of repression, including lethal force, are easier to use in authoritarian regimes (Tom Mertes, 2004: 245). Even so-called democratic regimes, however, engage in torture, extrajudicial killings, and other tactics that violate their espoused principles and their laws. Much depends upon the ability and willingness of the media to expose such crimes contemporaneously. The links between democratic and oppressive regimes are quite thick, with democratic states providing support to the latter and overthrowing democratic regimes to install more pliable authoritarian ones that guarantee raw materials (i.e., the transfer of carrying capacity such as oil, biofuels, and metals), cheap and obedient labor, and a sink for pollution (Richard Tucker, 2000; Michael Klare, 2001; Stephen Kinzer, 2006). It is not coincidental that in authoritarian regimes the damage to the biosphere is most pronounced (Judith Shapiro, 2001: 1), that is, where protest or other opposition is tolerated less and NGOs least developed.

Relations among states (and international regimes) can also work to constrain repression when NGOs are able to work these levers (McAdam and Sewall, 2001: 115; Clifford Bob, 2005); more on this below.

NGOs should assume that the lessons learned by one repressive agency quickly diffuse to counterparts around the globe. The militaries and clandestine services of various power blocs have long shared information and provided training and support for repression within their bloc (Jennifer Harbury, 2005). More recently police departments are increasingly sharing information about movement and NGO tactics within countries and transnationally (W. Lance Bennett, 2001: 221). This includes information on how best to disrupt movements and undermine their legitimacy. Police are also borrowing tactics from the military in countries where historically the two entities have not done so. Repression is also being contracted out more frequently; this provides governments and businesses with deniability and keeps formal police and troop levels looking lower.

Effective NGO responses to repression

The most effective NGO and movement responses to repression are strategic and three-pronged (Jeff Goodwin and Steven Pfaff, 2001: 284–5; Ruud Koopmans, 2005: 159–61). They counter repression's negative effects on movement participants psychologically and materially; they directly counter repressive actions; and they counter repressive action taken to influence other audiences in an effort to undermine their sympathy or support for the movement.

- *Countering fear.* Fear is a problematic emotion. When grounded and in the right intensity it prevents reckless behavior. When too intensely felt it can paralyze movement participants resulting in demobilization or cause costly mistakes. Directly addressing participants' fear is important, especially when widespread. Activists in the US civil rights movement and in the East German resistance mentioned fear more often than anger and outrage combined (Jeff Goodwin and Steven Pfaff, 2001: 285–7, 290–1, 293–4). These and other movements used many approaches to bolster participants, diminish fear, and sustain mobilization. They included fostering support by social networks so people did not feel alone; holding mass meetings and communal gatherings that included singing and passionate speeches reinforcing the movement's importance, its rightness and notions of its inevitable triumph; shaming defections; formal training in nonviolence; obtaining mass media attention; and organizing mass public declarations of support. Belief in divine protection was important to some. In the case of the US civil rights movement in which nonviolent public protest was central, participants kept firearms in their homes, making them feel they had something of a safe haven. Outrage – which can be cultivated – undercuts fear as does rejecting notions that repression can break the movement (Elisabeth Wood, 2001: 272, 277).

- Making participants feel safer also depends on actually making things safer for them by raising the costs of repression, enabling participants to counter it, and by establishing organizational actions to counter it.

- *Counter self-blame.* Conservationists should remind themselves that they need do nothing unlawful or wrong to be targets of repression. Repression results from the powerful seeking to protect their interests, and from repressive agencies' need to justify their existence or from their personal dislike of movements, participants, or goals.

- *Understanding repressive actors.* A good knowledge of those who engage in repressive activity – state, business, or countermovements – including their history, past campaigns of repression, repertoire of repressive tactics, favored tools, likely triggers, internal culture, and external constraints, can help NGOs and participants plan countermeasures and be prepared to respond. Institutional leaders can be as important as bureaucratic imperatives in this regard, for example, J. Edgar Hoover's paranoia and hypocrisy. NGO infiltration of police and military agencies is in some cases necessary to obtain good information, although sources include sympathetic agency personnel or their friends. Any information leaked directly or indirectly from repressive agencies should be evaluated for the likelihood it is being planted.

- *Educate movement participants about repressive tactics and effective responses.* Participant knowledge of tactics can blunt their ability to catch people off guard and makes rapid responses tenable. When participants know the methods used by infiltrators to entrap or encourage destructive behavior and the tools used to foment mistrust and discord, it renders them less effective. Forewarned is not always forearmed; knowing one is likely to be beaten if taken into custody may or may not alter a participant's response to a beating.

- NGOs and movements can only thrive when they are open and receptive to new participants, but this should not deter them from inquiring into the background of those who join. Many agents and many informers are very good at manipulation and almost invariably those they fool thought of themselves as excellent judges of character.

- *Invest in relationship building between and within NGOs.* Strong relationships, by encouraging people to go to each other to verify events rather than react to rumors or bogus claims or documents, make efforts to foment conflict less likely to take root.

- *Establish routines of watchfulness.* It is common practice in protest actions to have people look out for each other so they know when someone goes missing or gets into trouble. In some countries or in some circumstances in all countries, this needs to be extended to other NGO activities. The risks of traveling or undertaking tasks alone should be evaluated. Watchfulness and interactivity among participants helps ensure that individuals subject to repressive tactics have witnesses or others know about it quickly and can respond. Organizations forced underground by heavy repression must operate by different rules and this is not the appropriate place for such a discussion.

- *Secure information, offices, and other resources.* By routinely securing homes, offices, vehicles, computers, and information it is more difficult to gain access to them and easier to determine when unauthorized access has occurred. When intruders have entered a locked office with locked file cabinets or computers they usually leave signs. Copies of important records should be kept off-site or in some cases no written records should be kept.

- *Expect misrepresentation.* Secret police lore holds that one should never tell a friend what one would not tell an enemy. No healthy human society can operate this way (which partly explains why secrecy so quickly corrupts police and other political institutions). Nonetheless movement participants should understand that what they say, write, and E-mail, although perfectly legal, can easily be taken out of context and used to threaten, create bogus legal charges, and for attempted blackmail. Offhand comments about an official (e.g., someone ought to take care of that guy) or a simple placeholder comment (yeah) in response to an informer's carefully worded statement (It's important that *nothing* get in our way.) can be woven into an effective indictment.

- *Arrange on-call legal support.* This means little in many countries, but in those with somewhat effective judiciaries having established relationships with sympathetic and skilled attorneys can help blunt the use of some repressive tactics including use of the legal system.

- *Use legal proceedings to expose repression.* Trials and other legal proceedings can be used in some societies to gain media coverage of claims, show that the government or others are breaking the law, and to expose their use of the legal process to unfairly persecute groups because of their political goals. Success with this tool is highly dependent on media attention and framing. Using a trial to expose repression can put defendants at greater risk of conviction or otherwise compromise a purely legal strategy.

- Many countries have no meaningful judicial system. Resorting to aid from domestic human rights groups is often problematic because they are usually higher priority targets for repression. It often makes more sense to cultivate international legal connections that can assist with pursuing cases before various international adjudicatory bodies or to pressure treaty secretariats to act when they have authority (Linda Malone and Scott Pasternak, 2006). This path is a slow one.

- *Enlist broad community support against repression.* Strong ties to the community(ies) in which an NGO operates turn the eyes and ears of the community into those of the movement, making them witnesses to repressive activities by government and others. A community need not be sympathetic to an NGO's cause, only have an aversion to official lawbreaking. Agencies aware of these ties can be discouraged from acting in ways that if made public would reflect badly on them. These ties can also lead community members to warn NGOs of agents' presence or repressive activity afoot and to resist or expose repressive activities. Citizens of Chicago let demonstrators unknown to them into their houses to escape police violence. In another instance neighbors warned conservationists about housebreaking by agents.

- *Cultivate and enlist international political connections.* Conservationists are easy targets if not primary targets for repression in authoritarian countries or in countries with immature conservation NGOs. International NGO connections can help bring exposure to repressive actions, provide material assistance locally, or bring pressure through other countries, providing some protection (Margaret Keck and Kathryn Sikkirk, 1998: 121–63; Clifford Bob, 2005). There is a much greater need than international NGO resources can fill, however, so establishing relationships ahead of need can make a major difference. Reliance on external support can also generate justification for repression by making it easier for repressive actors to claim an NGO is disloyal or represents foreign interests.

- *Document encounters with authorities.* Photos, videos, witnesses, and contemporaneous written records are all useful tools in documenting encounters.
  Repressive actors know this and seize materials and intimidate witnesses when
  possible. Third-party documentation, for example, by the mainstream media,
  can be particularly effective in exposing or constraining repressive action. The
  media, however, are often unsympathetic and reporters enjoy no special immunity from repressive tactics such as beatings, tear-gassing, breaking cameras,
  or taking film. Media documentation of events has been exploited by officials
  to identify, harass, and prosecute movement participants.

- *Have strategies in place for dealing with infiltrators.* Infiltrators, informers,
  and provocateurs may be dealt with in a variety of ways. In some cases
  public exposure is most beneficial because it undermines their employer's
  credibility or reputation. Often infiltrators are criminals, have lied egregiously
  to create cover, grossly manipulated or taken advantage of people, broken
  the law, and possess other attributes that bring discredit when exposed. Legal
  action against an agency can keep embarrassing details of repressive activity
  before the public for a time. Widely disseminating photos of infiltrators
  throughout the movement can ensure they are less likely to play infiltrator in
  the future. Laws ban this in some countries but the internet makes it difficult
  to stop or prosecute. In other cases it makes sense to isolate infiltrators in order
  to make them ineffective and avoid public exposure if it would bring additional
  repression. In still other cases infiltrators can be used to feed disinformation
  to repressive agencies; this can make them appear effective and discourage
  additional infiltrators.

- *Prepare for attacks on NGO claims and integrity.* By anticipating how attacks
  on movement claims and credibility will be framed conservationists can
  be ready with compelling stories and analogies that appeal to deeply held
  notions of justice, values, and feelings. Many countries, for example, have a
  revolutionary heritage the elite pays lip service to, but the elite treats movement
  NGOs as the historical (and heroic) revolutionaries were once treated. Pointing
  out how the heirs to a revolutionary tradition have become the repressive
  authorities can cause important constituencies to question authorities' actions
  and motives.

- *Cultivate the media.* Movements depend on the media to get their story out
  and the media must choose between the competing stories offered by the
  movement and their foes, including repressive agencies. Good relationships
  with a wide range of media outlets can help persuade them to adopt more
  sympathetic frames and stories in portraying movement claims and actions or
  at least to question official and opponents' claims. Although media owners'
  interests may run counter to conservation interests thus blocking sympathetic

coverage, owners do not consistently interfere in news operations. Entertainment media in particular tend to overlook content if it is profitable. Even when
a media outlet is not sympathetic to a movement's claims, it may hold certain
repressive tactics to be wrong and cover them unsympathetically.

- *Cultivate connections with the famous and members of the elite.* Ties with
  sympathetic members of an elite can give some protection from repression or
  result in critical public attention being paid to repressive acts. Not all members
  of an elite are in a position to restrain coercive agents. Those with high profiles
  (writers, artists, scientists, etc.) often have good media access or access to those
  in other countries who can help publicize and criticize repressive actions. They
  are not immune from repression themselves, however, and these efforts can
  backfire if not thought through. Students, as the children of the elite in many
  countries, and religious leaders may also enjoy some immunity from repression
  or the ability to criticize it (Misagh Parsa, 2003: 82). Increasingly few outside
  the elite enjoy special protection as the assassination of bishops and writers
  suggests.

- *Plan transgressive action with repression in mind.* Transgressive action is
  usually essential to force significant change and therefore unavoidable. In
  societies where such action can result in death, long prison sentences, and
  other extreme measures, conservationists can rely on hit-and-run actions (rapid
  dispersal of leaflets, short street theater events, hanging a banner unexpectedly,
  disruption of official or opposition events by sabotaging or tapping into
  sound equipment), temporarily seizing events to communicate claims to those
  present and encourage them to join (this depends on participants in such
  events being sympathetic and giving cover), or actions aimed at satirizing
  the symbols of power (Hank Johnston, 2005: 122–5). Much of the guerrilla
  war literature, although focused on armed conflict, offers valuable lessons
  for nonviolent action in circumstances where significant power differentials
  exist. Ivan Arreguin-Toft's (2005) analysis of "asymmetric" conflicts over the
  last 200 years finds that weak actors win about two-thirds of wars fought
  against more powerful actors when they use nonconventional tactics and their
  stronger opponents do not match their tactics, usually because of stronger
  actors' political or military rigidity. The weak lose when they take a more
  powerful opponent head-on.

- *Political jujitsu.* Encounters with authorities seeking to repress an action or
  group can sometimes be turned against them, as when southern police turned
  fire hoses and dogs on scrupulously nonviolent civil rights demonstrators
  (including children) in front of national television cameras. Although civil
  rights activists could not count on the media exposure, their use of nonviolence
  against a more powerful enemy was not simply an ethical choice, but a practical

political one. After this experience others did plan on media coverage and were able to take advantage of police heavy-handedness.

- *Undermine the legitimacy of repressive actors.* The secrecy associated with repressive activities and agencies lends itself to corruption and lawbreaking, often on a very large scale (Chalmers Johnson, 2007). Exposing corruption, hypocrisy, and ineptitude can undermine support for repressive institutions especially when leaders are found to be personally profiting. Many of the most powerful opponents of conservation are businesses that profit from the degradation of Nature. Public officials often have close ties to such businesses and profit as well. Repressive agencies, the officials they report to, and officials' business constituents are typically involved with activities such as bribery, sweetheart deals, overcharging and underperforming on contracts, violating pollution laws, stealing resources, drug running, kidnapping, murder, and overthrowing governments (Stephen Kinzer, 2006; Chalmers Johnson, 2001, 2004, 2007). Such actions need not be associated with damaging Nature but they often are; they certainly injure many people. Exposing these activities makes official justifications for repression ring hollow and can even expose some actors, in rare circumstances, to prosecution. Many perpetrators are able to avoid personal responsibility by hiding behind the corporate veil or official immunity, but pursuing civil suits can keep the issues in the public eye. In authoritarian regimes those exposing such activities by the powerful can subject them to harsher repression.

- *Mass noncooperation.* Repression depends on the acquiescence of important sectors of the population. When they withdraw active support from a regime – by refusing to work, provide information, or buy certain products – authorities can be weakened. Such actions cause authorities to increase repression against high-profile noncooperators to force others back in line. Mass noncooperation can also give pause to authorities by reminding them that further repression will likely raise their costs dramatically. In some instances mass noncooperation – not an easy feat to organize – can grow into mass public demonstrations; it only takes about 4% of the population in the streets to overwhelm security forces (Ronald Francisco, 2005: 67).

- *Armed self-defense.* Even nonviolent US civil rights movement's participants often kept arms in their homes to discourage attacks by police and the Klan. Those involved in antipoaching operations or who face threats from the likes of Amazonian cattle ranchers, Indonesian logging companies, or oil companies operating in Nigeria, may find preparedness for self-defense necessary for survival and something audiences important to them easily understand. More typically conservation NGOs have helped to ensure law enforcement protecting reserves are adequately equipped rather than arming themselves.

Arming – especially formally and publicly and in opposition to the state or other powerful interests – usually opens a group to crushing repression because it provides an excuse (Jeff Goodwin and Steven Pfaff, 2001: 288–9). Today the terrorist label would quickly be applied to any armed group (as it is to many unarmed and even nonviolent groups). Long before the terrorist label was fashionable US security forces justified their violent repression of the Black Panther Party and American Indian Movement by pointing out they were armed. In many countries, of course, no such excuses are needed. Note that arming for self-defense is not the same as committing to armed struggle, such as guerrilla warfare. This is a distinction easy for the media, observers, and security forces to miss, especially if they want to.

- *Keep important audiences in mind.* Much repressive activity is carried out against movement NGOs to send a message to potential sympathizers and supporters: don't join (Ruud Koopmans, 2005: 160–1, 183). In deciding whether and how to act against an NGO, repressive forces look more to how the NGO is perceived by important constituencies or portrayed in the media. If the NGO is seen as making authorities appear weak or vulnerable, or as chipping away at their credibility, this is what triggers repression. If NGOs are at the same time disparaged by the media then repression can be carried out with little fear of criticism. Influencing how they are perceived, as every NGO knows, is at the heart of mobilizing support and realizing goals. The principles governing the crafting and conveying of mobilization messages are applicable to communicating about repression. Authoritarian settings and an unfriendly media create serious obstacles to successfully getting both types of messages out making it difficult for NGOs to influence the frame in which they are seen by important groups. Given these circumstances, NGOs must try to anticipate what perceptions are likely to trigger repression and to be prepared.

- Even when the circumstances are such that an NGO might avoid triggering repression by influencing its image, it may not be desirable to do so. Reaching important goals may require actions that unavoidably give rise to perceptions that trigger repression.

20 The life cycle of organizations

Organizations are living entities and have a life cycle during which much changes that bears on sustaining mobilization. Leadership becomes more routine and less charisma-dependent, organizational concerns temper mission, vision weakens and risk taking declines, formality increases, and as influence increases so all too often does timidity. But like individuals, nongovernmental organization (NGO) trajectories vary from the average. Some NGOs avoid senescence and timidity and reinvent themselves. Similar life-cycle changes do not always affect mobilization in the same way; much depends on political circumstances. When a government is hostile to conservation this may sustain mobilization even in dysfunctional NGOs as occurred in the United States during the Reagan and second Bush administrations.

Leadership trajectories and mobilization

Charismatic leaders often play a central role in the early life of an organization. They are usually among NGO founders because of their desire to create new and more effective organizations and for the recognition they get from leading and breaking ground. Their ability to excite others, think strategically, recognize opportunities, and their connections, makes them invaluable in early mobilization. Some of these attributes hurt mobilization as well. Charismatic leaders continue to produce new ideas which they often expect others to follow without question. The desire to control their creation or to treat it as something personal makes good decisions difficult and drives away others. Sustaining mobilization requires a range of leadership skills and abilities throughout an organization (Aldon Morris and Suzanne Staggenborg, 2004). To retain such people requires sharing the reins with them.

Founders, charismatic leaders, and self-aggrandizing leaders can interfere with sustaining mobilization in other ways. They contribute to contentious leadership transitions, which can destroy organizations. NGOs may fade away when a controlling leader leaves because no real organization was created. If financial support was based primarily on a leader's attractiveness – as it often is early on – their departure can leave an NGO facing an abyss.

On the other hand, if founders and charismatic leaders have the self-discipline and wisdom or can otherwise be compartmentalized, they can continue to play an important role in initiating and sustaining mobilization. They are often master storytellers and possess a contagious energy that can't be duplicated through training. People and media gravitate to them. Because most people find it easier to form bonds with a person than with a vision or organization, their ability to personify a vision or organization is helpful. Martin Luther King and Nelson Mandela are good examples of those who understood they needed others, needed to share power with them, and chose lieutenants well. King still plays an important role long after his death, as many charismatic leaders do. Self-aggrandizing leaders are ultimately destructive and should be removed or abandoned.

Charisma is not everyone's political pheromone. Issues are important to many, especially to those senior people in an NGO who plan and carry out programs. They are also important to many supporters, and to those who are strongly predisposed to the cognitive and analytical. These people often see the passion associated with charismatic leaders as unstable and a threat to political effectiveness. When charismatic leadership approaches demagoguery it adds to wariness and concern for organizational integrity.

Although most organizations grow beyond founders, charismatic or otherwise, the pitfalls associated with early leaders and leadership transitions can be avoided or better managed if understood. The same is true of problems associated with routinization of leadership and the loss of passion and vision.

Regardless of the legal structure, most NGOs begin with activist founders as governors; they often do much of the staff work as well. In time this circle is diversified to bring on those with broader connections and additional skills that enable the organization to become more effective. This change, along with more staff, allows the governing group to focus on setting overall direction and assisting staff with fund-raising and similar tasks. Some NGOs go further (or in a few cases start here) and the organization's formal governing group becomes primarily concerned with securing financial resources and rubber-stamping senior staff decisions about policy. In order to recruit and retain a governing group that can raise large sums an NGO must often refrain from conservation activity that runs contrary to interests. Such organizations can achieve much, but not profound policy change. This type of governance arrangement leaves little room for supporters to do more than write checks and read newsletters – not the sort of role that deepens commitment or makes people part of a conservation community on which the sustained political action needed for major reform is based.

Some NGOs bring allies into their governing bodies, which dilutes the NGO's conservation focus and can divide and weaken the organization, creating the same obstacles to needed action and deepening mobilization. There must be structures for alliances to work, but when a conservation NGO's governing body becomes that place it ceases to be a conservation NGO.

Keeping an NGO vigorous and on track requires a diverse leadership that combines as many desired attributes as attainable: vision, strategic good sense, organizational skills, boldness and intelligent risk taking, communications expertise, tactical insight and experience, access to insiders, and access to a variety of resources. There is no one right set of attributes for a leadership group because organizations' roles differ. There is one criterion, however, that every person in a leadership role *must* possess: a passionate commitment to conservation above all else. Leaders with divided, half-hearted or conditional commitments are the ruin of movements.

Organizational change, the iron law, and mobilization

Achieving their goals requires that organizations grow in effectiveness, which usually means more resources. As they grow NGOs are confronted with managerial, regulatory, and internal political challenges. More staff require new means of keeping focused and integrating the parts of the organization. Although no movement can succeed without the sacrifices of those who believe in it, Cesar Chavez argued, sacrifice is an exhaustible resource (Richard Jensen and John Hammerback, 2002: 65–6). Recruiting and retaining highly competent people with experience in wealthier countries entails reasonable pay and benefits. Volunteerism remains important and there will be times when sacrifice is required. In poorer countries conservation work often requires working in difficult conditions with low pay. In all countries conservation success will at times hinge on mass action that is by its nature unpaid and sometimes risky.

Organizational growth requires compliance with laws and creation of internal rules covering nonprogrammatic activities such as insurance, employment, taxation, and accounting. The increase in resources devoted to nonprogram operations can become a burden. Keeping the organization going becomes more of a focus and in some cases the primary focus. The mission remains, but the tail starts to wag the dog. Many smaller NGOs have the opposite problem: they ignore business operations causing them to lurch from crisis to crisis and this curtails their ability to carry out programs.

Organizational formality grows with size and explicit rules are necessary to ensure budgets are held to, programs carried out as planned, and staff treated equitably. Professional managers come to play a bigger role. If they lack a commitment to conservation values and do not understand the politics necessary to reach goals they often resist the sort type of mobilization that makes the movement strong because they experience it as interference. Memberships and contributions may go up under their direction, but not the type of mobilization that brings major policy change. New management techniques and periodic reorganization become the solution to every problem. It is unavoidable that tension exists between program and organizational requirements – it's even healthy – but if program is not dominant the organization is crippled. Too much tension between the two and valuable energy is drained. The best safeguard against this is a threshold filter in all hiring at senior levels: those hired must demonstrate a strong commitment to conservation and political understanding.

Two intertwining processes affecting organizational development are particularly corrosive of sustaining the kinds of mobilization required for significant change. These processes were described in the first part of the 20th century by Robert Michels (1962 [1915]) who saw them as inexorable; hence his "iron law of oligarchy."

As organizations become successful and gain a seat at the political table they become vested in staying at the table. To remain, they must play by the rules and refrain from throwing rocks or trying to overturn the table. Moreover, these organizations are for the first time in a position to seriously bargain and achieve some of what they want. NGOs that demanded a whole loaf from their picket line outside the store find that once inside the door they are within reach of only a *small piece of a slice;* to gain those meager rations they bind themselves to a set of rules that decrease their ability to get much more because they must behave themselves and be "reasonable." Thus do organizations by their success undergo moderation and often lose sight of their ultimate goals and where the best bargain lies. The ultimate goals become little more than empty rhetoric, still worthy of genuflection but relegated to realization in a distant future. Leaders become "practical," small thinkers, and the policies they pursue are constrained by the requirements of keeping a seat at the table. The clash between organizations within a movement over goals and appropriate compromises can often be explained by the difference between organizations who are "in the store" and those who are outside.

It is not just that keeping a seat at the table constrains a movement's leaders. Leaders themselves change with their experience as a political player – an experience that diverges from that of supporters and becomes more distant from those they represent. Leaders come to have more in common with other players in the political process and attached to the perquisites of leadership. Michels observed that as leaders' experiences diverge they seek to protect their position over the organization's mission. They become cautious and begin to speak of "political realism." The sad reality is that despite their willingness to comprises in exchange for insider status, their influence remains very limited absent considerable pressure from outsiders demanding much more.

This problem is compounded in the conservation movement because it is not just leaders speaking on behalf of supporters who differ from them, but of humans speaking on behalf of other species who cannot speak for themselves. Supporters can act as a check on leaders and in, for example, the labor or civil rights movements, the rank and file can rightly claim they know more about their daily lives and problems than leaders do. In the conservation movement supporters may defer to leaders as knowing more about what species and ecosystems need in the way of changes in human behavior and thus are less of a check on the tendency and temptation to bargain away the needs of other species and deny that this is what is being done.

Francis Piven and Richard Cloward (1977), who studied labor and economic movements in the United States, find substantiation for Michels's analysis. Others (David Meyer and Sidney Tarrow, 1998: 17; Christopher Rootes, 2004: 633) disagree, noting that Michels and Piven and Cloward studied centralized movements and the movements of the 1960s and others down to the present, including conservation, are not highly centralized (although some conservation NGOs are centralized). The applicability of the iron law depends on political context. Inclusion of conservation organizations by government in their decision processes does result in moderating goals and bureaucratization as does government funding of NGOs (John Dryzek et al., 2003: 99, 20). They do not observe, however, small radical NGOs evolving into bloated moderate organizations. For one thing, many organizations are not radical at their founding, but moderate. Some are even born with a seat at the table because members of the elite are involved in founding them. Many NGOs start and remain small and smallness does not equate with a radical program or feistiness. In many countries conservation NGOs are confined to an outsider strategy because the political system excludes nonelite players.

Many forces work against the conservation movement as a whole succumbing to the iron law, although clearly some large NGOs are oligarchical. The ongoing birth and rise of new organizations acts to partially discipline organizations susceptible to oligarchic pressures. More powerful organizations are seldom displaced by upstarts, nor do they easily admit aspiring organizations into their ranks, but the activity of hundreds and thousands of regionally or locally based grass-roots NGOs committed to their mission and vision above all else cannot be ignored. Smaller groups exert pressure on larger ones through the claims they make on government and business. When decision makers want a unified position from conservationists, including the ones in their home district, small NGO concerns have an effect. By publicly staking out positions smaller groups can embarrass larger groups for being too timid and willing to accept bad deals, or for spending a lot and getting little. Grass-roots groups are also the source of tremendous creative energy, most tactical and strategic innovation, and the transgressive politics that creates the new opportunities essential to sustaining mobilization and programmatic success.

Organizations that grow without abandoning a democratic structure are more susceptible to reform and reinvigoration from below. Leaders are also more likely to listen to members than simply issue edicts with the backing of a self-perpetuating board. Organizational democracy is not without its problems. Decisions may take longer, organizations can be pulled in many directions by supporter blocs, there is more chaos and sometimes bickering, more disruption by provocateurs and those with a personal agenda, and by those who want to exclude "radical factions." (This is why direct action organizations are small, tight, and informal.) Depite all of these difficulties more democratic organizations, such as the US Sierra Club, are more likely to remain vital and capable of reform than those that structurally resemble for-profit corporations. "More likely" does not mean easy. Internal change is difficult, Amatai Etzioni (1968: 395) observed – more so than changing policy.

Sustaining the kinds of mobilization that leads to major policy change requires organizations to avoid the reification that often accompanies age, bigness, and close relationships with power and wealth on their terms.

Funding

As organizations take on more they need more money to make it all run. This need increases the pressure to raise funds and can shift the focus of work to what is most easily fundable. Conservation relies heavily on gifts and gift giving is concentrated in a few countries. Giving to conservation attracts relatively little money compared to other causes where names can be chiseled into plaques adorning new hospital wings, symphony halls, and university buildings. (In the United States in 2006 only 2.2% of gifts went to conservation, environment, and animal welfare combined [Jim Holt, 2008: 11–2].). The money available is far less than what's needed, creating stiff competition. This in turn can diminish the willingness of NGOs to cooperate or to acknowledge the work of others. Receiving money from large conservative donors who are risk and controversy adverse (with notable exceptions), can make recipient organizations more timid. This is one reason some organizations achieve very little with tens of millions of dollars, whereas others achieve a great deal with thousands. Sometimes timidity is hidden behind a vigorous rhetoric, enabling more timid organizations to increase membership and participation while demanding little from policy makers and getting even less.

Dependence on charitable contributions limits the ability to sustain mobilization in other ways. Many countries forbid organizations receiving charitable gifts to engage in electoral politics. Such NGOs cannot make electoral campaigning a direct focus of their work, limiting not only program work (lobbying starts with electioneering) but the degree to which the energy generated by electoral politics can be used to sustain mobilization. It also limits the ability of NGOs to participate directly in electoral or ruling coalitions. Michael Shuman and Merrian Fuller (2005: 13–22) suggest that NGOs establish money-making enterprises in line with their goals in order to free themselves from reliance on donors. A concomitant advantage is that NGOs are creating economic institutions compatible with conservation, itself a major step in needed social change and a type of mobilization that is more and more necessary. Such enterprises can offer people conservation compatible livelihoods other than as activists or biologists. On the downside, if not properly structured such enterprises can divert leaders' attention away from programs and tie NGOs, like pension funds, to overall economic growth, that is, the transformation of Nature into commodities.

21 The need for many organizations

A range of nongovernmental organizations (NGOs) is important for reasons other than resisting reification. Individuals and cohorts are initially attracted to different types of organizations. Those groups conservationists must mobilize are at different levels of political development and are attracted to different types of action (lobbying, land acquisition, litigation, protest), looking for differing degrees of involvement (check writing, a staff position), have different levels of comfort with controversy and confrontation, and are passionate about different things (a place, a scale of concern, a species). The conservation movement must be able to tap into the diversity. A single organization cannot.

Over time people change and grow. Their political understanding and commitment deepens or wanes. They see the need for doing more or for doing things differently. In some cases the NGOs they belong to change with them or they are able to bring about change in them. Many individuals move to NGOs that better fit their concerns. Sustaining mobilization over the course of movement participants' lives requires that a range of NGOs exists to accommodate changes in supporters.

A single organization cannot be all things to all supporters. It is difficult for an NGO to accommodate administratively and politically more than one or two related types of action and do them well. (Larger organizations have some advantage in that they can shift resources between regions or take advantage of opportunities [Robert Repetto, 2006: 21], but this usually involves the same kind of efforts.) It is quite difficult to house and effectively manage all of the expertise needed to undertake many different types of activities. When organizations take on many types of tasks it's inevitable that important ones get short shrift and are compromised. Internal competition between programs saps energy. Potential constituents are more likely to be confused about what an organization does. A focus in one or a few linked program areas will result in better management and allow the organization to deepen its expertise and support.

Focus is not just an administrative issue, but a political one. It is difficult, for example, to sit down with business people and obtain sizable contributions with which to buy land, while at the same time directing boycotts against their businesses, suing them, or vigorously lobbying against some of their interests. Similarly, collaboration and confrontation are tools every movement needs, but they are not easily blended within the same organization. Some groups must be "conveners," able to reach out to wary constituencies and even opponents, in order to keep dialogue going and diffuse hostility or to explore potential common ground. Others need to be aggressive advocates that can put the fear in decision makers and opponents when appropriate. There are other reasons for a multiplicity of organizations. Competition among organizations – if there are not too many – enhances mobilization (Amatai Etzioni, 1968: 408–9). How many is too many? More organizations make repression more difficult (Doug McAdam et al., 2001: 157); it is more work infiltrating many groups than one. When there are many organizations the destruction of one does not destroy the movement. However, as the repressive activities against the New Left, Civil Rights, and Native American movements suggest, many groups can make exploiting the differences between them and between their leaders easier (see Chapter 19).

Making the best use of movement diversity requires cooperation and coordination beyond solidarity in the face of repression. Formal and informal coordination in support of a common vision – even if not shared in totality – with each group doing what it does best, adds enormous value. It helps ensure that places and species don't fall between the cracks, that opportunities aren't missed and threats are addressed rapidly. Coordination entails putting fears about competition for scarce funds on the agenda and addressing them. It requires that each group's sense of propriety – so essential to identity and defining what doing the right thing for Nature means – does not give way to self-righteousness and heretic burning. Those who fancy themselves hard-nosed pragmatists see purists as muddleheaded and threatening to tenuous public support and delicate deals. To those of more vision such pragmatists are sellouts and suck-ups who will do anything to be a player. Much of this mutual regard is self-indulgence and reflects political immaturity and the inability to put mission first. One would think even political novices would see the advantages of good-cop/bad-cop politics and insider/outsider strategies (John Dryzek et al., 2003: 161; Marco Giugni, 2004), not to mention the costs of infighting. The failure to act strategically prevents resources from being put to their best use, undercuts success, and thereby undercuts mobilization. Many models of cooperation are documented here. Networks provide a framework for cooperation, for sharing news and lessons learned, and provide political and psychological support. They are an incubator of community. Some transnational networks involve asymmetrical partnerships between larger international nongovernmental organizations (INGOs) and smaller local groups that depend on the former for many economic, political, and often scientific resources. Campaign coalitions and alliances are familiar models, with deep roots in social movements and other forms of politics. They are most successful when they include powerful allies that can bring pressure from within on elites and grass-roots NGOs that can bring mass pressure on elites from the outside. It is also true that grass-roots coalitions, with broader and intensely felt support, can temporarily overwhelm the more powerful in some fights.

More problematic is the relationship among movement, interest group, and electoral or regime politics. Influencing government starts in many societies with being part of the coalition that brings a government to power. In other societies being too closely allied with any party can create exclusion from influence when others are in power. At the same time social movement organizations or interest groups and parties operate differently and play different roles. Parties and their equivalents select leaders by various methods, aggregating support or gaining acquiescence from various organized and unorganized groups. They pursue definite policies, normally favoring the more powerful among their coalition, but are in the business of arranging compromises among elite factions and some nonelite groups in specific policy areas. Movement organizations and interest groups are advocates, though they too are in the business of cobbling together support for policies that involve compromise. They also are increasingly involved in leadership vetting or selection. In many societies movement NGOs are dependent on a tax status that keeps them out of electoral politics, at least formally, creating complex relationships between parties, candidates, and NGO programs. All of the functions performed by these three types of organizations are important to advancing conservation goals and they need to be coordinated.

Andy Kerr (2004) has a point when he says the US conservation movement could benefit from some mergers, where it seems "every man has his own NGO." In much of the developing world it is a different story. Conservation NGOs are sparse and face enormous obstacles even with help from the outside. Internal development pressures and globalized market pressures, including unequal terms of trade and political bullying, generate an intense demand for resources. When forests are protected in the developed world it doesn't mean wood consumption goes down – it means more trees are cut in poorer countries where conservationists are weaker and businesses often much stronger than government. A global priority for conservation must be more groups in poorer countries, more diversity among them, and more resources for them (including deft political assistance). Some smaller NGOs from the north have been particularly effective in assisting NGOs in the south to greatly enhance their effectiveness; larger international NGOs have played a major role, though often not very cost-effectively. They are often encumbered by internal bureaucracy and more frequently by reliance on entities such as the Global Environment Facility that combines the agility of an oil tanker with the commitment to conservation of the World Bank.

The conservation movement, for all of its ups and downs and the emergence and passing of NGOs, has demonstrated staying power. Its reason for being grows stronger as human destruction of life on Earth increases. Addressing and reversing human destructiveness toward the living systems on which humans depend requires all the perseverance, wisdom, diversity of approaches, expertise, creativity, organizational inventiveness, and toughness that can be brought to bear. All of these attributes are multiplied with coordination and cooperation and can be lost with bigness and centralization.

22 A final question

Mobilization is not the only factor constraining conservation effectiveness. Increasing the resources conservation commands will not automatically increase its influence. How well those resources are used counts for much. Expertise and experience – understanding the levers of politics and policy, knowing the sticks and carrots available and when and how to employ them, knowing what needs to be done – are equally critical to success. Conservationists have often failed because they used sufficient resources poorly, were surprised by opponents' lack of principle, or were otherwise outmaneuvered. Much depends on the political opportunities presented and on the ability to see them coming. Many opportunities are structural and cyclical, as when cumulative economic, technological, and social changes weaken the position of powerful groups and strengthen previously weaker groups. Recognizing opportunities contemporaneously and knowing how to use them play a huge role in success. In some circumstances opportunities can be made. Despite these many political variables, much of politics is a simple, crude, and ugly business. Effectiveness varies with the degree of power one holds: the control of institutions that make decisions, the ability to reward and punish decision makers more effectively than opponents – that is, out-mobilizing opponents. Greater resources (higher levels of mobilization) for conservationists depend on vibrant, well-coordinated nongovernmental organizations (NGOs) and powerful allies. These depend in turn on the movement's ability to reach out to new constituencies, involve people in activism (not just membership), nurture a conservation identity among participants, and sustain a deep and lasting commitment to the natural world, the movement and other participants. Crafting messages that resonate, using the right messengers and channels, developing and sustaining rituals and other elements of a rich movement culture, building a sense of community, engaging people in meaningful action, and demonstrating movement achievements are all elements in this. Conservation must become part of people's daily lives; only then can it become a priority in their lives. Conservation must be prepared to survive repression and periods of relative weakness.

The conservation movement can significantly increase its successes if it can substantially increase its capacity to reward and punish and use that capacity with intelligence and passion over time. But that won't be enough to halt the current loss of life on Earth. Arrayed against conservation is the very structure of the existing economic and social order: an order based on unending growth in material consumption and human population, on hierarchy, and on the mania for control.

Much more can be achieved within this framework than has been achieved, but existing structure and inertia limit conservation success to less than what is needed to halt the great extinction event underway. In the near term conservationists, by increased mobilization, can protect large parts of the planet, connect them, see they are well managed, and defend them from exploitation. But they cannot sustain such areas over the long term without basic social change – change that diminishes the overall human footprint. If human numbers continue to grow and human appetites remain insatiable then wild places and creatures will be eaten into oblivion. In short, conservationists must change what is possible even though they have not reached the limits of what is possible within the existing order of things. In seeking to bring about what is possible with enhanced mobilization, conservationists also start to shift the very structure of things, changing what is possible.

Bringing about the fundamental change needed to halt and reverse the impoverishment and degradation of the natural world will not be easy, but neither was bringing down apartheid or many other important transformations. Can people come to care as much about the wild as they do about injustice within human society? That is conservation's great challenge.

Bibliography

Abbey, E. (1982) *Down The River*. Dutton, New York.

Abernathy, D.B. (2000) *Global Dominance*. Yale University Press, New Haven, CT.

Adorno, T., Frenkel-Brunswik, E., Levinson, D.J. and Sanford, R.N. (1950) *The Authoritarian Personality*. Harper and Row, New York.

Ahern, T. (2002) Story telling for the media. In: Whybrow, H. (ed.) *The Story Handbook*. The Trust for Public Land, San Francisco, pp. 40–45.

Allen, M. (2005) Power is in the details: administrative technology and the growth of ancient near Eastern cores. In: Chase-Dunn, C. and Anderson, E.N. (eds) *The Historical Evolution of World Systems*. Palgrave Macmillan, New York, pp. 75–91.

Allen, T.F.H., Tainter, J.A. and Hoekstra, T.W. (2003) *Supply Side Sustainability*. Columbia University Press, New York.

Aminzade, R.R., Goldstone, J.A. and Perry, E.J. (2001) Leadership dynamics and the dynamics of contention. In: Aminzade, R.R., Goldstone, J.A., McAdam, D., Perry, E.J., Sewell, W.H., Tarrow, S. and Tilly, C. (eds) *Silence and Voice in the Study of Contentious Politics*. Cambridge University Press, Cambridge, UK, pp. 126–154.

Aminzade, R.R. and McAdam, D. (2001) Emotions and contentious politics. In: Aminzade, R.R., Goldstone, J.A., McAdam, D., Perry, E.J., Sewell, W.H., Tarrow, S. and Tilly, C. (eds) *Silence and Voice in the Study of Contentious Politics*. Cambridge University Press, Cambridge, UK, pp. 14–50.

Amnesty International.2007 Honduras: *Environmental Activists Killed in Olancho Department*. http://www.amnesty.org/en/library/asset/AMR37/001/2007/en/dom-MR370012007en.pdf

Anker, P. (2002) *Imperial Ecology*. Harvard University Press, Cambridge, MA.

Ansell, C. (2003) Community embeddedness and collaborative governance in the San Francisco Bay area environmental movement. In: Diani, M. and McAdam, D. (eds) *Social Movements and Networks*. Oxford University Press, Oxford, pp. 123–144.

Aronson, E. (1997) Back to the future: retrospective review of Leon Festinger's a theory of cognitive dissonance. *The American Journal of Psychology* 110(1):127–157.

Arreguin-Toft, I. (2005) *How the Weak Win Wars*. Cambridge University Press, Cambridge.

Asian Human Rights Commission (2005) Urgent Appeals. www.ahrchk.net/ua/mainfile.php/2005/1145, posted 30 June 2005; accessed 11th April 2008.

Babbitt, B. (2004) Personal Communication. At the High Desert Conference on 11th September 2004 I asked Babbitt if This story were true. He said "Close enough."

Babbitt, Bruce. (2005) *Cities in the Wilderness*. Island Press. Washington, DC.

Bagdikian, B.H. (2004) *The New Media Monopoly*. Beacon Press, Boston.

Battle for Wilderness, The (Televised Documentary Film). (1989) PBS-The American Experience/ Lawrence R. Hott Production, Boston.

Balmford, A., Manica, A., Airey, L., Birkin, L., Oliver, A. and Schleicher, J. (2004) Hollywood, climate change, and the public. *Science* 305:1713.

Barber, E.W. and Barber, P.T. (2004) *When they Severed Earth from Sky*. Princeton University Press, Princeton, NJ.

Barcott, B. (2001) For God So Loved the World, 26 Outside 3, pp. 84–126.

Barker, C. (2001) Fear, laughter, and collective power: the making of solidarity at the Lenin Shipyard in Gdansk, Poland, August 1980. In: Goodwin, J., Jasper, J.M. and Polletta, F. (eds) *Passionate Politics*. University of Chicago Press, Chicago, IL, pp. 175–194.

Barlow, C. (1997) *Green Space, Green Time*. Copernicus, New York.

Barnet, R.J. (1990) *The Rockets' Red Glare*. Simon and Schuster, New York.

Barnett, H.G. (1953) *Innovation: The Basis of Cultural Change*. McGraw-Hill, New York.

Bator, R.J. (2000). The application of persuasion theory to the development of effective proenvironmental public service announcements. *The Journal of Social Issues* **56**(3):527–541.

Baumgartner, F.R. (2006). Punctuated equilibrium theory and environmental policy. In: Repetto, R. (ed.) *Punctuated Equilibrium and the Dynamics of U.S. Environmental Policy*. Yale University Press, New Haven, CT, pp. 24–46.

Beck, R. and Kolankiewicz, L. (2000) The environmental movement's retreat from advocating U.S. Population stabilization(1970–1998): a first draft of history. *Journal of Policy History* **12**(1):123–156.

Becker, E. (1973) *The Denial of Death*. The Free Press, New York.

Beckwith, S.L. (2006) *Publicity for Nonprofits*. Kaplan, Chicago.

Beder, S. (2005). *Suiting Themselves*. Earthscan, London.

Beil, R. (2000) *The New Imperialism*. Zed Books, London.

Bell, C. (1997) *Ritual Perspectives and Dimensions*. Oxford University Press, New York.

Bennett, W.L. (1980). Myth, ritual and political control. *The Journal of Communication* **30**(4):166–179.

Bennett, G. (1995) *EECONET and the Wildlands Project*. Institute for European Environmental Policy, Arnhem, Netherlands.

Bennett, W.L. (2002) *News: The Politics of Illusion*, 5th edn., Longman, New York.

Bennett, W.L. (2005) Social movements beyond borders: understanding two eras of transnational activism. In: Donatella, D.P. and Tarrow, S. (eds) *Transnational Protest and Global Activism*. Rowman and Littlefield, Lanham, MD, pp. 203–226.

Bennett, W.L., Lawrence, R. and Livingston, S. (2007) *When The Press Fails*. University of Chicago Press, Chicago.

Bennett, G. and Mulongoy, K.J. (2006) *Review of the Experience with Ecological Networks, Corridors and Buffer Zones. CBD Technical Series 23*. Secretariat of the Convention on Biodiversity, Montreal.

Bergesen, A.J. and Bartley, T. (2005) World system and ecosystem. In: Hall, T.D. (ed.) *A World Systems Reader*. Rowman & Littlefield, Lanham, MD, pp. 307–324.

Berman, M. (2000) *Wandering God*. State University of New York Press, Albany, NY.

Bernard, C. (1865) *An Introduction to the Study of Experimental Medicine*. Dover, New York.

Bernays, E.L. (1928) *Propaganda*. Liveright, New York.

Bernays, E.L. (1947) The engineering of consent. *The Annals of the American Academy of Political and Social Science* **250**:113–120.

Bernays, E.L., (ed.) (1955) *The Engineering of Consent*. University of Oklahoma Press, Norman.

Berry, B.J.L. (1991) *Long Wave Rhythms in Economic Development and Political Behavior*. Johns Hopkins University Press, Baltimore.

Berry, B.J., Elliot, E., Harpham, E.J. and Kim, H. (1998) *The Rhythms of American Politics*. University Press of America. Lanham, MD.

Berry, W. (1993) *Sex, Economy, Community and Freedom*. Pantheon, New York.

Bettig, R.V. and Hall, J.L. (2003). *Big Media, Big Money*. Rowman & Littlefield, Lanham, MD.

Bighorn Association (1870) Editorial. *Cheyenne Daily Leader*, 13 March.

Biodiversity Project Staff (2002) *Ethics for a Small Planet*. Biodiversity Project, Madison, WI.

Bittman, M. (2008) Rethinking the meat guzzler. *The New York Times*, 27 January.

Blake, W. (1977, 1790). The marriage of heaven and hell. In: Blake, W. (ed.) *The Complete Poems*. Penguin Press, Harmondsworth, UK, pp. 180–194.

Blakeslee, S. (2003) Humanity? Maybe It's in the Wiring. *The New York Times*, 9 December.

Bob, C. (2005) *Marketing Rebellion*. Cambridge University Press, Cambridge, UK.

Boehm, C. (1999) *Hierarchy in the Forest*. Harvard University Press, Cambridge, MA.

Bohannon, J. (2006). Tracking people's electronic footprints. *Science* **314**:914–916.

Boudreau, V. (2005) Precarious regimes and matchup problems in the explanation of repressive policy. In: Davenport, C., Johnson, H. and Mueller, C. (eds) *Repression and Mobilization*. University of Minnesota Press, Minneapolis, pp. 33–57.

Boykoff, J. (2007) *Beyond Bullets*. AK Press, Oakland, CA.

Bracht, N. (2001) Community partnership strategies in health campaigns. In: Rice, R.E. and Atkin, C.K. (eds) 1989 *Public Communication Campaigns*, 2nd edn. Sage Publications, Newbury, pp. 323–342.

Brader, T. (2005) *Campaigning for Hearts and Minds*. University of Chicago Press, Chicago.

Brecht, B. (1926, 1977). A Man's a Man. *Brecht Collected Plays*, Vol. 2. Vintage, New York, pp. 1–70.

Broadbent, J. (2003) Movement in context: thick network and Japanese environmental protest. In: Diani, M., McAdam, D. (eds) *Social Movements and Networks*. Oxford University Press, Oxford, pp. 204–229.

Brock, W.A. (2006) Tipping points, abrupt opinion changes, and punctuated policy change. In: Repetto, R. (ed.) *Punctuated Equilibrium and the Dynamics of U.S. Environmental Policy*. Yale University Press, New Haven, pp. 47–77.

Brody, J.E. (2004). TV's toll on young minds and bodies. *The New York Times*, 3 August.

Brooks, D. (2005) Meet the poor republicans. *The New York Times*, 15 May.

Brown, N.O. (1959) *Life Against Death*. Wesleyan University Press, Middletown, CT.

Bruner, J.S. (1960) Myth and identity. In: Murray, H.A. (ed.) *Myth and Mythmaking*. Braziller, New York, pp. 276–287.

Bruntland Commission (World Commission on Environment and Development) (1987) *Our Common Future*. Oxford University Press, Oxford.

Bryner, G.C. (2001) *Gaia's Wager*. Rowman & Littlefield, Lanham, MD.

Bryner, G.C. (2007) Congress and clean air policy. In: Kraft, M.E. and Kamieniecki, S. (eds)
  Business and Environmental Policy. MIT Press, Cambridge, MA, pp. 127–151.

Buchanan, M. (2002) *Nexus*. WW Norton, New York.

Burton, L. (2002) *Worship and Wilderness: Culture, Religion and Law in Public Lands
  Management*. University of Wisconsin Press, Madison.

Butler, M. (2001) America's sacred cow. In: Rice, R.E. and Atkin, C.K. (eds) *Public
  Communication Campaigns*, 3rd edn. Sage Publications, Thousand Oaks, CA,
  pp. 309–314.

Butler, R. (1992) Parrots, pressures, people, and pride. In: Beissinger, S.R. and
  Snyder, N.F.R. (eds) *New World Parrots in Crisis*. Smithsonian Press, Washington, DC,
  pp. 25–46.

Buying the War (television documentary film) (2007) (U.S.) Public Broadcasting System.
  Films for the Humanities and Social Science, Princeton, NJ, Washington, DC.

Caldini, R.B. (2001) Littering, when every litter bit hurts. In: Rice, R.E. and Atkin, C.K. (eds)
  Public Communication Campaigns, 3rd edn. Sage Publications, Thousand Oaks, CA,
  pp. 280–282.

Campaign for America's Wilderness (2003) *A Mandate to Protect America's Wilderness*.
  Campaign for America's Wilderness, Washington, DC.

Campbell, J. (1959) *The Masks of God*, Vol. 1, *Primitive Mythology*. Viking, New York.

Campbell, J. (1988) *The Power of Myth*. Doubleday, New York.

Carroll, C., Noss R., and Paquet P. (2001) Carnivores as Focal Species for Conservation
  Planning in the Rocky Mountain Region. *Ecological Applications* 4(11):961–980.

Carroll, C., Noss, R. and Paquet, P. (2004). A proposed wildlands network for carnivore
  conservation in the rocky mountains. *Wild Earth* 14(1–2):28–33.

Carson, R. (1984) *Sense of Wonder*. Harpers, New York.

Carter, N. (2007) *The Politics of the Environment*, 2nd edn. Cambridge University Press,
  Cambridge, UK.

Cashore, B. and Howlett, M. (2006) Behavioral thresholds and institutional rigidities as
  explanations of punctuated equilibrium processes in Pacific Northwest forest policy
  dynamics. In: Repetto, R. (ed.) *Punctuated Equilibrium and the Dynamics of U.S.
  Environmental Policy*. Yale University Press, New Haven, CT, pp. 137–161.

Cassen, B. (2004) Inventing ATTAC. In: Mertes, T. (ed.) *A Movement of Movements*. Verso,
  London, pp. 152–174.

Center for Responsive Politics (2008) US Election Will Cost 5.3 Billion.
  http://www.opensecrets.org/news/2008/10/us-election-will-cost-53-billi.html
  (accessed 22 October 2008).

Chafe, W.H. (2005) *Private Lives/Public Consequences*. Harvard University Press,
  Cambridge, MA.

Chaudhuri, A. (2006) *Emotion and Reason in Consumer Behavior*. Elsevier, Amsterdam.

Chernus, I. (2004) Presidential Fiction, The Story Behind the Debate. Posted at
  www.tomdispatch.com.

Chester, C. (2006) *Conservation Across Borders*. Island Press, Washington, DC.

Chew, S.C. (2001) *World Ecological Degradation*. AltaMira Press, Walnut Creek, CA.

Chew, S.C. (2005) From Harappa to Mesopotamia and Egypt to Mycenae (2200–700BC). In: Chase-Dunn, C. and Anderson, E.N. (eds) *The Historical Evolution of World Systems.* Palgrave Macmillan, New York, pp. 52–74.

Churchill, W. (1948) *The Gathering Storm, Vol. 1 The Second World War.* Houghton Mifflin, Boston, MA.

Churchill, W. and Vander Wall, J. (1990) *Agents of Repression.* South End Press, Boston.

Churchill, W. and Vander Wall, J. (1900b) *Agents of Repression,* 2nd edn. South End Press, Boston.

Clark, G. (1986) *Symbols of Excellence.* Cambridge University Press, Cambridge, UK.

Clarke, T.W., Curlee, P., Minta, S.C. and Kareiva, P. (eds) (1999) *Carnivores in Ecosystems.* Yale University Press, New Haven, CT.

Clayton, S. (2000). Models of justice in the environmental debate. *The Journal of Social Issues* 56(3):459–474.

Cogkianese, G. (2007) Business interests and information in environmental rulemaking. In: Kraft, M.E. and Kamieniecki, S. (eds) *Business and Environmental Policy.* MIT Press, Cambridge, MA, pp. 185–210.

Cohen, P.S. (1969) Theories of myth. *Man* 4(3):337–353.

Coles, R. (1964) Social struggle and weariness *Psychiatry* 27:305–315.

Coles, R. (1986) *The Political Life of Children.* Atlantic Monthly Press, Boston.

Collier, P. and Hoeffler A. (1999) *Justice- Seeking and Loot-Seeking in Civil War.* World Bank. Washington, DC.

Collins, R. (2001) Social movements and the focus of attention. In: Goodwin, J., Jasper, J.M. and Polletta, F. (eds) *Passionate Politics.* University of Chicago Press, Chicago, pp. 27–44.

Combs, J.E. and Nimmo, D. (1993) *The New Propaganda: The Dictatorship of Palaver in Contemporary Politics.* Longman, New York.

Commission on Population and the American Future (1972) *Report of the Commission on Population and the American Future.* U.S. Government Printing Office, Washington, DC.

Conca, K. (1993) Environmental change and the deep structure of world politics. In: Lipschutz, R.D. and Conca, K. (eds) *The State and Social Power in Global Environmental Politics.* Columbia University Press, New York, pp. 306–326.

Conca, K. and Lipschutz, R.D. (1993) A tale of two forests. In: Lipschutz, R.D. and Conca, K. (eds) *The State and Social Power in Global Environmental Politics.* Columbia University Press, New York, pp. 1–18.

Controlling Interest (film) (1978). California Newsreel, San Francisco.

Cornog, E. (2004) *The Power and the Story.* Penguin Press, New York.

Cox, R. (2006) *Environmental Communication and the Public Sphere.* Sage, Thousand Oaks, CA.

Crosby, A.W. (1986) *Ecological Imperialism.* Cambridge University Press, Cambridge, UK.

Cunningham, D. (2004) *There's Something Happening Here.* University of California Press, Berkeley.

Czech, B. (2000) *Shoveling Fuel for a Runaway Train.* University of California Press, Berkeley.

Daly, H.E. and Farley, J. (2004) *Ecological Economics: Principles and Applications*. Island Press, Washington, DC.

Damasio, A. (1994) *Descartes' Error*. Grosset/Putnam, New York.

Damasio, A. (1999) *The Feeling of What Happens*. Harcourt, Brace, New York.

DaMatta, R. (1977) Constraint and license: a preliminary study of two Brazilian National rituals. In: Moore, S.F. and Myerhoff, B.G. (eds) *Secular Ritual*. VanGorcum, Assen, Netherlands, pp. 244–264.

Danaher, K. and Mark, J. (2003) *Insurrection*. Routledge, New York.

D'Aquila, E. and Laughlin, C.D. (1975) The biophysiological determinants of religious ritual behavior. *Zygon* **10**:32–57.

D'Aquila, E. and Laughlin, C.D. (1979) The neurobiology of myth and ritual. In: d'Aqila, E. Eugene G, Charles D Laughlin Jr and John McManus (eds) *The Spectrum of Ritual*. Columbia University Press, New York.

Dasgupta, P. (2003) Externalities of population change. In: Demeny, P. and McNicoll, G. (eds) *Encyclopedia of Population*. Thomson-Gale, New York, pp. 336–340.

Johns, D. (2002) Wilderness and Energy, *14 Wild Earth* 3:12–13 (Fall).

Davis, C. (2006) The politics of grazing on federal lands. In: Repetto, R. (ed.) *Punctuated Equilibrium and the Dynamics of U.S. Environmental Policy*. Yale University Press, New Haven, CT, pp. 232–252.

Dearing, J.W. (2001) The Cumulative Community Response to AIDS in San Francisco. In:Rice, R.E. and Atkin, C.K. (eds) 1989 *Public Communication Campaigns*, 2nd edn. Sage Publications, Newbury Park, CA, pp. 305–308.

Decision Research (2002) *Poll Conducted for MacWilliams Robinson and Partners*. MacWilliams, Robinson and Partners, Washington, DC.

Dempsey, M. (1989) Quatsi means life. *Film Quarterly* **42**(3):2–12.

Dennis, M. (2004) The invention of Martin Luther King's birthday. In: Etzioni, A. and Bloom, J. (eds) *We Are What We Celebrate*. New York University Press, New York, pp. 178–193.

DePalma, A. (1998) 19 Nations see U.S. as threat to cultures. *The New York Times*, 1 July.

de Quervain, D.J.-F., Fischbacher, U., Treyer, V., Schellhammer, M., Schnyder, U., Buck, A. and Fehr, E. (2004) The neural basis of altruistic punishment. *Science* **305**:1254–1258.

Deudney, D. (1993) Global environmental rescue and the emergence of world domestic politics. In: Lipschutz, R.D. and Conca, K. (eds) *The State and Social Power in Global Environmental Politics*. Columbia University Press, New York, pp. 281–305.

DeYoung, R. (2000) Expanding and evaluating motives for environmentally responsible behavior. *The Journal of Social Issues* **56**(3):509–526.

Dezenhall, E. (1999) *Nail 'Em*. Prometheus books, Amherst, NY.

Diamond, J. (2005) *Collapse*. Viking, New York.

Diani, M. (2003a) Introduction: social movements, contentious actions, and social networks: 'From metaphor to substance'. In: Diani, M. and McAdam, D. (eds) *Social Movements and Networks*. Oxford University Press, Oxford, pp. 1–18.

Diani, M. (2003b) 'Leaders' or brokers? Positions and influence in social movement networks. In: Diani, M. and McAdam, D. (eds) *Social Movements and Networks*. Oxford University Press, Oxford, pp. 105–122.

Dillon, M. and Paul W. (2007) *In the Course of a Lifetime*. University of California Press, Berkeley.

Dinnerstein, D. (1976) *The Mermaid and the Minotaur*. Harper and Row, New York.

Dobbin, F. (2001) The Business of Social Movements. In Goodwin, J., Jasper, J.M. and Polletta, F. (eds) *Passionate Politics*. University of Chicago Press, pp. 74–80.

Domhoff, G.W. (1998) *Who Rules America?* 3rd edn. Mayfield Publishing, Mountain View, CA.

Donatella, D.P. and Tarrow, S. (2005) Preface. In: Donatella, D.P. and Tarrow, S. (eds) *Transnational Protest and Global Activism*. Rowman and Littlefield, Lanham, MD, pp. 1–17.

Doniger, W. (1996) Minimyths and maximyths and political points of view. In: Patton, L.L. and Doniger, W. (eds) *Myth and Method*. University Press of Virginia, Charlottesville, pp. 109–127.

Donner, F. (1990) *Protectors of Privilege*. University of California Press, Berkeley.

Doob, L.W. (1950) Goebbels's principles of propaganda. *Public Opinion Quarterly* 14:419–442.

Douglass, F. (1985, 1857). The significance of emancipation in the West Indies. In: Blassingame, J.W. (ed.) *The Frederick Douglass Papers, Series One: Speeches, Debates, and Interviews. Volume 3: 1855–63*. Yale University Press, New Haven, CT.

Doyle, J. (2000) *Taken for a Ride: Detroit's Big Three and the Politics of Pollution*. Four Walls Eight Windows, New York.

Dozier, D.M., Grunig, L.A. and Grunig, J.E. (2001) Public relations as communication campaign. In: Rice, R.E. and Atkin, C.K. (eds) *Public Communication Campaigns*, 3rd edn. Sage Publications, Thousand Oaks, CA, pp. 231–248.

Drohan, M. (2003) *Making a Killing*. Random House Canada, Toronto.

Dryzek, J.S., Downes, D., Hunold, C. and Schlosberg, D. (2003) *Green States and Social Movements*. Oxford University Press, Oxford.

Dudley, N., Higgins-Zogib, L. and Mansourian, S. (2005) *Beyond Belief: Linking Faiths and Protected Areas to Support Conservation*. WWF International and ARC, Gland, Switzerland.

Duffy, R.J. (2007) Business, elections and the environment. In: Kraft, M.E. and Kamieniecki, S. (eds) *Business and Environmental Policy*, MIT Press, Cambridge, MA, pp. 61–90.

Dunbar, R.I.M. (2001) Brains on two legs: group size and the evolution of intelligence. In: deWaal, F.B.M. (ed.) *Tree of Origin*. Harvard University Press, Cambridge, MA, pp. 173–191.

Dunlap, R., Van Liere, K., Mertig, A. and Jones, R.E. (2000) Measuring endorsement of the new ecological paradigm: a revised NEP scale. *The Journal of Social Issues* 56(3):425–442.

Durant, R.F., Fiorino, D.J. and O'Leary, R. (2004) Conclusion. In: Durant, R.F., Fiorino, D.J. and O'Leary, R. (eds) *Environmental Governance Reconsidered*. MIT Press, Cambridge, MA, pp. 483–525.

Durkheim, E. (1995, 1912) *The Elementary Forms of the Religious Life*. Free Press, New York.

Durning, A.T. and Crowther, C.D. (1997) *Blaming the Victim*. Northwest Environmental Watch, Seattle, WA.

Dwyer, J. (2006) Police files say arrest tactics calmed protest. *The New York Times*, 25 April.

Dwyer, J. (2007) At the protest, a civics lesson gets a twist. *The New York Times*, 17 March.

Dye, T.R. (1995) *Who's Running America?* 6th edn. Prentice Hall, Englewood Cliffs, NJ.

Dye, T.R. (2002) *Who's Running America?* 7th edn. Prentice Hall, Upper Saddle River, NJ.

Edelman, M. (1964) *The Symbolic uses of Politics*. University of Illinois, Urbana, IL.

Edelman, M. (1988) *Constructing the Political Spectacle*. University of Chicago Press, Chicago.

Edelman, M. (1995) *From Art to Politics*. University of Chicago Press, Chicago.

Ehrenfeld, D. (1979) *The Arrogance of Humanism*. Oxford University Press, New York.

Ehrlich, P. and Anne. (1997) *The Betrayal of Science and Reason*. Island Press, Washington, DC.

Eliade, M. (1991) Toward a definition of myth. In: Bonnefoy, Y. (ed.) *Mythologies*. University of Chicago Press, Chicago, pp. 3–5.

Ellul, J. (1972) *Propaganda*. Knopf, New York.

Environmental Working Group (2003) Web Publication of An Excerpt from a Scanned Document Developed by "The Luntz Research Companies – Straight Talk", pp. 131–146, www.ewg.org/briefings/luntzmemo, 14 November 2003.

Erikson, E.H. (1968) *Identity, Youth and Crisis*. Norton, New York.

Estes, J.A., DeMaster, D.P., Doak, D.F., Williams, T.M. and Brownell Jr, R.L. (eds) (2007) *Whales, Whaling and Ocean Ecosystems*. University of California, Berkeley.

Etzioni, A. (1968) *The Active Society*. Free Press, New York.

Etzioni, A. (2004) Holidays and rituals. In: Etzioni, A. and Bloom, J. (eds) *We Are What We Celebrate*. New York University Press, New York, pp. 3–40.

Evans, D. (2001) *Emotion, the Science of Sentiment*. Oxford University Press, Oxford.

Evernden, N. (1985) *The Natural Alien*. University of Toronto Press, Toronto.

Ewen, S. (1996) *PR! A Social History of Spin*. Basic Books, New York.

Fairbank, Maslin, Maullin and Associates (2004) Lessons Learned Regarding the Language of Conservation. Memo of 1 June 2004 to TNC/Trust for Public Land, Santa Monica, CA.

Feder, B.J. (2004) Report backs regulating utility emissions. *The New York Times*, 14 April.

Fellowes, M.C. and Wolf, P.J. (2004) Funding mechanisms and policy instruments: how business campaign contributions influence congressional votes. *Political Research Quarterly* **57**(2):315–324.

Ferree, M.M. (2005) Soft repression. In: Davenport, C., Johnson, H. and Mueller, C. (eds) *Repression and Mobilization*. University of Minnesota Press, Minneapolis, pp. 138–155.

Festinger, L. (1957) *A Theory of Cognitive Dissonance*. Stanford University Press, Stanford, CA.

Festinger, L., Riecken, H.W. and Schachter, S. (1956) *When Prophecy Fails*. University of Minnesota Press, Minneapolis.

Fischer, D.H. (1996) *The Great Wave*. Oxford University Press, New York.

Fleischman, D.E. and Cutler, H.W. (1955) Themes and symbols. In: Bernays, E. (ed.) *The Engineering of Consent*. University of Oklahoma Press, Norman, pp. 138–155.

Flock of Dodos (film) (2006). Prairie Starfish Productions, Los Angeles.

Ford, J. and Dennis, M. (eds) (2000) Special section on "Traditional ecological knowledge and wisdom". *Ecological Applications* 10:1249–1340.

Foreman, D. (1998–1999) Around the campfire: the river Wild. *Wild Earth* 8(4).

Foreman, D. (2004) *Rewilding North America*. Island Press, Washington, DC.

Forgas, J. (ed.) (2000) *Feeling and Thinking*. Cambridge University Press, Cambridge, MA.

Forgas, J. and Cromer, M. (2004) On being sad and evasive: affective influences on verbal communication strategies in conflict situations. *Journal of Experimental Social Psychology* 40:511–518.

Forgas, J. P., Williams K. D., and von Hippel W. (2003) Responding to the Social World. In: Forgas, J. P., Williams K. D., and von Hippel W. (eds). *Social Judgments*. Cambridge University Press, Cambridge, UK pp. 1–20.

Foster, R. and Kreitzman, L. (2004) *Rhythms of Life*. Profile, London.

Francisco, R.A. (2005) The dictator's dilemma. In: Davenport, C., Johnson, H. and Mueller, C. (eds) *Repression and Mobilization*. University of Minnesota Press, Minneapolis, pp. 58–81.

Frank, A.G. (1993) *The World System: Five Hundred Years or Five Thousand?* Routledge, London.

Frank, T. (2004) *What's the Matter With Kansas*. Holt, New York.

Franklin, B. (1987, [1732]) *Poor Richard's Almanac*. Blackwell North America, Lake Oswego, OR.

Freddi, C. (1983) *Pork*. RKP, London.

Free Speech for Sale (Television Documentary Film) (1999) (U.S.) Public Broadcasting System, Washington, DC (Bill Moyers' Productions, New York).

Freeman, A. and Forcese, C. (1994) Get Tough on Corporate Crime. The Toronto Star, 17 November.

Freilich, M. (1972) Manufacturing culture: man the scientist. In: Freilich, M. (ed.) *The Meaning of Culture*. Xerox College Publishing, Lexington, MA, pp. 267–325.

Freud, S. (1959, 1922) *Group Psychology and the Analysis of the Ego*. Norton, New York.

Freud, S. (1961, 1930) *Civilization and Its Discontents*. Norton, New York.

Frye, N. (1960) New directions from old. In: Murray, H.A. (ed.) *Myth and Mythmaking*. Braziller, New York, pp. 115–131.

GBOP (2007) *North Cascades Grizzly Bear Outreach Project* (brochure). Grizzly Bear Outreach Project, Bellingham, WA.

Gebhard, U., Nevers, P. and Billman-Machecha, E. (2003) Moralizing trees: anthropomorphism and identity in children's relationships to nature. In: Clayton, S. and Opotow, S. (eds) *Identity and the Natural Environment*. MIT Press, Cambridge, MA, pp. 91–112.

Gigerenzer, G. (2007) *Gut Feelings*. Viking, New York.

Gilbert, G. (1947) *Nuremberg Diary*. Farrar, Straus and Company, New York.

Ginsburgh, A.R. (1955) The tactics of public relations. In: Bernays, E.L. (ed.) *The Engineering of Consent*. University of Oklahoma Press, Norman, pp. 214–236.

Giugni, M. (2004) *Social Protest and Policy Change*. Rowman and Littlefield, Lanham, MD.

Goldman, E. (1934) *Living my Life*. Knopf, New York.

Goldman, M. (2005) *Imperial Nature.* Yale University Press, New Haven, CT.

Goldstein, J.S. (1988) *Long Cycles.* Yale University Press, New Haven, CT.

Goldstone, J.A. (1980) The weakness of organization. *The American Journal of Sociology* **85(5)**:1017–1042.

Goldstone, J.A. and Charles, T. (2001) Threat (and opportunity): popular action and state response in the dynamics of contentious action. In: Aminzade, R.R., Goldstone, J.A., McAdam, D., Perry, E.J., Sewell, W.H., Tarrow, S. and Charles, T. (eds) *Silence and Voice in the Study of Contentious Politics.* Cambridge University Press, Cambridge UK, pp. 179–194.

Goldstone, J.A. and McAdam, D. (2001) Contention in demographic and life-course context. In: Aminzade, R.R., Goldstone, J.A., McAdam, D., Perry, E.J., Sewell, W.H., Tarrow, S. and Charles T (eds) *Silence and Voice in the Study of Contentious Politics.* Cambridge University Press, Cambridge, UK, pp. 195–221.

Goodwin, J., Jasper, J.M. and Polletta, F. (eds) (2001) Introduction. *Passionate Politics.* University of Chicago Press, Chicago, pp. 1–24.

Goldstone, J.A. (2003) Introduction: bridging institutionalized and noninstitutionalized politics. In: Goldstone, J.A. (ed.) *States, Parties, and Social Movements.* Cambridge University Press, Cambridge, MA, pp. 1–24.

Gonzalez, G.A. (2001) *Corporate Power and the Environment.* Rowman & Littlefield, Lanham, MD.

Goodwin, J. and Pfaff, S. (2001) Emotion work in high-risk social movement: managing fear in the US and East German civil rights movements. In: Goodwin, J., Jasper, J.M. and Polletta, F. (eds) *Passionate Politics.* University of Chicago Press, Chicago, pp. 282–302.

Gorz, A. (1980) *Ecology as Politics.* South End Press, Boston.

Gould, R.V. (2003) Why do networks matters? Rationalist and structuralist interpretations. In: Diani, M. and McAdam, D. (eds) *Social Movements and Networks.* Oxford University Press, Oxford, pp. 233–257.

Grant, S.-M. (1997) Making history: myth and the construction of American nationhood. In: Hosking, G. and Schopflin, G. (eds) *Myths and Nationhood.* Routledge, New York, pp. 88–106.

Green, D.P. and Gerber, A.S. (2004) *Get Out the Vote!* Brookings Institution, Washington, DC.

Greider, W. (1981) The education of David Stockman. *The Atlantic Monthly* **248(6)**:27–54.

Grevin Jr, P.J. (1977) *The Protestant Temperament.* Knopf, New York.

Grimm, K. (ed.) (2006) *Discovering the Activation Point.* Communications Leadership Institute, San Francisco.

Grover, W.F. (1989) *The President as Prisoner.* State University Press of New York, Albany, NY. Quoting Kennedy at a 1961 press conference.

Guber, D.L. (2003) *The Grassroots of a Green Revolution.* MIT Press, Cambridge, MA.

Guigni, M. (1999) Introduction: how social movements matters: past research, present problems, future developments. In: Giugni, M., McAdam, D., and Tilly, C. (eds) *How Social Movements Matter.* University of Minnesota Press, Minneapolis, pp. 13–33.

Gunter, T.R. (2001) The Steens Mountain Divide: Beyond Compromise in Oregon's High Desert. MS Thesis, University of Montana, Missoula.

Gunter, M.M. (2004) *Building the Next Ark*. Dartmouth University Press, Hanover, NH.

Gurr, T.R. (1970) *Why Men Rebel*. Princeton University Press, Princeton, NJ.

Haidt, J. (2001) The emotional dog and its rational tail. *Psychological Review* **108(4)**:814–834.

Hall, T.D. (2000) World-systems analysis. In: Hall, T.D. (ed.) *A World Systems Reader*. Rowman and Littlefield, Lanham, MD, pp. 3–27.

Halpern, B.S., Shaun Walbridge, Kimberly A. Selkoe, Carrie V. Kappel, Fiorenza Micheli, Caterina D'Agrosa, et al. (2008) A global map of human impact on marine ecosystems. *Science* **319**:948–952.

Hammer, O. and Lewis, J.R. (2007) Introduction. In: Lewis, J.R. and Hammer, O. (eds) *The Invention of Sacred Tradition*. Cambridge University Press, Cambridge, MA, pp. 1–17.

Harbury, J.K. (2005) *Truth, Torture, and the American Way*. Beacon, Boston.

Harmon-Jones, E. and Mills, J. (eds) (1999) *Cognitive Dissonance*. American Psychological Association, Washington, DC.

Harris, M. (1974) *Cows, Pigs, Wars, and Witches*. Random House, New York.

Harris, M. (1975) *Culture, People, Nature*, 2nd edn. Crowell, New York.

Harris, M. (1977) *Cannibals and Kings*. Random House, New York.

Hart, K. (1999) Foreword. In: Rappaport, R.A. (ed.) *Ritual and Religion in the Making of Humanity*. Cambridge University Press, Cambridge, MA, pp. xiv–xix.

Hawken, P. (1993) *The Ecology of Commerce*. Harper Business, New York.

Harvest of Shame (1960) (broadcast). CBS News/CBS Reports, New York.

Heimert, A. and Delbanco, A. eds (1985) Excerpt from cotton Mather's reserved memorials. In: *The Puritans in America*, Harvard University Press, Cambridge, MA.

Helvarg, D. (1994) *The War Against the Greens*. Sierra Club Books, San Francisco.

Helvarg, D. (2004) *The War Against the Greens*, 2nd edn. Johnson, Boulder, CO.

Heuer, K. (2004). Personal communication at IUCN mountain meeting.

Hightower, J. (1995–6) Get the hogs out of the creek. *Earth Island Journal* **11(1)**:32.

Hironaka, A. (2005) *Neverending Wars*. Harvard University Press. Cambridge, MA.

Hobsbawm, E. and Terence R. (1984 [1983]) *The Invention of Tradition*. Cambridge University Press, Cambridge, UK.

Holt, J. (2008) Good instincts. *The New York Times Magazine*, 9 March.

Hornick, R.C. (1989) Channel effectiveness in development communication programs. In: Rice, R.E. and Atkin, C.K. (eds) *Public Communication Campaigns*, 2nd edn. Sage, Newbury Park, CA, pp. 309–330.

Hsu, Ming, Anen C. and Quartz S.R. (2008). The Right and the Good: Distributive Justice and Neural Encoding of Equity and efficiency. Science 320 :1092-1095.

Hughes, R.T. (2003) *Myths America Lives By*. University of Illinois Press, Urbana, IL.

Illyn, P. (2001). Lecture at Portland State University (6 November).

Ingram, H. and Fraser, L. (2006) Path dependency and adroit innovation: the case of California water. In: Repetto, R. (ed.) *Punctuated Equilibrium and the Dynamics of U.S. Environmental Policy*. Yale University Press, New Haven, CT, pp. 78–109.

International Labor Organization (United Nations) (1997) *World Labor Report 1997–98. Industrial Relations, Democracy and Social Stability*. International Labor Office, Geneva.

Isikoff, M. (2005). Profiling: how the FBI tracks eco-terror suspects. *Newsweek* **146**:121.

Jacobson, S. (1999) *Communication Skills for Environmental Professionals*. Island Press, Washington, DC.

Jacobson, S. (2006) *Conservation Education and Outreach Techniques*. Oxford University Press, New York.

Jasper, J.M. (1998) *The Art of Moral Protest*. University of Chicago Press, Chicago.

Jasper, J.M. (1999) Recruiting intimates, recruiting strangers: building the contemporary animal rights movement. In: Freeman, J. and Johnson, V. (eds) *Waves of Protest*. Rowman and Littlefield, Lanham, MD, pp. 65–82.

Jasper, J.M. and Poulsen, J.D. (1997, 1993). Fighting back: vulnerabilities, blunders, and countermobilization by targets in three animal rights campaigns. In: McAdam, D. and Snow, D.A. (eds) *Social Movements: Readings on Their Emergence, Mobilization, and Dynamics*. Roxbury Publishing, Los Angeles, pp. 397–406.

Jeffers, R. (1965, 1925). Boats in a fog. In: *Selected Poems*. Vintage, New York.

Jefferson, T. (1955, 1787). Letter to William S Stephens (13 November). In: Boyd, J.P. (ed.) *Jefferson Papers* Vol 12. Princeton University Press, Princeton, p. 356.

Jefferson, T. (1992, 1817). Letter to Horatio Spafford (17 March). In: Boyd, J.P. (ed.) *Jefferson Papers* Vol 14. Princeton University Press, Princeton, p. 221.

Jensen, R.J. and Hammerback, J.C. (2002, 1974). *The Words of Cesar Chavez*. Texas A & M University Press, College Station.

Johns, D. (1999) Biological science in conservation. In: McCool, S.F. et al. (ed.) *Wilderness Science in a Time of Change* Vol. 2, USDA Forest Service, Rocky Mountain Research Station, Ogden UT pp. 223–229.

Johns, D. (2003) The wildlands project outside North America. In: Watson, Alan and Janet Sproull, (comps). *Seventh World Wilderness Congress Symposium: Science and Stewardship to Protect and Sustain Wilderness Values*. Forest Service, Rocky Mountain Research Station Proceedings RMRS-P-000. U.S. Department of Agriculture, Ogden, UT, pp. 114–120.

Johnson, A.W. and Earle, T. (2001) *The Evolution of Human Societies*, 2nd edn. Stanford University Press, Stanford CA.

Johnson, S.M. (1985) *Characterological Transformation*. Norton, New York.

Johnson, S.M. (1987) *Humanizing the Narcissistic Style*. Norton, New York.

Johnson, C. (1982) *Revolutionary Change*, 2nd edn. Stanford University Press, Stanford, CA.

Johnson, C. (2001) *Blowback*. Henry Holt, New York.

Johnson, M. (2003) *Street Justice*. Beacon Press, Boston.

Johnson, C. (2004) *The Sorrows of Empire*. Henry Holt, New York.

Johnson, C. (2007) *Nemesis*. Henry Holt, New York.

Johnson, A.W. and Price-Williams, D. (1996) *Oedipus Ubiquitous: The Family Complex in World Folk Literature*. Stanford University Press, Stanford, CA.

Johnson-Cartee, K.S. and Copeland, G.A. (1997) *Inside Political Campaigns*. Praeger, Westport, CT.

Johnson-Cartee, K.S. and Copeland, G.A. (2004) *Strategic Political Communication.* Rowman and Littlefield, Lanham, MD.

Johnston, H. (2005) Talking the Walk. In: Davenport, C., Johnson, H. and Mueller, C. (eds) *Repression and Mobilization.* University of Minnesota Press, Minneapolis, pp. 108–137.

Jost, J.T., Glaser, J., Kruglanski, A.W. and Sulloway, F. (2003) Conservatism as motivated social cognition. *Psychological Bulletin* **129**(3):339–375.

Juergensmeyer, M. (2008) *Global Rebellion.* University of California Press, Berkeley.

Kahn, P.H. (1999) *The Human Relationship with Nature.* MIT Press, Cambridge, MA.

Kahn, P.H. (2002) Children's affiliations with nature. In: Kahn, P.H. and Kellert, S.R. (eds) *Children and Nature.* MIT Press, Cambridge, MA, pp. 93–116.

Kairys, D. (ed.) (1998) *The Politics of Law,* 3rd edn. Basic Books, New York.

Kals, E. and Ittner, H. (2003) Children's environmental identity: indicators and behavior impact. In: Clayton, S. and Opotow, S. (eds) *Identity and the Natural Environment.* MIT Press, Cambridge, MA, pp. 135–157.

Kamieniecki, S. (2006) *Corporate America and Environmental Policy.* Stanford University Press, Stanford, CA.

Kaplan, S. (2000) Human nature and environmentally responsible behavior. *The Journal of Social Issues* **56**(3):491–508.

Kareiva, P. and Marvier, M. (2007) Conservation for the people. *Scientific American* **297**(4):50–57.

Kasser, T., Ryan, R.M., Couchman, C.E. and Sheldon, K.M. (2003) Materialistic values: their causes and consequences. In: Kasser, T., Kanner, A.D. (eds) *Psychology and Consumer Culture.* American Psychological Association, Washington, DC, pp. 11–28.

Katzenstein, M.F. (1998) Stepsisters: feminist movement activism in different institutional spaces. In: Meyer, D.S. and Tarrow, S. (eds) *The Social Movement Society.* Rowman and Littlefield, Lanham, MD, pp. 195–216.

Kauffman, S.A. (1993) *The Origins of Order.* Oxford University Press, New York.

Kaye, R. (2006) *Last Great Wilderness.* University of Alaska Press, Fairbanks.

Keck, M.E. and Sikkink, K. (1998) Transnational advocacy networks in the movement society. In: Meyer, D.S. and Tarrow, S. (eds) *The Social Movement Society.* Rowman and Littlefield, Lanham, MD, pp. 217–238.

Keleman, D. (1999) Beliefs about purpose: on the origins of teleological thought. In: Corballis, M.C. and Lea, S.E.G. (eds) *The Descent of Mind.* Oxford University Press, Oxford, pp. 278–294.

Kellert, S.R. (2002) Experiencing nature. In: Kahn, P.H and Stephen R.K. (eds) *Children and Nature.* MIT Press, Cambridge, pp. 17–51.

Kempton, W. and Holland, D.C. (2003) Identity and sustained environmental practice. In: Clayton, S. and Opotow, S. (eds) *Identity and the Natural Environment.* MIT Press, Cambridge, MA, pp. 317–341.

Kerr, A. (2004) Mergers, acquisitions, diversifications, restructurings, and/or die-offs in the conservation movement. *Wild Earth* **14**(1–2):44–51.

Kertzer, D.I. (1988) *Ritual, Politics and Power.* Yale University Press, New Haven.

Kinzer, S. (2006) *Overthrow.* Times Books, New York.

Kirk, G.S. (1970) *Myth, Its Meaning and Function in Ancient and Other Cultures*. University of California, Berkeley.

Klandermans, B., Roefs, M. and Olivier, J. (1998) A movement takes office. In: Meyer, D.S. and Tarrow, S. (eds) *The Social Movement Society*. Rowman and Littlefield, Lanham, MD, pp. 173–194.

Klare, M. (2001) *Resource Wars*. Henry Holt, New York.

Klein, N. (2004) Reclaiming the commons. In: Mertes, T. (ed.) *A Movement of Movements*. Verso, London, pp. 219–229.

Knutson, B. (2004) Sweet revenge? *Science* **305**:1246–1247.

Koehler, B. (2005) *Stand by Your Land*. The Wilderness Society Wilderness Support Center, Durango, CO.

Koopmans, R. (2004) Protest in Time and Space. In: Snow D.A., Soulé S.A. and Kriesi H. (eds). *The Blackwell Companion to Social Movements*. Blackwell, Malden, MA, pp 19–46.

Koopmans, R. (2005) Repression and the public sphere. In: Davenport, C., Johnson, H. and Mueller, C. (eds) *Repression and Mobilization*. University of Minnesota Press, Minneapolis, pp. 159–188.

Kraft, M.E. and Kamieniecki, S. (2007) Analyzing the role of business in environmental policy. In: Kraft, M.E. and Kamieniecki, S. (eds) *Business and Environmental Policy*. MIT Press, Cambridge, MA, pp. 3–31.

Krauss, C. (2008) Commodities relentless surge. *The New York Times*, 15 January.

Kreisi, H. (2004) Political Context and Opportunity. In: Snow D.A., Soulé S.A. and Kriesi H. (eds). *The Blackwell Companion to Social Movements*. Blackwell. Malden, MA, pp 67–90.

Kull, S. (2003) *Misperceptions, the Media, and the Iraq War*. Program on International Policy Attitudes, University of Maryland, Washington, DC.

Kull, S. (2004) *Americans and Iraq on the Eve of the Presidential Election*. Program on International Policy Attitudes, University of Maryland, Washington, DC.

Kull, S. (2006) *Americans on Iraq Three Years On*. Program on International Policy Attitudes, University of Maryland, Washington, DC.

Lakoff, G. (1996) *Moral Politics*. University of Chicago Press, Chicago.

Lange, D. (1995). Personal Communication. Unpublished interview conducted with the former Prime Minister of New Zealand conducted at Portland State University, Portland, OR.

Larson, G. (1998) *There's A Hair in My Dirt*. Harper Collins, New York.

Lasch, C. (1978) *The Culture of Narcissism*. Norton, New York.

Lawrence, D.H. (1936, 1930) The grand inquisitor by F.M. Dostoievsky. In: *Phoenix*. Viking, New York, pp. 283–291.

Layzer, J.A. (2006) *The Environmental Case*, 2nd edn. Congressional Quarterly Press, Washington, DC.

Layzer, J. (2007) Deep freeze: how business has shaped the global warming debate in congress. In: Kraft, M.E. and Kamieniecki, S. (eds) *Business and Environmental Policy*. MIT Press, Cambridge, MA, pp. 93–125.

Lee, M.F. (1995) *Earth First! Environmental Apocalypse*. Syracuse University Press, Syracuse, NY.

Lee, R.B. (1972) The intensification of social life among the !Kung Bushmen. In: Spooner, B. (ed.) *Population Growth: Anthropological Implications*. MIT Press, Cambridge, MA, pp. 343–350.

Leeming, D.A. (2002) Myth: *A Biography of Belief*. Oxford University Press, New York.

Leopold, Aldo. 1987, (1949) *A Sand county Almanac and Sketches Here and There*. Oxford University Press, New York.

Lewin, T. (2003) Catholics liberalize in college, survey finds. *The New York Times*, 5 March.

Lian, P.K., Dunn, R.L., Sodhi, N.S., Colwell, R.K., Proctor, H.C. and Smith, V.S. (2004) Species coextinctions and the biodiversity crisis. *Science* **305**:1632–1634.

Libby, R.T. (1999) *Eco-Wars: Campaigns and Social Movements*. Columbia University Press, New York.

Life of Brian, The (film) (1979). Python Pictures-Handmade Films, London.

Lincoln, A. (1953–1955, 1864) Address at sanitary fair. In: Basler, R.P. (ed.) *The Collected Works of Lincoln*, Rutgers University Press, New Brunswick, NJ, Volume 7, pp. 301–303.

Lincoln, B. (1996) Mythic narrative and cultural diversity in American society. In: Patton, L.L. and Doniger, W. (eds) *Myth and Method*. University Press of Virginia, Charlottesville, pp. 163–176.

Lindblom, C.E. (1977) *Politics and Markets*. Basic Books, New York.

Linden, E. (2006) *The Winds of Change*. Simon and Schuster, New York.

Lipschutz, R.D. and Mayer, J. (1993) Not seeing the forest for the trees: rights, rules, and the renegotiation of resource management regimes. In: Lipschutz, R.D. and Conca, K. (eds) *The State and Social Power in Global Environmental Politics*. Columbia University Press, New York, pp. 246–273.

Lipschutz, R.D and Ken, C. (eds) (1993) The implications of global ecological interdependence. *The State and Social Power in Global Environmental Politics*. Columbia University Press, New York, pp. 327–343.

Liptak, A. (2006a) In leak cases, new pressure on journalists. *The New York Times*, 30 April.

Liptak, A. (2006b) Gonzales says prosecutions of journalists are possible. *The New York Times*, 22 May.

Liptak, A. (2008) A corporate view of Mafia tactics: protesting, lobbying and citing Upton Sinclair. *The New York Times*.

Locke, H. and Dearden, P. (2005) Rethinking protected area categories and the new paradigm. *Environmental Conservation* **32**(1):1–10.

Locke, J. (2003 [1689]) *Two Treatises of Government and a Letter Concerning Toleration*. Yale University Press, New Haven, CT.

Long, B. (2002). Presentation. *Wild Idaho North Conference*. Bonner's Ferry, ID. (19 Oct).

Lopez, B.H. (1978) *Of Wolves and Men*. Scribners, New York.

Louv, R. (2005) *Last Child in the Woods*. Chapel Hill, Algonquin.

Lowen, A. (1975) *Bioenergetics*. Coward, McCann and Geoghegan, New York.

Lucero, L.J. (2003) The politics of ritual. *Current Anthropology* **44**(4):523–558.

McAdam, D. and Sewell, W.H. (2001) Its about time: temporality in the study of social movements and revolutions. In: Aminzade, R.R, Goldstone, J.A, McAdam, D., Perry, E.J., Sewell, W.H., Tarrow, S. and Tilly, C. (eds) *Silence and Voice in the Study of Contentious Politics*. Cambridge University Press, Cambridge, UK, pp. 89–125.

McAdams, D.P. (1993) *The Stories We Live By*. Morrow, New York.

Machiavelli, N. (1984, 1532). *The Prince*. Oxford University Press, Oxford, UK.

MacLeod, M. (2002). *D'oh! Homer's Powerful Nuclear Lesson for Scots*, in Scotland on Sunday (Edinburgh), 14 July: 1.

MacWilliams, Robinson and Partners (2002) Y2YCI Message Framework. MacWilliam, Robinson and Partners, Washington, DC.

Madison, J. (1961, 1788) Federalist No. 10. In: *The Federalist*. Wesleyan University Press, Middletown, CT, pp. 56–65.

Mahler, J. (2005) With Jesus as our connector. *The New York Times* 30–57, 27 March.

Malone, L.A. and Pasternack, S. (2006) *Defending the Environment*, 2nd edn. Island Press, Washington, DC.

Marcus, J.T. (1960) The world impact of the West: the mystique and the sense of participation in history. In: Murray, H.A. (ed.) *Myth and Mythmaking*. Braziller, New York, pp. 221–239.

Marcuse, H. (1964) *One Dimensional Man*. Beacon Press, Boston.

Margoluis, R. and Salalfsky, N. (1998) *Measures of Success*. Island Press, Washington, DC.

Marker, L. (2006). Interview Conducted by Author. DATE. Portland, OR.

Markowitz, G. and Rosner, D. (2002) *Deceit and Denial*. University of California Press, Berkeley.

Marmot, M. (2004) *The Status Syndrome*. Times Books, New York.

Marx, K. (1976, 1896) *The Poverty of Philosophy. Karl Marx and Frederick Engels Collected Works*, 3rd edn. Progress Publishers, Moscow, Volume 6.

Marx, K.-B. (1979, 1869) *The 18th Brumaire of Louis Bonaparte. Karl Marx and Frederick Engels Collected Works*, 2nd edn. Progress Publishers, Moscow, Volume 11.

Marx, A.W. (1992) *Lessons of Struggle*. Oxford University Press, New York.

Marx, K. and Engels, F. (1975, 1844). *The Holy Family*. Progress, Moscow.

Marx, K. and Engels, F. (1998, 1848). *The Communist Manifesto*.

Maslow, A. (1968) *Toward a Psychology of Being*, 2nd edn. Van Nostrand, New York.

Maslow, A. (1970) *Motivation and Personality*, 2nd edn. Harper and Row, New York.

Masterson, J.F. (1985) *The Real Self*. Brunner/Mazel, New York.

Mazlish, B. (1990) *The Leader, the Led, and the Psyche*. Wesleyan University Press, Hanover, NH.

Mazur, L.A. (ed.) (1994) *Beyond the Numbers*. Island Press, Washington, DC.

McAdam, D. (1997, 1983) Tactical innovation and the pace of insurgency. In: McAdam, D. and Snow, D.A. (eds) *Social Movements: Readings on Their Emergence, Mobilization, and Dyanmics*. Roxbury Publishing, Los Angeles, pp. 340–356.

McAdam, D., Tarrow, S. and Tilly, C. (2001) *Dynamics of Contention*. Cambridge University Press, Cambridge, UK.

McAdam, D. (2003) Beyond structural analysis: toward a more dynamic understanding of social movements. In: Mario, D. and McAdam, D. (eds) *Social Movements and Networks*. Oxford University Press, Oxford, pp. 281–298.

McCarthy, J.D. and McPhail, C. (1998) The institutionalization of protest in the United States. In: Meyer, D.S. and Tarrow, S. (eds) *The Social Movement Society*. Rowman and Littlefield, Lanham, MD, pp. 83–110.

McDonald, P. (2003) Family policy. In: Demeny, P. and McNicoll, G. (eds) *Encyclopedia of Population*. Thomson-Gale, New York, pp. 371–374.

McGuire, W.J. (1989) Theoretical foundations of campaigns. In: Rice, R.E. and Atkin, C.K. (eds) *Public Communication Campaigns*, 2nd edn. Sage Publications, Newbury Park, CA, pp. 43–65.

McGuire, W.J. (2001) Input and output variables currently promising for constructing persuasive communications. In: Rice, R.E. and Atkin, C.K. (eds) *Public Communication Campaigns*, 3rd edn. Sage Publications, Thousand Oaks, CA, pp. 22–48.

McIver, J., Muttillinja, S., Pickering, D. and VanBuskirk, R. (1991) *Population Dynamics and Habitat selection of the Oregon silverspot butterfly (Speyeria zerene hippolyta): a comparative study at four primary sites in Oregon. Report to the U. S. Dept. of Agriculture, Forest Service, Siuslaw National Forest*. USDA Forest Service, Corvallis, OR.

McKenzie-Mohr, D. (2000) Promoting sustainable behavior: an introduction to community-based social marketing. *The Journal of Social Issues* 56(3):543–554.

McKibben, B. (2007) *Deep Economy*. Henry Holt, New York.

McLuhan, M. (1964) *Understanding Media*. McGraw-Hill, New York.

McNeill, W.H. (1995) *Keeping Together in Time*. Harvard University Press, Cambridge, MA.

Meadow, R.G. (1989) Political campaigns. In: Rice, R.E. and Atkin, C.K. (eds) *Public Communication Campaigns*, 2nd edn. Sage Publications, Newbury Park, CA, pp. 253–272.

Meine, C., Soulé, M. and Noss, R.F. (2006) A mission driven discipline: the growth of conservation biology. *Conservation Biology* 20:631–651.

Melson, G.F. (2001) *Why the Wild Things Are*. Harvard University Press, Cambridge.

Merchant, C. (1980) *The Death of Nature*. Harper and Row, New York.

Mertes, T. (2004) Grass-roots globalism. In: Mertes, T. (ed.) *A Movement of Movements*. Verso, London, pp. 235–247.

Meyer, D.S. and Tarrow, S. (1998) A movement society: contentious politics for a new society. In: Meyer, D.S. and Tarrow, S. (eds) *The Social Movement Society*. Rowman and Littlefield, Lanham, MD, pp. 1–28.

Michels, R. (1962, 1915) *Political Parties*. Free Press, Glencoe, IL.

Micklethwait, J. and Wooldridge, A. (2004) *The Right Nation*. Penguin Press, New York.

Milburn, M.A. and Conrad, S.D. (1996) *The Politics of Denial*. MIT Press, Cambridge, MA.

Miller, A. (1985) *Thou Shalt Not Be Aware*. Farrar, Strauss and Giroux, New York.

Miller, A. (2001) *Banished Knowledge*. Doubleday, New York.

Miller, B., Reading, R., Strittholt, J., Carroll, C., Noss, R., Soulé, M., Sanchez, O., Terborgh, J., Brightsmith, D., Cheeseman, T. and Foreman, D. (1998–9). Using focal species in the design of nature reserve networks. *Wild Earth* 8(4):81–92.

Miroff, B., Seidelman, R. and Swanstrom, T. (2002) *The Democratic Debate*, 3rd edn. Houghton Mifflin, Boston.

Mische, A. (2003) Cross-talk in movements: reconceiving the culture-network link. In: Giugni, M., McAdam, D. and Tilly, C. (eds) *How Social Movements Matter*. University of Minnesota Press, Minneapolis, pp. 258–280.

Mooney, C. (2005) *The Republican War on Science*. Basic Books, New York.

Moore, B. (1966) *The Social Origins of Dictatorship and Democracy*. Beacon Press, Boston.

Moore, S.F. (1977) Political meetings and the simulation of unanimity, Kilimanjaro 1973. In: Moore, S.F. and Myerhoff, B.G. (eds) *Secular Ritual*. VanGorcum, Assen, Netherlands, pp. 151–172.

Moore, S.F. and Myerhoff, B.G. (1977) Introduction: secular ritual: forms and meanings. In: Moore, S.F. and Myerhoff, B.G. (eds) *Secular Ritual*. Van Gorcum, Assen, Netherlands, pp. 3–24.

Morgan, C.P., Davis, J., Ford, T. and Laney, N. (2004) Promoting understanding: the approach of the North Cascades Grizzly Bear Outreach Project. *Ursus* 15(Suppl. 1): 137–141.

Morris, E. (1979) *Rise of Theodore Roosevelt*. Putnam, New York.

Morris, A.D. and Staggenborg, S. (2004) Leadership in social movements. In: Snow, D., Soulé, S. and Kriesi, H. (eds) *The Blackwell Companion to Social Movements*. Blackwell, Oxford, pp. 171–196.

Muir, D. (2004) Proclaiming thanksgiving throughout the land. In: Etzioni, A. and Bloom, J. (eds) *We Are What We Celebrate*. New York University Press, New York, pp. 194–212.

Murray, H.A. (1960) The possible nature of a "Mythology" to come. In: Murray, H.A. (ed.) *Myth and Mythmaking*. Braziller, New York, pp. 300–352.

Myers, O.E. and Saunders C.D. (2002) Animals as links toward developing caring relationships with the natural world. In: Kahn, P. H. and Kellert S.R. (eds). *Children and Nature*. MIT Press, Cambridge pp. 153–178.

Myers, G. (2007) *The Significance of Children and Animals*, 2nd edn. Purdue University Press, West Lafayette, IN.

Nadeau, R.L. (2006) *The Environmental Endgame*. Rutgers University Press, New Brunswick, NJ.

Nash, R. (1982) *Wilderness and the American Mind*, 3rd edn. Yale University Press, New Haven, CT.

Nepstad, S.E. and Christian, S. (2001) The social structure of moral outrage in recruitment to the US Central America peace movement. In: Goodwin, J., Jasper, J.M., Polletta, F. (eds) *Passionate politics*. University of Chicago Press, Chicago, pp. 158–174.

Nicholsen, S.W. (2002) *The Love of Nature and the End of the World*. MIT Press, Cambridge.

Niemi, E., Courant, P. and Whitelaw, E. (1997) *The Ecosystem-economy Relationship: Insights from Six Forested LTER Sites, Report to the National Science Foundation*. ECONorthwest, Portland, Oregon.

Niemi, E. and Gall, M. (1998) *The Economics of ICBEMP: An Initial Assessment of the Draft Environmental Impact Statement for the Interior Columbia Basin Ecosystem Management Project*. ECONorthwest, Portland, Oregon.

Niemi, E., MacMullan, E., Whitlaw, E. and Taylor, D. (1996) *The Potential Economic Consequences of Designating Critical Habitat for the Marbled Murrelet, Final Report to the U.S. Fish and Wildlife Service*. ECONorthwest, Portland, Oregon.

Niemi, E., Whitlaw, E. and Johnson, A. (1999a) *The Sky did not Fall: The Pacific Northwest's Response to Logging Reductions*. ECONorthwest, Portland, Oregon.

Niemi, E., Gall, M. and Johnston, A. (1999b) *An Economy in Transition: The Klamath-Siskiyou Ecoregion*. ECONorthwest, Portland, Oregon.

Nissenbaum, S. (1996) *The Battle for Christmas*. Knopf, New York.

Noble, D.F. (1997) *The Religion of Technology*. Knopf, New york.

Noelle-Newman, E. (1973) Return to the concept of powerful mass media. *Studies of Broadcasting* 9:67–112.

Norris, P. and Inglehart, R. (2004) *Sacred and Secular*. Cambridge University Press, Cambridge, UK.

Noss, R. (1993) The wildlands project land conservation strategy. *Wild Earth* (Special Issue 1):10–25.

Noss, R. and Cooperrider, A. (1994) *Saving Nature's Legacy*. Island Press, Washington, DC.

Oates, J.F. (1999) *Myth and Reality in the Rainforest: How Conservation Strategies are Failing in West Africa*. University of California, Berkeley.

Obach, B.K. (2004) *Labor and the Environmental Movement*. MIT Press, Cambridge, MA.

O'Keefe, G.J. and Reid-Martinez, K. (2001) The McGruff crime prevention campaign. In: Rice, R.E. and Atkin, C.K. (eds) *Public Communication Campaigns*, 3rd edn. Sage Publications, Thousand Oaks, CA, pp. 273–275.

Oliver, P.E. and Meyers, D.J. (2003) Networks, diffusion and cycles of collective action. In: Diani, M. and McAdam, D. (eds) *Social Movements and Networks*. Oxford University Press, Oxford, pp. 173–203.

Orwell, G. (1968, 1946) Politics and the English language. In: Orwell, G. (ed.) *Collected Essays, Journalism and Letters Volume 4, In Front of Your Nose*. Harcourt, Brace and World, New York, pp. 127–139.

Overing, J. (1997) The role of myth: an anthropological perspective. In: Hosking, G. and Schopflin, G. (eds) *Myths and Nationhood*. Routledge, New York, pp. 1–18.

Paisley, W.J. (2001) Public communication campaigns, the American experience. In: Rice, R.E. and Atkin, C.K. (eds) Public Communication Campaigns, 3rd edn. Sage Publications, Thousand Oaks, CA, pp. 3–21.

Paletz, D.L. and Entman, R.M. (1981) *Media Power and Politics*. Free Press, New York.

Parsa, M. (2003) Will democratization and globalization make revolutions obsolete? In: Foran, J. (ed.) *The Future of Revolutions*. Zed Books, London, pp. 73–82.

Passy, F. (2003) Social networks matter. But how? In: Diani, M. and McAdam, D. (eds) *Social Movements and Networks*. Oxford University Press, Oxford, pp. 21–48.

Patterson, T.E. (2002) *The Vanishing Voter*. Knopf, New York.

Peck, M.S. (1978) *The Road Less Traveled*. Simon and Schuster, New York.

Peet, J. (1992) *The Ecological Economics of Sustainability*. Island Press, Washington, DC.

Phillips, K. (2006) *American Theocracy*. Viking, New York.

Pimm, S.L. (2001) *The World According to Pimm*. McGraw-Hill, New York.

Piven, F.F. and Cloward, R. (1977) *Poor People's Movements*. Pantheon, New York City.

Polletta, F. (2004) Can you celebrate dissent? In: Etzioni, A. and Bloom, J. (eds) *We Are What We Celebrate*. New York University Press, New York, pp. 151–177.

Polletta, F. (2006) *It Was Like a Fever*. University of Chicago Press, Chicago.

Powell, J.W. (1961, 1875). *Exploring the Colorado River and Its Canyons*. Dover, New York.

Powers, T.M. (1996) *Lost Landscapes and Failed Economies*. Island Press, Washington, DC.

Powers, T.M. and Barrett, R.N. (2001) *Post Cowboy Economics*. Island Press, Washington, DC.

Rambo, L.R. (1993) *Understanding Conversion*. Yale University Press, New Haven, CT.

Rappaport, R.A. (1974) Sanctity and adaptation. *The Coevolution Quarterly* **Summer**(2), 54–68.

Rappaport, R.A. (1969 [1974]) Sanctity and adaptation. Presented at the Wenner-Gren Foundation Conference "The Moral and Esthetic Structure of Human Adaptation", *July CoEvolution Quarterly* 2:54–68. (Summer)

Rappaport, R.A. (1976) Adaptations and maladaptations in social systems. In: Hill., I. (ed.) *The Ethical Basis of Economic Freedom*. American Viewpoint, Chapel Hill, NC, pp. 39–79.

Rappaport, R.A. (1999) *Ritual and Religion in the Making of Humanity*. Cambridge University Press, Cambridge.

Rasker, R. and Alexander, B. (1997) *The New Challenge: People, Commerce and the Environment in the Yellowstone to Yukon Region*. The Wilderness Society, Washington, DC.

Redford, K.H. and Richter, B.D. (1999) Conservation of biodiversity in a world of use. *Conservation Biology* 13(6):1246–1256.

Reich, W. (1970, 1946) *The Mass Psychology of Fascism*. Orgone Institute Press, New York. (A much more available 1st English language edition, newly translated, is available from Noonday/Farrar, Strauss and Giroux, New York, 1970.)

Reich, W. (1961, 1949) *Character Analysis*, 3rd edn. Orgone Institute Press, New York. (A much more available 3d Edition is available from Noonday/Farrar, Strauss and Giroux, New York, 1961.)

Reich, W. (1972, 1934). What is class consciousness. In: *Sex-Pol: Essays 1929–1934*. Random House, New York, pp. 277–358.

Reichart, T. and Lambiase, J. (eds) (2002) *Sex in Advertising*. Lawrence Erlbaum Associates, Mahwah, NJ.

Reichart, T. and Lambiase, J. (eds) (2005) *Sex in Consumer Culture*. Lawrence Erlbaum Associates, Mahwah, NJ.

Repetto, R. (2006) Introduction. In: Repetto, R. (ed.) *Punctuated Equilibrium and the Dynamics of U.S. Environmental Policy*. Yale University Press, New Haven, pp. 1–23.

Rich, A. (2005) *Think Tanks, Public Policy, and the Politics of Expertise*. Cambridge University Press, Cambridge, UK.

Robinson, J. (2004) Parks, people and pipelines. *Conservation Biology* 18(3):607–608.

Rodgers, D.T. (1987) *Contested Truths*. Basic Books, New York.

Rohter, L. (2005) Brazil, bowing to protests, reopens logging in Amazon. *The New York Times*, 13 February.

Romano, L. (2005) Literacy of college graduates is on decline. *Washington Post*.

Roosevelt, T. (1911) *The New Nationalism*. speech delivered 31 August 1910 in Osawatomie, Kansas, Outlook, New York, pp. 3–33.

Rootes, C. (2004) Environmental movements. In: Snow, D., Soulé, S. and Kriesi, H. (eds) *The Blackwell Companion to Social Movements*. Blackwell, Oxford, pp. 608–640.

Rosenbaum, W.A. (1998) *Environmental Politics and Policy*, 4th edn. Congressional Quarterly Press, Washington, DC.

Rosenblatt, R. (ed.) (1999) *Consuming Desires: Consumption, Culture, and the Pursuit of Happiness*. Island Press, Washington, DC.

Rosnick, D. and Weisbrot, M. (2006) *Are Shorter Work Hours Good for the Environment?* Center for Economic and Policy Research, Washington, DC.

Rucht, D. (1999) The impact of environmental movements in western societies. In: Giugni, M., McAdam, D. and Tilly, C. (eds) *How Social Movements Matter*. University of Minnesota Press, Minneapolis, pp. 204–224.

Russell, C. (1923) Speech to the great falls, Montana booster club. Cited at p. 89 in William Kittredge. 1987. *Owning It All*. Graywolf Press, St Paul, MN.

Sahlins, M. (1972) *Stone Age Economics*. Aldine, Chicago.

Sale, K. (1973) *SDS*. Random House, New York.

Sanderson, E.W., Jaiteh, M., Levy, M.A., Redford, K.H., Wannebo, A.V. and Woolmer, G. (2002) The human footprint on the last of the wild. *Bioscience* **52**:891–904.

Sartre, J.P. (1947) *No Exit and The Flies*. Knopf. New York.

Saunders, F.S. (1999) *Who Paid the Piper? The CIA and the Cultural Cold War*. Granta, London.

Scahill, J. (2007) *Blackwater*. Perseus, New York.

Schlosser, E. (2001) *Fast Food Nation*. Houghton Mifflin, New York.

Schopflin, G. (1997) Functions of myth and a taxonomy of myth. In: Hosking, G. and Schopflin, G. (eds) *Myths and Nationhood*. Routledge, New York, pp. 19–35.

Schultz, B. and Schultz R. (2001) *The Price of Dissent*. University of California Press, Berkeley.

Schwartz, J. (2003) Choosing whether to cover-up or come clean. *The New York Times*, 1 July.

Scott, J.C. (1990) *Domination and the Arts of Resistance*. Yale University Press, New Haven, CT.

Scott, J.C. (1998) *Seeing Like a State*. Yale University Press, New Haven, CT.

Searles, H.F. (1960) *The Non-human Environment in Normal Development and Schizophrenia*. International Universities Press, New York.

Searles, H.F. (1979) Unconscious processes in relation to the environmental crisis. In: *Countertransference*. International Universities Press, New York, pp. 228–242.

See It Now (broadcast). (1954) CBS News, New York. (9 March).

Sellars, J. (2004) Raising a ruckus. In: Mertes, T. (ed.) *A Movement of Movements*. Verso, London, pp. 175–191.

Servheen, C., Herrero, S. and Payton, B. (1999) *Bears: Status Survey and Conservation Action Plan*. IUCN Press, Gland, Switzerland.

Shaiko, R.G. (1999) *Voices and Echoes for the Environment*. Columbia University Press, New York.

Shapiro, J. (2001) *Mao's War Against Nature*. Cambridge University Press, Cambridge, UK.

Shepard, P. (1973) *The Sacred Game and the Tender Carnivore*. Scribners, New York.

Shepard, P. (1990) *Nature and Madness*. Sierra Club Books, San Francisco.

Shepard, P. (1995) Virtually hunting reality in the forests of simulacra. In: Soulé, M.E. and Lease, G. (eds) *Reinventing Nature?* Island Press, Washington, DC, pp. 17–30.

Shuman, M. and Fuller, M. (2005) Profits for justice. *Nation* **280**(3):13–22, 24 January.

Singer, P.W. (2008) *Corporate Warriors*. Cornell University Press, Ithaca, NY.

Singhal, A. and Rogers, E.M. (2001) The entertainment-education strategy in communication campaigns. In: Rice, R.E. and Atkin, C.K. (eds) *Public Communication Campaigns*, 3rd edn. Sage Publications, Thousand Oaks, CA, pp. 343–356.

Smith, A. (1976, 1759) *The Theory of Moral Sentiments*. Clarendon, Oxford University Press, Oxford.

Snyder, G. (1990) *Practice of the Wild*. North Point Press. San Francisco.

Smith, R.M. (2003) *Stories of Peoplehood*. Cambridge University Press, Cambridge, UK.

Snow, D.A. and McAdam, D. (2000) Identity work processes in the context of social movements: clarifying the identity/movement nexus. In: Sheldon, S., Owens, T.J. and White, R.W. (eds) *Self, Identity, and Social Movements*. University of Minnesota, Minneapolis, MN, pp. 41–67.

Solomon, S., Greenberg, J.L. and Pyszczynski, T. (2004) Lethal consumption: death denying materialism. In: Kasser, T. and Kramer, A.D. (eds) *Psychology and Consumer Culture*. American Psychological Association, Washington, DC, pp. 127–146.

Sorel, G. (1961, 1907). *Reflections on Violence*. Collier Books, New York.

Soulé, M. (1995) The social siege of nature. In: Soulé, M.E. and Lease, G. (eds) *Reinventing Nature?* Island Press, Washington, DC, pp. 131–170.

Soulé, M. and Noss, R. (1998) Rewilding and biodiversity. *Wild Earth* 8(3):19–28.

Soulé, M. and Terborgh, J. (eds) (1999) *Continental Conservation*. Island Press, Washington, DC.

Spradley, J.P. and McCurdy, D.M. (2004) *The Cultural Experience*, 2nd edn. Waveland Press, Long Grove, IL.

Staggenborg, S. (1997, 1988) The consequences of professionalization and formalization in the pro-choice movement. In: McAdam, D. and Snow, D.A. (eds) *Social Movements: Readings on Their Emergence, Mobilization, and Dyanmics*. Roxbury Publishing, Los Angeles, pp. 421–439.

Stauber, J.C. and Rampton, S. (1995) *Toxic Sludge is Good for You*. Common Courage Press, Monroe, ME.

Steckel, R.H. and Rose, J.C. (eds) (2002) *Backbone of History*. Cambridge University Press, Cambridge, UK.

Stedile, J.P. (2004) Brazil's landless battalions. In: Mertes, T. (ed.) *A Movement of Movements*. Verso, London, pp. 17–48.

Stein, A. (2001) Revenge of the shamed: the Christian right's emotional cultural war. In: Jeff, G., Jasper, J.M. and Polletta, F. (eds) *Passionate Politics*. University of Chicago Press, Chicago, pp. 115–131.

Stern, P. (2000) Toward a coherent theory of environmentally significant behavior. *The Journal of Social Issues* 56(3):407–424.

Stryker, S., Owens, T.J. and White, R.W. (2000) Social Psychology and social movements: cloudy past and bright future. In: Sheldon, S., Owens, T.J. and White, R.W. (eds) *Self, Identity, and Social Movements*. University of Minnesota, Minneapolis, MN, pp. 1–15.

Tainter, J. (1988) *The Collapse of Complex Systems*. Cambridge University Press, Cambridge, UK.

Tarrow, S. (2005) *The New Transnational Activism*. Cambridge University Press, Cambridge, UK.

Tarrow, S. and McAdam, D. (2005) Scale shift in transnational contention. In: Donatella, D.P. and Tarrow, S. (eds) *Transnational Protest and Global Activism*. Rowman and Littlefield, Lanham, MD, pp. 121–147.

Tavris, C. and Aronson, E. (2007) *Mistakes Were Made (but not by me)*. Harcourt, New York.

Taylor, V. (1989) Social movement continuity: the women's movement in abeyance. *American Sociological Review* **54**(5):761–775.

Teicher, M.H. (2002a) The neurobiology of child abuse. *Scientific American* **286**(3):68–75.

Teicher, M.H. (2002b) Letter to the editor response. *Scientific American* **287**(1):16.

Terborgh, J. (1999) Requiem for Nature. Island Press. Washington, DC.

Tetlock, P.E. (2005) *Expert Political Judgment. How Good Is It? How Can We Know?* Princeton University Press, Princeton, NJ.

Theoharis, A. (1978) *Spying on Americans*. Temple University Press, Philadelphia.

Theoharis, A. (2004) *The FBI and American Democracy*. University Press of Kansas, Lawrence.

Thomas, C. (2003) *Bureaucratic Landscapes*. MIT Press, Cambridge, MA.

Thompson, W.R. (2000). K-waves, leadership cycles, and global war. In: Hall, T.D. (ed.) *A World Systems Reader*. Rowman and Littlefield, Lanham, MD, pp. 83–104.

Thompson, C. (2003) There's a sucker born in every medial prefrontal cortex. *The New York Times* 54–57, 26 October.

Thompson, W.R. (2005) Eurasian C-wave crises in the first millennium B.C. In: Chase-Dunn, C. and Anderson, E.N. (eds) *The Historical Evolution of World Systems*, Palgrave Macmillan, New York, pp. 20–51.

Thoreau, H.D. (1964, 1854) Walden. In: Bode, C. (ed.) *The Portable Thoreau*. Viking, New York, pp. 258–572.

Tierney, J. (2006) The SWAT syndrome. *The New York Times*.

Tilly, C. (1999) From interactions to outcomes in social movements. In: Giugni, M., McAdam, D. and Tilly, C. (eds) *How Social Movements Matter*. University of Minnesota Press, Minneapolis, pp. 253–270.

Tilly, C. (2002) *Stories, Identities, and Political Change*. Rowman and Littlefield, Lanham, MD.

Tilly, C. (2004) *Social Movements 1768–2004*. Paradigm, Boulder, CO.

Tilly, C. (2005) Repression, mobilization, and explanation. In: Davenport, C., Johnson, H. and Mueller, C. (eds) *Repression and Mobilization*. University of Minnesota Press, Minneapolis, pp. 211–226.

Tilly, C. (2006) *Regimes and Repertoires*. University of Chicago Press, Chicago.

Tompkins, D. (1998) Personal communication.

Torgerson, D. (2005) Obsolescent leviathan. In: Paehlke, R. and Torgerson, D. (eds) *Managing Leviathan*, 2nd edn. Broadview, Peterborough, ONT, pp. 11–24.

Torgerson, D. and Paehlke, R. (2005) Environmental administration. In: Paehlke, R. and Torgerson, D. (eds) *Managing Leviathan*, 2nd edn. Broadview, Peterborough, ONT, pp. 3–10.

Trent, J.S. and Friedenberg, R.V. (2004) *Political Campaign Communication*, 5th edn. Rowman and Littlefield, Latham, MD.

Tucker, R.P. (2000) *Insatiable Appetite*. University of California Press, Berkeley.

Turner, T.S. (1977a) Transformation, hierarchy and transcendence: a reformulation of VanGennep's model of the structure of Rites de Passage. In: Moore, S.F. and Myerhoff, B.G. (eds) *Secular Ritual*. VanGorcum, Assen, Netherlands, pp. 53–70.

Turner, V. (1977b) Variations on a theme of liminality. In: Moore, S.F. and Myerhoff, B.G. (eds) *Secular Ritual*. Van Gorcum, Assen, Netherlands, pp. 36–52.

USA Foundation (2007) *Giving USA*. USA Foundation, Glenview, IL.

U.S. Congress Joint Economic Committee (1976) *Report to the Joint Economic Committee: Estimating the Social Costs of National Economic Policy*. USGPO, Washington, DC.

U.S. Congress Joint Economic Committee (1984). *Report to the Joint Economic Committee: Estimating the Effects of Economic Change on National Social Well Being*. USGPO, Washington, DC.

U.S. Declaration of Independence (1776) *(Second Continental Congress)*.

U.S. General Accountability Office (2003) *Youth Illicit Drug Use Prevention: DARE Long-Term Evaluations and Federal Efforts to Identify Effective Programs* (GAO-03-172-R), Washington, DC.

Valery, P. (1919) *The Crisis of the Mind*. Athenaeum, London 11 April and 2 May. (Originally published as "La Crise de l'esprit" in *La Nouvelle Revue Française,* August 1919.

Van Praag, B.M.S. (2003) Cost of children. In: Demeny, P. and McNicoll, G. (eds) *Encyclopedia of Population*. Thomson-Gale, New York, pp. 177–180.

Vaughn, J. and Cortner, H.J. (2005) *George W Bush's Healthy Forests*. University of Colorado Press, Boulder, CO.

Viguerie, R.A. and Franke, D. (2004) *America's Right Turn*. Bonus Books, Chicago.

Vygotsky, L.S. (1962, 1934) *Thought and Language*. MIT Press/Wiley, Cambridge MA.

Waldau, P. and Patton, K. (eds) (2006) *A Communion of Subjects*. Columbia University Press, New York.

Walker, R. (2004) The hidden (in plain sight) persuaders. *The New York Times* 69–131.

Wallace, A.F.C. (1956) Revitalization movements. *American Anthropologist* **58**: 264–281.

Wallace, A.F.C. (1970a) *Culture and Personality*, 2nd edn. Random House, New York.

Wallack, L. and Dorfman, L. (2001) Putting policy into health communication. In: Rice, R.E. and Atkin, C.K. (eds) *Public Communication Campaigns*, 3rd edn. Sage Publications, Thousand Oaks, CA, pp. 389–401.

Wallerstein, I. (1976) *The Modern World System I*. Academic Press, New York.

Wallerstein, I. (2004) New revolts against the system. In: Mertes, T. (ed.) *A Movement of Movements*. Verso, London, pp. 262–273.

Weber, M. (1947) *The Theory of Social and Economic Organization*. Oxford University Press, Oxford.

Weiss, P. (1965) *Marat/Sade*. Atheneum, New York.

Weiss Reid, J. and Beazley, K. (2004) Public preferences for wildlife as a focus for biodiversity conservation. In: Munro, N.W.P., Herman, T.B., Beazley, K. and Dearden, P. (eds) *Making Ecosystem-based Management Work. Proceedings of the Fifth*

International Conference on Science and Management of Protected Areas, Victoria, BC, May, 2003. Science and Management of Protected Areas Association, Wolfville, Nova Scotia.

Wheat, F. (1999) *California Desert Miracle*. Sunbelt Publications, San Diego.

White, L. (1967) The historical roots of our ecological crisis. *Science* **155**:1203–1207.

White, L. (1987, 1975) The energy theory of cultural development. In: *Ethnological Essays*. University of New Mexico Press, Albuquerque.

Wilde, A.D. (2004) Mainstreaming kwanzaa. In: Etzioni, A. and Bloom, J. (eds) *We Are What We Celebrate*. New York University Press, New York, pp. 120–130.

The Wildlands Project (2000) *Sky Islands Wildlands Network Conservation Plan*. The Wildlands Project, Tucson, AZ (now Titusville, FL)

The Wildlands Project (2003) *New Mexico Highlands Wildlands Network Vision*. The Wildlands Project, Titusville, FL.

The Wildlands Project (2004) *Heart of the West Wildlands Network Vision*. The Wildlands Project, Titusville, FL.

The Wildlands Project, Southern Rockies Ecosystem Project and Zoo, D. (2003). *Southern Rockies Wildlands Network Vision*. Colorado Mountain Club Press, Golden, CO.

The Wildlands Project (2006) From the Adirondaks to Arcadia. Wildlands Project Special Paper No. 7. Titusville, FL.

Wilkinson, T. (1998) *Science Under Siege*. Johnson Books, Boulder, CO.

Williams, T.T. (1991) *Refuge*. Pantheon, New York.

Williams, M. (2002) *Deforesting the Earth*. University of Chicago Press, Chicago.

Williams, T. (2002) Lynx, lies, and media hype. *Audubon* **104**(3):24–33.

Williams, A. (2004) The alchemy of a political slogan. *The New York Times*, 22 August.

Wilson, E.O. (1984) *Biophilia*. Harvard University Press, Cambridge , MA.

Wilson, D.S. (2002) *Darwin's Cathedral*. University of Chicago, Chicago.

Winsten, J.A. and DeJong, W. (2001) The designated driver campaign. In: Rice, R.E. and Atkin, C.K. (eds) *Public Communication Campaigns*, 3rd edn. Sage Publications, Thousand Oaks, CA, pp. 290–294.

Wood, E.J. (2000) *Forging Democracy from Below*. Cambridge University Press.

Wood, E.J. (2001). The emotional benefits of insurgency in El Salvador. In: Goodwin, J., Jasper, J.M. and Polletta, F. (eds) *Passionate Politics*. University of Chicago Press, Chicago, pp. 267–281.

Wright, R. (2004) *A Short History of Progress*. Anansi, Toronto.

Wring, D. (1996) Political marketing and party development in Britain: a "Secret" history. *European Journal of Marketing* **30**(10/11):92–103.

Young, H.P. (1998) *Individual Strategy and Social Structure*. Princeton University Press, Princeton.

Young, M.P. (2001) A revolution of the soul: transformative experiences and immediate abolition. In: Goodwin, J., Jasper, J.M. and Polletta, F. (eds) *Passionate Politics*. University of Chicago Press, Chicago, pp. 99–114.

Young, J. (2003) (Read all about it, or not). Cox News Service. 22 October.

Zakin, S. (1993) *Coyotes and Town Dogs*. Viking Press, New York.

Zelezny, L.C., Chua, P.-P. and Aldrich, C. (2000) Elaborating on gender differences in environmentalism. *The Journal of Social Issues* **56**(3):443–458.

Zuckerman, L. (2000) How the C.I.A. played dirty tricks with culture. *The New York Times*, 18 March.

Zwerman, G. and Steinhoff, P. (2005) When activists ask for trouble. In: Davenport, C., Johnson, H. and Mueller, C. (eds) *Repression and Mobilization*. University of Minnesota Press, Minneapolis, pp. 85–107.

Index